NATURAL HISTORY
UNIVERSAL LIBRARY

西方博物学大系

主编：江晓原

THE
MAMMALS OF AUSTRALIA
澳大利亚哺乳动物志

[英] 约翰·古尔德 著

华东师范大学出版社

图书在版编目(CIP)数据

澳大利亚哺乳动物志 = The Mammals of Australia : 英文 /
(英) 约翰·古尔德 (John Gould) 著. — 上海 : 华东师范大学
出版社, 2019
(寰宇文献)
ISBN 978-7-5675-9162-2

Ⅰ.①澳… Ⅱ.①约… Ⅲ.①哺乳动物纲–动物志–澳大利
亚–英文 Ⅳ.①Q959.808

中国版本图书馆CIP数据核字(2019)第085282号

澳大利亚哺乳动物志
The Mammals of Australia
(英) 约翰·古尔德 (John Gould)

特约策划　黄曙辉　徐　辰
责任编辑　庞　坚
特约编辑　许　倩
装帧设计　刘怡霖

出版发行　华东师范大学出版社
社　　址　上海市中山北路3663号　邮编 200062
网　　址　www.ecnupress.com.cn
电　　话　021-60821666　行政传真　021-62572105
客服电话　021-62865537
门市(邮购)电话　021-62869887
地　　址　上海市中山北路3663号华东师范大学校内先锋路口
网　　店　http://hdsdcbs.tmall.com/

印　刷　者　虎彩印艺股份有限公司
开　　本　787×1092 16开
印　　张　30
版　　次　2019年6月第1版
印　　次　2019年6月第1次
书　　号　ISBN 978-7-5675-9162-2
定　　价　598.00元(精装全一册)

出 版 人　王　焰

(如发现本版图书有印订质量问题，请寄回本社客服中心调换或电话021-62865537联系)

总　目

《西方博物学大系总序》（江晓原）　　　　1

出版说明　　　　1

The Mammals of Australia VOL.I　　　　1

The Mammals of Australia VOL.II　　　　187

The Mammals of Australia VOL.III　　　　349

《西方博物学大系》总序

江晓原

　　《西方博物学大系》收录博物学著作超过一百种，时间跨度为 15 世纪至 1919 年，作者分布于 16 个国家，写作语种有英语、法语、拉丁语、德语、弗莱芒语等，涉及对象包括植物、昆虫、软体动物、两栖动物、爬行动物、哺乳动物、鸟类和人类等，西方博物学史上的经典著作大备于此编。

中西方"博物"传统及观念之异同

　　今天中文里的"博物学"一词，学者们认为对应的英语词汇是 Natural History，考其本义，在中国传统文化中并无现成对应词汇。在中国传统文化中原有"博物"一词，与"自然史"当然并不精确相同，甚至还有着相当大的区别，但是在"搜集自然界的物品"这种最原始的意义上，两者确实也大有相通之处，故以"博物学"对译 Natural History 一词，大体仍属可取，而且已被广泛接受。

　　已故科学史前辈刘祖慰教授尝言：古代中国人处理知识，如开中药铺，有数十上百小抽屉，将百药分门别类放入其中，即心安矣。刘教授言此，其辞若有憾焉——认为中国人不致力于寻求世界"所以然之理"，故不如西方之分析传统优越。然而古代中国人这种处理知识的风格，正与西方的博物学相通。

　　与此相对，西方的分析传统致力于探求各种现象和物体之间的相互关系，试图以此解释宇宙运行的原因。自古希腊开始，西方哲人即孜孜不倦建构各种几何模型，欲用以说明宇宙如何运行，其中最典型的代表，即为托勒密（Ptolemy）的宇宙体系。

　　比较两者，差别即在于：古代中国人主要关心外部世界"如何"运行，而以希腊为源头的西方知识传统（西方并非没有别的知识传统，只是未能光大而已）更关心世界"为何"如此运行。在线性发展无限进步的科学主义观念体系中，我们习惯于认为"为何"是在解决了"如何"之后的更高境界，故西方的分析传统比中国的传统更高明。

然而考之古代实际情形，如此简单的优劣结论未必能够成立。例如以天文学言之，古代东西方世界天文学的终极问题是共同的：给定任意地点和时刻，计算出太阳、月亮和五大行星（七政）的位置。古代中国人虽不致力于建立几何模型去解释七政"为何"如此运行，但他们用抽象的周期叠加（古代巴比伦也使用类似方法），同样能在足够高的精度上计算并预报任意给定地点和时刻的七政位置。而通过持续观察天象变化以统计、收集各种天象周期，同样可视之为富有博物学色彩的活动。

还有一点需要注意：虽然我们已经接受了用"博物学"来对译 Natural History，但中国的博物传统，确实和西方的博物学有一个重大差别——即中国的博物传统是可以容纳怪力乱神的，而西方的博物学基本上没有怪力乱神的位置。

古代中国人的博物传统不限于"多识于鸟兽草木之名"。体现此种传统的典型著作，首推晋代张华《博物志》一书。书名"博物"，其义尽显。此书从内容到分类，无不充分体现它作为中国博物传统的代表资格。

《博物志》中内容，大致可分为五类：一、山川地理知识；二、奇禽异兽描述；三、古代神话材料；四、历史人物传说；五、神仙方伎故事。这五大类，完全符合中国文化中的博物传统，深合中国古代博物传统之旨。第一类，其中涉及宇宙学说，甚至还有"地动"思想，故为科学史家所重视。第二类，其中甚至出现了中国古代长期流传的"守宫砂"传说的早期文献：相传守宫砂点在处女胳膊上，永不褪色，只有性交之后才会自动消失。第三类，古代神话传说，其中甚至包括可猜想为现代"连体人"的记载。第四类，各种著名历史人物，比如三位著名刺客的传说，此三名刺客及所刺对象，历史上皆实有其人。第五类，包括各种古代方术传说，比如中国古代房中养生学说，房中术史上的传说人物之一"青牛道士封君达"等等。前两类与西方的博物学较为接近，但每一类都会带怪力乱神色彩。

"所有的科学不是物理学就是集邮"

在许多人心目中，画画花草图案，做做昆虫标本，拍拍植物照片，这类博物学活动，和精密的数理科学，比如天文学、物理学等等，那是无法同日而语的。博物学显得那么的初级、简单，甚至幼稚。这种观念，实际上是将"数理程度"作为唯一的标尺，用来衡量一切知识。但凡能够使用数学工具来描述的，或能够进行物理实验的，那就是"硬"科学。使用的数学工具越高深越复杂，似乎就越"硬"；物理实验设备越庞大，花费的金钱越多，似乎就越"高端"、越"先进"……

这样的观念，当然带着浓厚的"物理学沙文主义"色彩，在很多情况下是不正确的。而实际上，即使我们暂且同意上述"物理学沙文主义"的观念，博物学的"科学地位"也仍然可以保住。作为一个学天体物理专业出身，因而经常徜徉在"物理学沙文主义"幻影之下的人，我很乐意指出这样一个事实：现代天文学家们的研究工作中，仍然有绘制星图，编制星表，以及为此进行的巡天观测等等活动，这些活动和博物学家"寻花问柳"，绘制植物或昆虫图谱，本质上是完全一致的。

这里我们不妨重温物理学家卢瑟福(Ernest Rutherford)的金句："所有的科学不是物理学就是集邮(All science is either physics or stamp collecting)。"卢瑟福的这个金句堪称"物理学沙文主义"的极致，连天文学也没被他放在眼里。不过，按照中国传统的"博物"理念，集邮毫无疑问应该是博物学的一部分——尽管古代并没有邮票。卢瑟福的金句也可以从另一个角度来解读：既然在卢瑟福眼里天文学和博物学都只是"集邮"，那岂不就可以将博物学和天文学相提并论了？

如果我们摆脱了科学主义的语境，则西方模式的优越性将进一步被消解。例如，按照霍金(Stephen Hawking)在《大设计》(*The Grand Design*)中的意见，他所认同的是一种"依赖模型的实在论(model-dependent realism)"，即"不存在与图像或理论无关的实在性概念(There is no picture- or theory-independent concept of reality)"。在这样的认识中，我们以前所坚信的外部世界的客观性，已经不复存在。既然几何模型只不过是对外部世界图像的人为建构，则古代中国人干脆放弃这种建构直奔应用（毕竟在实际应用中我们只需要知道七政"如何"运行），又有何不可？

传说中的"神农尝百草"故事，也可以在类似意义下得到新的解读："尝百草"当然是富有博物学色彩的活动，神农通过这一活动，得知哪些草能够治病，哪些不能，然而在这个传说中，神农显然没有致力于解释"为何"某些草能够治病而另一些则不能，更不会去建立"模型"以说明之。

"帝国科学"的原罪

今日学者有倡言"博物学复兴"者，用意可有多种，诸如缓解压力、亲近自然、保护环境、绿色生活、可持续发展、科学主义解毒剂等等，皆属美善。编印《西方博物学大系》也是意欲为"博物学复兴"添一助力。

然而，对于这些博物学著作，有一点似乎从未见学者指出过，而鄙意以为，当我们披阅把玩欣赏这些著作时，意识到这一点是必须的。

这百余种著作的时间跨度为 15 世纪至 1919 年，注意这个时间跨度，正是西方列强"帝国科学"大行其道的时代。遥想当年，帝国的科学家们乘上帝国的军舰——达尔文在皇家海军"小猎犬号"上就是这样的场景之一，前往那些已经成为帝国的殖民地或还未成为殖民地的"未开化"的遥远地方，通常都是踌躇满志、充满优越感的。

作为一个典型的例子，英国学者法拉在（Patricia Fara）《性、植物学与帝国：林奈与班克斯》（*Sex, Botany and Empire, The Story of Carl Linnaeus and Joseph Banks*）一书中讲述了英国植物学家班克斯（Joseph Banks）的故事。1768 年 8 月 15 日，班克斯告别未婚妻，登上了澳大利亚军舰"奋进号"。此次"奋进号"的远航是受英国海军部和皇家学会资助，目的是前往南太平洋的塔希提岛（Tahiti，法属海外自治领，另一个常见的译名是"大溪地"）观测一次比较罕见的金星凌日。舰长库克（James Cook）是西方殖民史上最著名的舰长之一，多次远航探险，开拓海外殖民地。他还被认为是澳大利亚和夏威夷群岛的"发现"者，如今以他命名的群岛、海峡、山峰等不胜枚举。

当"奋进号"停靠塔希提岛时，班克斯一下就被当地美丽的土著女性迷昏了，他在她们的温柔乡里纵情狂欢，连库克舰长都看不下去了，"道德愤怒情绪偷偷溜进了他的日志当中，他发现自己根本不可能不去批评所见到的滥交行为"，而班克斯纵欲到了"连嫖妓都毫无激情"的地步——这是别人讽刺班克斯的说法，因为对于那时常年航行于茫茫大海上的男性来说，上岸嫖妓通常是一项能够唤起"激情"的活动。

而在"帝国科学"的宏大叙事中，科学家的私德是无关紧要的，人们关注的是科学家做出的科学发现。所以，尽管一面是班克斯在塔希提岛纵欲滥交，一面是他留在故乡的未婚妻正泪眼婆娑地"为远去的心上人绣织背心"，这样典型的"渣男"行径要是放在今天，非被互联网上的口水淹死不可，但是"班克斯很快从他们的分离之苦中走了出来，在外近三年，他活得倒十分滋润"。

法拉不无讽刺地指出了"帝国科学"的实质："班克斯接管了当地的女性和植物，而库克则保护了大英帝国在太平洋上的殖民地。"甚至对班克斯的植物学本身也调侃了一番："即使是植物学方面的科学术语也充满了性指涉。……这个体系主要依靠花朵之中雌雄生殖器官的数量来进行分类。"据说"要保护年轻妇女不受植物学教育的浸染，他们严令禁止各种各样的植物采集探险活动"。这简直就是将植物学看成一种"涉黄"的淫秽色情活动了。

在意识形态强烈影响着我们学术话语的时代，上面的故事通常是这样被描述的：库克舰长的"奋进号"军舰对殖民地和尚未成为殖民地的那些地方的所谓"访问"，其实是殖民者耀武扬威的侵略，搭载着达尔文的"小猎犬号"军舰也是同样行径；班克斯和当地女性的纵欲狂欢，当然是殖民者对土著妇女令人发指的蹂躏；即使是他采集当地植物标本的"科学考察"，也可以视为殖民者"窃取当地经济情报"的罪恶行为。

后来改革开放，上面那种意识形态话语被抛弃了，但似乎又走向了另一个极端，完全忘记或有意回避殖民者和帝国主义这个层面，只歌颂这些军舰上的科学家的伟大发现和成就，例如达尔文随着"小猎犬号"的航行，早已成为一曲祥和优美的科学颂歌。

其实达尔文也未能免俗，他在远航中也乐意与土著女性打打交道，当然他没有像班克斯那样滥情纵欲。在达尔文为"小猎犬号"远航写的《环球游记》中，我们读到："回程途中我们遇到一群黑人姑娘在聚会，……我们笑着看了很久，还给了她们一些钱，这着实令她们欣喜一番，拿着钱尖声大笑起来，很远还能听到那愉悦的笑声。"

有趣的是，在班克斯在塔希提岛纵欲六十多年后，达尔文随着"小猎犬号"也来到了塔希提岛，岛上的土著女性同样引起了达尔文的注意，在《环球游记》中他写道："我对这里妇女的外貌感到有些失望，然而她们却很爱美，把一朵白花或者红花戴在脑后的发髻上……"接着他以居高临下的笔调描述了当地女性的几种发饰。

用今天的眼光来看，这些在别的民族土地上采集植物动物标本、测量地质水文数据等等的"科学考察"行为，有没有合法性问题？有没有侵犯主权的问题？这些行为得到当地人的同意了吗？当地人知道这些行为的性质和意义吗？他们有知情权吗？……这些问题，在今天的国际交往中，确实都是存在的。

也许有人会为这些帝国科学家辩解说：那时当地土著尚在未开化或半开化状态中，他们哪有"国家主权"的意识啊？他们也没有制止帝国科学家的考察活动啊。但是，这样的辩解是无法成立的。

姑不论当地土著当时究竟有没有试图制止帝国科学家的"科学考察"行为，现在早已不得而知，只要殖民者没有记录下来，我们通常就无法知道。况且殖民者有军舰有枪炮，土著就是想制止也无能为力。正如法拉所描述的："在几个塔希提人被杀之后，一套行之有效的易货贸易体制建立了起来。"

即使土著因为无知而没有制止帝国科学家的"科学考察"行为，这事也很像一个成年人闯进别人的家，难道因为那家只有不懂事的小孩子，闯入者就可以随便打探那家的隐私、拿走那家的东西、甚至将那家的房屋土地据为己有吗？事实上，很多情况下殖民者就是这样干的。所以，所谓的"帝国科学"，其实是有着原罪的。

如果沿用上述比喻，现在的局面是，家家户户都不会只有不懂事的孩子了，所以任何外来者要想进行"科学探索"，他也得和这家主人达成共识，得到这家主人的允许才能够进行。即使这种共识的达成依赖于利益的交换，至少也不能单方面强加于人。

博物学在今日中国

博物学在今日中国之复兴，北京大学刘华杰教授提倡之功殊不可没。自刘教授大力提倡之后，各界人士纷纷跟进，仿佛昔日蔡锷在云南起兵反袁之"滇黔首义，薄海同钦，一檄遥传，景从恐后"光景，这当然是和博物学本身特点密切相关的。

无论在西方还是在中国，无论在过去还是在当下，为何博物学在它繁荣时尚的阶段，就会应者云集？深究起来，恐怕和博物学本身的特点有关。博物学没有复杂的理论结构，它的专业训练也相对容易，至少没有天文学、物理学那样的数理"门槛"，所以和一些数理学科相比，博物学可以有更多的自学成才者。这次编印的《西方博物学大系》，卷帙浩繁，蔚为大观，同样说明了这一点。

最后，还有一点明显的差别必须在此处强调指出：用刘华杰教授喜欢的术语来说，《西方博物学大系》所收入的百余种著作，绝大部分属于"一阶"性质的工作，即直接对博物学作出了贡献的著作。事实上，这也是它们被收入《西方博物学大系》的主要理由之一。而在中国国内目前已经相当热的博物学时尚潮流中，绝大部分已经出版的书籍，不是属于"二阶"性质（比如介绍西方的博物学成就），就是文学性的吟风咏月野草闲花。

要寻找中国当代学者在博物学方面的"一阶"著作，如果有之，以笔者之孤陋寡闻，唯有刘华杰教授的《檀岛花事——夏威夷植物日记》三卷，可以当之。这是刘教授在夏威夷群岛实地考察当地植物的成果，不仅属于直接对博物学作出贡献之作，而且至少在形式上将昔日"帝国科学"的逻辑反其道而用之，岂不快哉！

2018 年 6 月 5 日
于上海交通大学
科学史与科学文化研究院

约翰·古尔德（John Gould），英国鸟类学家兼鸟类画家。生于多塞特郡，其父是温莎城堡御花园的园丁主管。由于家庭关系，他幼年常往来于御花园，十四岁时正式成为那里里的园丁学徒。这份工作培养了他观察自然界生物的本领和剥制标本的手艺。1827年，他受聘为伦敦的动物学会博物馆首席标本剥制师。新的职务给了古尔德接触国内顶尖博物学家的大好机会，送到动物学会的鸟类标本，也大都由他第一时间查验过目，其中很多种类都是英国人前所未见。达尔文完成第二次小猎犬号科考航行之后，也将收集到的鸟类样本交给古尔德研究。古尔德一生共留下近三千幅博物学绘画，其中最被人称道的是鸟类画。

约翰·古尔德
（1804-1881）

《澳大利亚哺乳动物志》是古尔德的博物学著作之一。1838年，古尔德对澳大利亚及南太平洋诸岛的鸟类进行了实地研究。翌年2月，古尔德再次搭船前往悉尼进行了两个月的研究。同年5月，又和探险家查尔斯·斯图尔特一起到阿德莱德旅行，1840年5月才回到英国。在此期间，除鸟类外，澳大利亚本土的哺乳动物也是古尔德兴趣所在。自1845年起，他开始出版《澳大利亚哺乳动物志》，至1863年出齐三卷，共800余页，配有他和妻子伊丽莎白共同创作的约200幅精致动物画。今据外文原版影印。

THE

MAMMALS OF AUSTRALIA.

BY

JOHN GOULD, F.R.S.,

F.L.S., F.Z.S., M.E.S., F.R.GEOG.S., M.RAY S.: HON. MEMB. OF THE ROYAL ACADEMY OF SCIENCES OF TURIN; OF THE ROYAL
ZOOL. SOC. OF IRELAND; OF THE PENZANCE NAT. HIST. SOC.; OF THE WORCESTER NAT. HIST. SOC.; OF THE
NORTHUMBERLAND, DURHAM, AND NEWCASTLE NAT. HIST. SOC.; OF THE NAT. HIST. SOC.
OF DARMSTADT; OF THE TASMANIAN SOC. OF VAN DIEMEN'S LAND; OF THE NAT.
HIST. SOC. OF STRASBOURG; OF THE NAT. HIST. SOC. OF IPSWICH; AND
CORR. MEMB. SOC. OF NAT. HIST. OF WÜRTEMBERG.

IN THREE VOLUMES.

VOL. I.

LONDON:

PRINTED BY TAYLOR AND FRANCIS, RED LION COURT, FLEET STREET.

PUBLISHED BY THE AUTHOR, 26 CHARLOTTE STREET, BEDFORD SQUARE.

1863.

TO

HIS ROYAL HIGHNESS

THE PRINCE CONSORT

THIS WORK

ON THE

MAMMALS OF AUSTRALIA

IS,

WITH HIS ROYAL HIGHNESS'S PERMISSION,

DEDICATED

BY HIS MOST OBLIGED AND FAITHFUL SERVANT

JOHN GOULD.

HAVING been permitted to dedicate my work on the "Birds of Australia" to Her Most Gracious Majesty Queen Victoria, I was naturally desirous of dedicating the companion-work on the Mammals of the same country to Her Majesty's most enlightened and accomplished Consort; and the required permission was readily and graciously granted me. The dispensation which has deprived Her Majesty and the Prince's adopted country of one whose untimely loss we all deplore, still leaves me the privilege of that permission, and my work will continue to have the honour of being inscribed to His Royal Highness. It is with a melancholy satisfaction that I accordingly retain that Dedication, which, should it meet my Sovereign's eye, will, I think, only recall to her that love which the whole country entertains for his cherished memory. I feel that nothing I can say respecting the admirable qualities of this most enlightened Prince can in any way add to the deservedly high reputation of one whose great learning and manifold virtues, while he was among us, did so much for Science and Art, and whose example, we trust, will influence generations yet unborn.

LIST OF SUBSCRIBERS.

HER MOST GRACIOUS MAJESTY QUEEN VICTORIA.

HIS MAJESTY THE KING OF HANOVER.
HIS MAJESTY THE KING OF THE BELGIANS.
HIS ROYAL HIGHNESS THE PRINCE CONSORT.

ABERDEEN, the University and King's College.
Allport, M., Esq. *Tasmania.*
Amsterdam, the Royal Zoological Society of Natura Artis Magistra at.
Angas, the Hon. George Fife. *Prospect Hall, North Adelaide, South Australia.*
Astor Library, the, *New York.*
Athenæum, the Library of the, *Pall Mall.*
Australian Library, the, *Sydney, New South Wales.*
Australian Museum, the, *Sydney, New South Wales.*
Aylesford, the Right Hon. the Earl of. *Packington Hall, Coventry, Warwickshire; and Aylesford, Maidstone, Kent.*
Baker, T. B. L., Esq., F.G.S., &c. *Hardwicke Court, Gloucester.*
Barclay, J. G., Esq. *Lombard Street; and Leyton, Essex.*
Barthés and Lowell, Messrs. *Great Marlborough Street.*
Bennett, Dr. *Sydney, New South Wales.*
Berlin, the Royal Library of.
Berlin, the Royal Museum of.
Bodleian Library, the, *Oxford.*
Boston Natural History Society, the.
Brassey, T., Esq. *Lowndes Square.*
British Museum, the Library of the.
Buccleuch and Queensbury, His Grace the Duke of, K.G., P.C., F.R.S., F.L.S., D.C.L., &c. *Whitehall Gardens; Broughton House, Kettering, Northamptonshire; Richmond, Surrey; Dalkeith Palace, Edinburgh; Drumlanrig Castle, and Langholm Lodge, Dumfriesshire; and Bowhill, Selkirk, N. B.*
Butler, C., Esq. *Sussex Square, Hyde Park.*
Cabbell, B. B., Esq., F.R.S., F.R.G.S., F.S.A., F.R.S.L., &c. *Brick Court, Temple; Portland Place; Aldwick, Sussex; and Cromer, Norfolk.*
Cambridge, the University of.
Campbell, A., Esq. *Seamore Place, Mayfair; and Blythswood, Renfrewshire, N. B.*
Campbell, R., Esq. *Sydney, New South Wales.*
Canada, the Library of the Parliament of.
Canning, the Rev. W. *The Cloisters, Windsor Castle; and Old Windsor.*
Christiania, the Museum of, *Norway.*
Classen's Library, *Copenhagen, Denmark.*
Coulon, Mons. Louis. *Neufchâtel, Switzerland.*
Craven, the Right Hon. the Earl of. *Charles Street, Berkeley Square; Coombe Abbey, Coventry, Warwickshire; Hampstead Marshall, Newbury; and Ashdown Park, Lambourn, Berkshire.*
Crowley, C. S., Esq. *Cavendish Place, Cavendish Square.* 2 Copies.
Currer, Miss. *Eshton Hall, Gargrave, Yorkshire.*

Devonshire, His Grace the Duke of, K.G., M.A., F.R.S., F.R.G.S., F.G.S., F.R.S.L. *Devonshire House, Piccadilly; Chatsworth, and Hardwicke Hall, Derbyshire; Bolton Abbey, Yorkshire; Holkar Hall, Newton in Cartmel, Lancashire; Compton Place, Eastbourne, Sussex; and Lismore Castle, Waterford, Ireland.*
Drummond, R., Esq. *Park Lane; and Ardvorlich, Lochearn-head, Crieff, N. B.*
East India Office, the Library of the.
Edinburgh, the Library of the Faculty of Advocates.
Edinburgh, the University of.
Eyton, T. C., Esq. *Eyton, Wellington, Shropshire.*
Fitzwilliam, the Right Hon. Earl. *Grosvenor Square; Wentworth Woodhouse, Rotherham, Yorkshire; and Coollatin House, Wexford, Ireland.*
Folliott, G., Esq. *Vicar's Cross, near Chester.*
Frank, M. *Amsterdam.*
Gibson, W. G., Esq. *Saffron Walden, Essex.*
Glasgow University, the.
Gosford, the Right Hon. the Earl of, K.P. *Grosvenor Street; Gosford Castle, Armagh, Ireland; and Worlingham Hall, Beccles, Suffolk.*
Gott, W., Esq. *Leeds, Yorkshire.*
Grey, His Excellency Sir George, Governor of New Zealand.
Gunn, R. C., Esq. *Launceston, Tasmania.*
Gurney, H. E., Esq. *Nutfield, Surrey.*
Gurney, J. H., Esq., M.P. *Catton Hall, near Norwich.*
Hale, R. B., Esq. *Alderley Park, Wootton-under-edge, Gloucestershire.*
Hamilton and Brandon, His Grace the Duke of. *Arlington Street, Piccadilly; Hamilton Palace, Lanarkshire; Brodick Castle, Isle of Arran; Kinniel House, Linlithgowshire; and Easton Park, Wickham Market, Suffolk.*
Hartree, Mrs. *Lewisham Road, Greenwich, Kent.*
Hewson, John, Esq. *Newlands, Lincoln.*
Hill, the Right Hon. the Viscount, F.G.S. *Hawkstone, near Shrewsbury; and Hardwicke Grange, Shropshire.*
Hull Subscription Library, the.
Huth, L., Esq. *Upper Harley Street, Cavendish Square.*
Ingram, Mrs. *Kensington Palace Gardens.*
Jardin des Plantes, la Bibliothèque de le.
Jourdan, M., Directeur du Muséum d'Histoire Naturelle à Lyon.
Kelk, J., Esq. *Eaton Square; and The Priory, Stanmore, Middlesex.*
Larking, J. W., Esq. *The Firs, Lee, Kent.*
Legh, G. C., Esq., M.P., F.G.S., &c. *Eaton Place; and High Legh, Warrington, Cheshire.*
Lilford, the Right Hon. Lord. *Lilford Hall, Oundle, Northampton; and Bewsay Hall, Warrington, Lancashire.*

Little and Brown, Messrs. *Boston, North America.*

Liverpool, the Free Public Library.

Liverpool Library, the.

Liverpool, the Royal Institution of.

Llewelyn, J. D., Esq. *Pennlergare, Swansea, South Wales.*

London Institution, the, *Finsbury Circus.*

Lucas, T., Esq. *Hyde Park Gardens; and Lowestoft, Suffolk.*

Melbourne Public Library, the, *Victoria.*

Milner, Sir W. M. E., Bart. *Nunappleton, Tadcaster, Yorkshire.*

Mitford, Admiral. *Hummanby Hall, Scarborough, Yorkshire.*

Muquardt, Mons. C. *Brussels.*

Naylor, J., Esq. *Liverpool; and Leighton Hall, Welchpool, Montgomeryshire.*

Newcastle, His Grace the Duke of, K.G., P.C. *Clumber Park, Worksop, Nottinghamshire.*

Northumberland, His Grace the Duke of, K.G. *Northumberland House, Charing Cross; Alnwick and Keilder Castles, Northumberland; Sion House, Middlesex; Werrington Park, Cornwall; and Stanwich, Darlington, Yorkshire.*

Nutt, Mr. D. *Strand.*

Owen, Professor. *British Museum; and Richmond Park, Surrey.*

Palatine Library, the, *Florence.*

Peckover, A., Esq. *Wisbech, Cambridgeshire.*

Percy, Dr. *Campden Hill, Bayswater.*

Peto, Sir S. Morton, Bart. *Kensington Palace Gardens.*

Philadelphia, the Academy of Natural Sciences of, *North America.*

Portland, His Grace the Duke of. *Hyde Park Gardens; Welbeck Abbey, Worksop, Nottinghamshire; and Fullarton House, Ayrshire, N. B.*

Powerscourt, the Right Hon. Viscount. *Powerscourt House, Enniskerry, Ireland.*

Radcliffe Library, the, *Oxford.*

Reeves, J. R., Esq. *Woodhays, Wimbledon, Surrey.*

Rolle, the Right Hon. Lady. *Upper Grosvenor Street; Stevenston near Torrington, Bicton near Exeter, and Bovey near Axminster, Devonshire.*

Rouen, the Public Library of.

Royal Institution of Great Britain, the, *Albemarle Street.*

Royal Society, the, *Burlington House, Piccadilly.*

Royal Society of Tasmania, the, *Hobart Town.*

Rucker, S., Esq. *Wandsworth.*

St. Andrew's, the University of.

St. Petersburg, the Library of the Imperial Academy of Sciences of.

Sanford, W. A., Esq. *Nynehead Court, Wellington, Somersetshire.*

Schlegel, Dr. H., Directeur du Musée Royale des Pays Bas, *Leyden.*

Senckenbergian Society, the, *Frankfort on the Maine.*

Shuttleworth, R. J., Esq., Director of the Zoological Department of the Museum of Berne, *Switzerland.*

Skaife, John, Esq. *Union Street, Blackburn, Lancashire.*

South Australian Institute, the, *Adelaide, South Australia.*

Staniforth, Rev. T. *Bolton Rectory, Clitheroe, Lancashire; and Storr's Hall, Windermere, Westmoreland.*

Strasbourg, le Musée d'Histoire Naturelle de.

Strickland, A., Esq. *Bridlington Quay, Yorkshire.*

Stuart, R. L., Esq. *Greenwich Street, New York.*

Surgeons of England, the Royal College of, *Lincoln's Inn Fields.*

Sutherland, His Grace the Duke of. *Stafford House, St. James's; Trentham, Staffordshire; Tarbet House, Park Hill, Ross-shire; Dunrobin Castle, and the House of Tongue, Sutherlandshire.*

Teylerian Library, the, *Haarlem.*

Tooth, R., Esq. *Sydney, New South Wales.*

Trinity College, *Dublin.*

Van den Hoeck and Ruprecht, Messrs. *Göttingen.*

Victoria, the National Museum of, *Australia.*

Vienna, the Imperial Library of.

Vrolik, Professor W. *Amsterdam.*

Walker, Mrs. *Southgate, Middlesex.*

Weigel, Mons. T. O. *Leipsic.*

Wellington, His Grace the Duke of, K.G., P.C., F.R.G.S. *Apsley House, Piccadilly; Strathfieldsaye, Hampshire; Thetford Lodge, Clermont, near Walton, and Hillborough Hall, Brandon, Norfolk.*

Wenlock, the Right Hon. Lord. *Escrick Park, near York.*

Zoological Society of London, the.

PREFACE.

IN the Preface to the 'Birds of Australia,' which has now been fifteen years before the public, I stated that, "Having in the summer of 1837 brought my work on the 'Birds of Europe' to a successful termination, I was naturally desirous of turning my attention to the Ornithology of some other region; and a variety of opportune and concurring circumstances induced me to select that of Australia, the birds of which country, although invested with the highest degree of interest, had been almost entirely neglected." But if the Birds of Australia had not received that degree of attention from the scientific ornithologist which their interest demanded, I can assert, without fear of contradiction, that its highly curious and interesting Mammals had been still less investigated. It was not, however, until I arrived in the country, and found myself surrounded by objects as strange as if I had been transported to another planet, that I conceived the idea of devoting a portion of my attention to the mammalian class of its extraordinary fauna.

The native black, while conducting me through the forest or among the park-like trees of the open plains, would often point out the pricking of an Opossum's nails on the bark of a *Eucalyptus* or other tree, and indicate by his actions that in yonder hole, high up, was sleeping an Opossum, a *Phalangista*, or a Flying *Petaurus*. Even the objects brought to our bush-fires were enough to incite a desire for a more extended knowledge of Australia's Mammals; for numerous were the species of Kangaroos and Opossums that were nightly roasted and eaten by these children of nature. Perchance a half-charred log, or the heated hollow branch of a *Eucalyptus*, would send forth into the lap of one or other of the surrounding guests the *Acrobates pygmæus*, the white-footed *Hapalotis*, or

b

other small quadruped. Tired by a long and laborious day's walk under a burning sun, I
frequently encamped for the night by the side of a river, a natural pond, or a water-
hole, and before retiring to rest not unfrequently stretched my weary body on the river's
bank; while thus reposing, the surface of the water was often disturbed by the little
concentric circles formed by the *Ornithorhynchus*, or perhaps an *Echidna* came trotting
up towards me.   With such scenes as these continually around me, is it surprising that
I should have entertained the idea of collecting examples of the indigenous Mammals of
a country whose ornithological productions I had gone out expressly to investigate?   To
have attempted to acquire a knowledge of more than the Birds and Mammals would
have been unwise; still I was not insensible to the interest which attaches to its insects
and to its wonderful botanical productions.   The *Eucalypti*, the *Banksiæ*, the *Casuarinæ*,
the native Cedar- and the Fig-trees will ever stand forth prominently in my memory.
While in the interior of the country, I formed the intention of publishing a monograph
of the great family of Kangaroos; but soon after my return to England I determined to
attempt a more extended work, under the title of the 'Mammals of Australia.'

It will always be a source of pleasure to me to remember that I was the first to
describe and figure the Great Black and Red Wallaroos (*Osphranter robustus* and *O. antilopinus*), the three species of *Onychogalea*, several of the equally singular *Lagorchestes*, and
many other new species of Kangaroos.   Mounted examples of all these animals, whether
discovered by myself or by others, are now contained in the national collection of this
country; but I regret to say that their colours are very different from what they were
while the animals were living, the continuous exposure to light, consequent upon their
being placed in a museum, causing their evanescent colouring rapidly to fade, both here
and in the collections of every other country.   Those who have seen the living *Osphranter
rufus* at the Zoological Gardens could scarcely for a moment suppose that the Museum
specimen of the same animal had ever been dressed in such glowing tints.   To see the
Kangaroos in all their glory, their native country must be visited; their beauty would
then be at once apparent, and their various specific distinctions easily recognizable.

The exploration of every new district has afforded ample proof of the existence of species
in every department of zoology with which we were previously unacquainted.   Under
these circumstances, I do not consider my work to be in any way complete, or that it
comprises nearly the whole of the Mammals of a country of which so much has yet to
be traversed; but I bring it to a close after an interval of eighteen years since its commencement, during which constant attention has been given to the subject, as treating
upon the genera and species known up to the present time.   If my life be prolonged,
and the blessing of health be continued to me, I propose, as in the case of the 'Birds

of Australia,' to keep the subject complete, by issuing a supplementary part, from time to time, should sufficient new materials be acquired to enable me so to do.

As with regard to my other publications, so also with this, I have to offer my best thanks to many persons for the kind and friendly assistance they have rendered me in prosecuting my labours on the 'Mammals of Australia.' I cannot, therefore, close these remarks without recording my obligations to Professor Owen, Dr. Gray, and G. R. Waterhouse, Esq., of the British Museum; to Ronald C. Gunn, Esq., of Launceston; the Rev. T. J. Ewing and Dr. Milligan of Hobart Town; to Dr. Bennett, W. S. Mac-Leay, Esq., Gerard Krefft, Esq., the late Dr. Ludwig Becker, W. S. Wall, Esq., the authorities of the Australian Museum, and the late Frederick Strange, of New South Wales; to Charles Coxen, Esq., of Queensland; John Macgillivray, Esq.; the late Commander J. M. R. Ince, R.N.; to His Excellency Sir George Grey, formerly Governor of South Australia, and now of New Zealand; the late John Gilbert; Professor M'Coy, of Melbourne; George French Angas, Esq., of Angaston, South Australia; W. Ogilby, Esq., formerly Secretary of the Zoological Society of London; Dr. Sclater, its present Secretary; R. F. Tomes, Esq.; M. Jules Verreaux, of Paris; Dr. W. Peters, of the Royal Museum of Berlin; and lastly, my son, Mr. Charles Gould, the Geological Surveyor of Tasmania. I believe I have here enumerated the names of all who have favoured me with specimens or with the benefit of their opinions, in reference to the subjects of the present work. To have omitted the name of one friend would be a source of much vexation to me; but if such should unfortunately have been done, I trust it will be considered the result of inadvertence, and not of intentional neglect.

To my artist, Mr. Richter, I consider (and I have no doubt my readers will concur in my opinion) that much credit is due for the manner in which he has executed the drawings, both from the dead as well as from the living examples from which they were taken. My Secretary, Mr. Prince, has also discharged the same praiseworthy services as heretofore.

It will be observed that, in mentioning the localities frequented by the various species, I have mostly employed the term Van Diemen's Land to designate the large island lying off the south coast of Australia; there is now, however, a very general desire that it should be called Tasmania—in honour of Tasman, its original discoverer; this term has, therefore, also been used, and hence has arisen the discrepancy of employing two names for one island. Even since the commencement of the work, new colonies have sprung up, or the older ones have been divided; thus the country now known as Queensland was formerly part of New South Wales, and Victoria, until lately, was known as Port Phillip.

INTRODUCTION.

In the foregoing Preface I have glanced at the principal groups of Mammals inhabiting the great country of Australia. It will now, however, be necessary to enter into greater detail respecting this division of its fauna; and I conceive that it will not be out of place if I commence with a retrospective view of the gradual discovery of countries and their zoological productions from the earliest historic times. Such a retrospect will not, I think, be deemed unnecessary, especially since my intention is to show to the general reader, rather than to the scientific naturalist, that each great division of the globe has its own peculiar forms of animal life, and that the fauna of Australia is widely different from that of every other part of the world. By a mere glance at the zoological features of the globe as at present existing, it will be perceived with what precision the animal life of each country has been adapted to its physical character; the absence of certain great families of birds and quadrupeds in some countries will also be apparent. To account for this on any scientific principle would be very difficult, when we cannot say why the Nightingale is not a summer visitant to Devonshire, or why the Grouse is not found south of Wales; why the aërial Swifts, Swallows, and Martins are numerous in Australia, and absent in New Zealand; or why Woodpeckers, which occur in nearly every other part of the globe, are not found in Australia, New Guinea, or any of the Polynesian Islands.

The ancient Egyptians appear to have been little acquainted with the natural productions of any other country than their own,—at least, we have no evidence that they were; for neither so conspicuous a bird as the Peacock, nor even the Common Fowl, are represented on their lasting monuments. Of the eastern countries Alexander's expedition doubtless greatly increased the knowledge of the Greeks, furnishing materials for the philosophic mind of Aristotle, and certainly extending the knowledge of Pliny, as is evidenced by his 'Historia Naturalis,' the only work which has come down to us of the latter great naturalist. Pliny, standing out as a bright star in zoological science at the period he lived, was doubtless tolerably acquainted with the natural productions of Eastern Europe, Arabia, North-eastern Africa, slightly with those of Persia, and still less so with those of India.

It may be fairly said, that the earliest dawn of natural history commenced with the Christian era,— Aristotle living just before, and Pliny soon after, the advent of our Saviour. This early dawn, however, was for a long period obscured by the dark ages which succeeded; for it was not until the commencement of the

c

17th century that Aldrovandus, Piso, Marcgrave, and Willughby wrote their works on this branch of science. At this comparatively late period, the productions of Europe were better known; Africa had been for a long time circumnavigated, and its southern fauna partially brought to light; India also in like manner furnished her quota, though sparingly, to the stock of human knowledge. What Alexander's celebrated expedition did for the naturalists Aristotle and Pliny, the discoveries of Columbus did by shedding a new light upon zoological science, and furnishing fresh food to the modern writers above mentioned. Linnæus, the greatest of all systematists, had a very extended knowledge of the natural productions of the globe, and the information this great man has left behind him in his numerous writings is considerable. Still, the southern land which we designate Australia (the mammalian products of which this work is intended to illustrate) was a sealed book to him. As regards this great country, it may be said that its most highly organized animals, if we except the Seals, are the various species of Rodents, and the equally numerous insectivorous and frugivorous Bats, both of which rank among the lowest of the *Placentals*. In America the *Marsupialia* are but feebly represented; in Africa and India none of this form exist. On the other hand, Australia is the great country of these pouched animals; they are universally distributed throughout its entire extent, from north to south, and from east to west; and they are not even absent from the neighbouring islands. Their presence in Tasmania on the south, and New Guinea on the north, testifies that these countries were formerly united to the mainland, and constituted a great natural division of the globe, characterized by a similar fauna and flora. It will be unnecessary for me to state that none of the *Quadrumana*, or Monkeys, are found in Australia; and that neither the Lion, the Tiger, the Leopard, nor any other of the *Felinæ*, roam among its forests, to disturb the harmony of its generally peaceful quadrupeds.

The great groups of the *Bovinæ*, or Oxen, the *Equinæ*, or Horses and Zebras, the stately Elephant, the huge Rhinoceros, as well as the *Cervidæ*, or Deer-kind, and the Antelopes, are totally unknown in Australia; yet the great grassy plains and other physical features of the country would appear to be well adapted for them and also for the smaller herbivorous quadrupeds, such as the Hare, the Rabbit, &c. Why there should occur so great a difference between the animals of Australia and those of the other countries of the world it is not for me to say. But I may ask, has creation been arrested in this strange land? and, if not, why are these higher types denied to it? Whatever opinion may be formed on this interesting subject, it is generally believed that no more highly organized animals than those which are now found there ever roamed over her plains or tenanted her luxuriant brushes. At the same time, the partially fossilized remains of distinct species of Kangaroos which have been discovered in her stalactitic caves, and the huge skeletons, or parts of skeletons, which have been exhumed from her alluvial beds, testify that Australia must be of remote origin. It is scarcely necessary to remark that all these remains belong to Marsupial animals; nor must it be imagined that I am oblivious of the fact that the remains of members of this group have been found in the older tertiary and secondary strata of Europe. I merely glance at these things, and leave their consideration to those who pay special attention to the sister science of geology.

Although the more highly organized animals do not inhabit, and seem never to have inhabited Australia, it is not a little interesting to observe how completely the law of representation is manifested among her mammals—how one family typifies another in the higher groups of the *Placentalia*; or, to be more explicit, to note how the *Herbivora* are represented by the Kangaroos, the *Felinæ* by the *Dasyures*, the Jerboas by the *Hapalotides*, &c. When speaking of the wonderful fossil *Diprotodon*, in his work on Palæontology,

Professor Owen states—" Australia yields evidence of an analogous correspondence between its last extinct and its present aboriginal mammalian fauna, which is the more interesting on account of the very peculiar organization of most of the native quadrupeds of that division of the globe. That the *Marsupialia* form one great natural group is now generally admitted by zoologists; the representatives in that group of many of the orders of the more exclusive Placental subclass of the Mammalia of the larger continents have also been recognized in the existing genera and species: the *Dasyures*, for example, play the parts of the *Carnivora*; the Bandicoots (*Perameles*), of the *Insectivora*; the Phalangers, of the *Quadrumana*; the Wombat, of the *Rodentia*; and the Kangaroos, in a remoter degree, of the *Ruminantia*. The first collection of mammalian fossils from the ossiferous caves of Australia brought to light the former existence on that continent of larger species of the same peculiar marsupial genera: some, as the Thylacine, and the Dasyurine subgenus represented by the *D. ursinus*, are now extinct on the Australian continent; but one species of each still exists on the adjacent island of Tasmania; the rest were extinct Wombats, Phalangers, Potoroos, and Kangaroos—some of the latter (*Macropus Atlas, M. Titan*) being of great stature. A single tooth, in the same collection of fossils, gave the first indication of the former existence of a type of the Marsupial group, which represented the Pachyderms of the larger continents, and which seems now to have disappeared from the face of the Australian earth,—of the great quadruped, so indicated under the name of *Diprotodon* in 1838; and successive subsequent acquisitions have established the true marsupial character and the near affinities of the genus to the Kangaroo (*Macropus*), but with an osculant relationship with the herbivorous Wombat. The entire skull of the *Diprotodon*, lately acquired by the British Museum, shows *in situ* the tooth on which the genus was founded. This skull measures 3 feet in length, and exemplifies by its size the huge dimensions of the primeval Kangaroo. Like the contemporary gigantic Sloth in South America, the *Diprotodon* of Australia, while retaining the dental formula of its living homologue, shows great and remarkable modifications of its limbs. The hind pair were much shortened and strengthened compared with those of the Kangaroo; the fore pair were lengthened, as well as strengthened. Yet, as in the case of the *Megatherium*, the ulna and radius were maintained free, and so articulated as to give the fore paw the rotatory actions. These, in *Diprotodon*, would be needed, as in the herbivorous Kangaroo, by the economy of the marsupial pouch. The dental formula of *Diprotodon* was the same as in *Macropus major*: the first of the grinding series was soon shed, but the other four two-ridged teeth were longer retained; and the front upper incisor was very large and scalpriform, as in the Wombat. The zygomatic arch sent down a process for augmenting the origin of the masseter muscle, as in the Kangaroo. The foregoing skull, with parts of the skeleton of the *Diprotodon australis*, were discovered in a lacustrine deposit, probably pleistocene, intersected by creeks, in the plains of Darling Downs, Australia.

" The same formation has yielded evidence of a somewhat smaller extinct herbivorous genus (*Nototherium*), combining, with essential affinities to *Macropus*, some of the characters of the Koala (*Phascolarctos*). The writer has recently communicated descriptions and figures of the entire skull of the *Nototherium Mitchelli* to the Geological Society of London. The genus *Phascolomys* was at the same period represented by a Wombat (*P. gigas*) of the magnitude of a Tapir. The pleistocene marsupial Carnivora presented the usual relations of size and power to the Herbivora whose undue increase they had to check."

In another work, Prof. Owen represents an almost entire skull, with part of the lower jaw, of an animal (*Thylacoleo*) rivalling the Lion in size, the marsupial character of which is demonstrated by the position of the lacrymal foramen in front of the orbit, by the palatal vacuity, by the loose tympanic bone, by the development of the tympanic bulla in the alisphenoid, by the very small relative size of the brain, and other characters. "The carnassial tooth is 2 inches 3 lines in longitudinal extent, or nearly double the size of that in the Lion. The upper tubercular tooth resembles, in its smallness and position, that in the placental Felines. But in the lower jaw the carnassial is succeeded by two very small tubercular teeth, as in *Plagiaulax*; and there is a socket close to the symphysis of the lower jaw of *Thylacoleo*, which indicates that the canine may have terminated the dental series there, and have afforded an additional feature of resemblance to the *Plagiaulax*."

As might naturally be expected, the climate of a country which extends over more than 30 degrees of latitude is very much diversified. Cape York and Arnheim's Land are as near 11° south as possible, while Wilson's Promontory, in Victoria, reaches 39°, and the southern part of Tasmania 44¼°. The parts of Australia approaching the Tropic differ very considerably from its southern portions; for, lying more to the north, the latter are under the influence of monsoons, and rains more or less regular occur in their proper seasons. Speaking generally, however, Australia may be characterized as one of the driest and most heated countries of our globe; for, although an island in the strictest sense of the word, it is so extensive that the surrounding seas have little influence upon the distant interior, which must still be regarded as a great sterile waste, destitute of mountains sufficient to attract the moisture requisite to form navigable or other rivers. In writing this in 1863, when travellers have crossed the country and so many valuable discoveries have lately been made, I am willing to admit that this great desert is here and there relieved by higher lands which will ultimately become useful to the enterprising settler, and that, in all probability, many fine and extensive oases have yet to be brought to light; but, at the same time, I believe there will always be considerable uncertainty in the seasons of the interior of this great land. In southern latitudes we know that this is the case, while in the north a wet or a dry monsoon greatly alters the face of the country, and exerts a powerful influence on animal and vegetable life. Hence it is that the scanty fauna of this part of Australia is so organized that it is able to exist without water: the various species of Rodents, such as the members of the genera *Mus* and *Hapalotis*, and the Wombats, Lagorchestes, and Bettongias, and other Kangaroos, are thus constituted; and it will be recollected that, when speaking of the Halcyons and other large Kingfishers in the 'Birds of Australia,' I stated that I believed they never partook of this element, their food consisting of lizards and insects, to which, in like manner, it was not essential. The Australian mammals must, however, be put to severe straits occasionally, not from the want, but from the superabundance of water,—a wet monsoon in the north, and the heavy rains which occasionally occur in the south, deluging the basin-like surface of the interior and rendering it untenable, and obliging them to retire to the higher ridges until the drought, which generally ensues, has restored it to its normal condition. The districts, or countries as I may call them, which constitute the other portions of Australia are very different, indeed completely opposite in character; I mean the rich lands which surround nearly the whole of the sterile centre. The mountain-ranges, of no very great elevation it is true, exert much influence upon the face of nature, constantly attracting rains, which, pouring down their sides, deposit a rich alluvial soil, favourable to the growth of gigantic trees and the most luxuriant vegetation. The forests of Palms which there occur are scarcely inferior to those of any other country, while the stately native Cedars and Fig-trees are wonders to every traveller. These giants of the forest are scarcely ever to be found in the interior; sterility is not suited to their existence; they do not occur in company with the *Banksiæ*, the *Hakeæ*, or the *Casuarinæ*, most of which are characteristics of land wherein the settler would not choose to risk his fortune.   The great physical features of Australia, then, as a whole, are the absence of high mountains and navigable rivers, its heated interior, its vast grassy plains, and its luxuriant brushes, particularly on its southern and south-eastern coasts.   Over the whole of this extensive country, with its ever-varying climate, certain groups of animals are universally spread, while others, particularly the more isolated forms, are strictly confined to their own districts, each adapted for some special end and purpose,—as much as the long bill of the Humming-bird (*Docimastes ensiferus*) is evidently formed for exploring the lengthened tubular corollas of the *Brugmansiæ*, or the greatly curved bill of two species of the same family of birds (the *Eutoxeres Aquila* and *E. Condaminei*) is for insertion into the honey-cups of the *Coryanthes speciosa* and its allies,—or, to take a more striking instance, as the brush-like tongues of the numerous honey-feeding Parrakeets and Honey-eaters of Australia are constituted for obtaining the nectar from the flowers of the universally spread and equally numerous *Eucalypti* which form so prominent a feature in the flora of that country.

I will now give, as far as my knowledge of the subject will permit, an enumeration of Australian mammals, the extent of their range, &c.   In doing this, I shall commence with the Monotrematous section of the *Marsupiata*, which includes the *Ornithorhynchus* and two species of *Echidna*; I shall then proceed to the genera *Myrmecobius, Tarsipes, Chœropus, Peragalea, Perameles, Phascolarctos, Phalangista, Cuscus, Petaurista, Belideus, Phascogale, Sarcophilus, Dasyurus, Thylacinus,* and *Phascolomys*; and these will be followed by the great family of Kangaroos, with remarks upon their structural differences and the especial object for which these appear to have been designed; next we shall come to the feebly represented Placentals, the Seals, and Rodents; and lastly, to the species of *Pteropus* and other Bats.

I have considered that, in a large illustrated work like the 'Mammals of Australia,' it would be out of place to enter into the anatomy of the objects I have represented.   I have therefore omitted all details of this kind; neither have I included therein a repetition of the generic characters and Latin descriptions which have appeared in general works on Mammalogy, where they may be easily referred to.   Those who wish to enter more fully into the generic characters of the Australian mammals will find all the information they can wish for in Mr. Waterhouse's valuable work, entitled 'A Natural History of the Mammalia,' a publication of such great promise and merit, that it becomes a matter of surprise and regret to all interested in this branch of science that the publisher decided upon not continuing it to its completion.

It will be observed that I have entirely omitted the Whales, Porpesses, and Dugong, my reason for so doing being that I had not sufficient opportunities for studying those animals in a state of nature, and therefore have not attempted that which I did not understand, and consequently could not have accomplished in a satisfactory manner.   With regard to the Dugong, I must not omit thanking my relative, Charles Coxen, Esq., of Queensland, for his attention in sending me a skin and part of the skeleton of this animal; but even with these materials I found I could not produce an accurate representation of it in the living state.   Although I do not inflict upon my readers the characters and distinctions of genera, I must not pass over unnoticed the principal features which distinguish the *Marsupiata* from the Placental Mammalia.   In the first place, the former are considered to be much less highly organized than the latter: according to Professor Owen, the brain is deficient in both the corpus callosum and the septum lucidum; the cerebrum is small in proportion

d

to the animal, contracted in front, and its surface is smooth, or presents but few convolutions; the cerebellum is entirely exposed, and has a vermiform process large in proportion to the lateral lobes; the olfactory lobes are large. Two venæ cavæ enter the heart; "the right auricle has no trace of a fossa ovalis." In point of fact, the main characteristic of the Marsupials, as distinguished from the Placentals, is that much of the embryotic life in the former is carried on in what may be called a sort of external uterus.

On my return from Australia, the venerable Geoffroy St.-Hilaire put the following question to me, "Does the Ornithorhynchus lay eggs?" and when I answered in the negative, that fine old gentleman and eminent naturalist appeared somewhat disconcerted. Now, this oviparous notion was nearly in accordance with the true state of things—somewhat akin to what is actually the case; and I consider the most striking peculiarity of this singular animal, and indeed of all the *Mursupiata*, to be the imperfectly formed state in which their young are born. The Kangaroo at its birth is not larger than a baby's little finger, and not very unlike it in shape: in this extremely helpless state, the mother, by some means at present unknown, places this vermiform object to one of the nipples within her pouch or marsupium; by some equally unknown process, the little creature becomes attached by its imperfectly formed mouth to the nipple, and there remains dangling for days, and even weeks, during which it gradually assumes the likeness and structure of its parents; at length it drops from this lacteal attachment into the pouch, re-attaches itself when hunger prompts it so to do, and as often again tumbles off when its wants have been supplied. It is scarcely necessary to say that, after gaining sufficient strength, it leaves this natural pocket of the mother, leaps into the open air and sports about the plains or the forest, as the case may be, and returns again to its warm home, until at length the wearied mother denies it this indulgence and proceeds again to comply with the law which governs all creatures, that of reproduction. This is a very low form of animal life, indeed the lowest among the Mammalia, and exhibits the first stage beyond the development of the bird.

This description has reference not only to the Kangaroos, which mostly have but one young at a time, but is equally descriptive of the other members of this group, some of which have two, while others have three or four, and others, the *Phascogalæ* for instance, eight or nine at a birth; but in all cases, even with these large numbers, the young hang to the mammæ in the way I have described.

Independently of the low structure of the brain and the low form of reproduction of the Kangaroos, I ought to mention that two little bones have been expressly provided for the support of the marsupium; there is also a considerable difference in the dentition, as well as in the form of the lower jaw, by which this group of animals may at all times be distinguished. I have not failed to notice much disparity in size in the *Marsupiata*; they seem to be always growing; for the males get larger and still larger for years, even long after they have commenced the duty of reproduction, and hence individuals of all sizes occur, and occasionally one extraordinarily large may be met with. I have observed this to occur with all the Marsupials, but particularly among the Kangaroos. The great herds of the grey species, *Macropus major*, are frequently headed by an enormous male, or Boomer as he is called. Like the "rogue Elephants" of Ceylon, these patriarchs are often solitary, and are generally very savage.

Commencing with the most lowly organized of the Australian mammals, I may state that the *Ornitho-rhynchus* has a very limited range, as is shown by its not being found either in Western or Northern Australia—the south-eastern portions of the continent and Van Diemen's Land being the localities to which it is confined.

The spiny *Echidna hystrix* has not yet been found to the northward of Moreton Bay on the east coast, and, except in New South Wales and the islands in Bass's Straits, it is very rare—so rare indeed, that I have never seen a specimen from South Australia; yet in all probability it will be found there, since Mr. Gilbert obtained an example at Swan River; this individual, however, did not come under my notice, and I am therefore unable to say if it were a true *E. hystrix*, or a western representative of that species.

The more hairy *Echidna setosa* is confined to Van Diemen's Land; but it is questionable whether it be really distinct from *E. hystrix*; the more southern position and colder climate of that island may have had the effect of giving it a warmer coat, whiter spines, and of altering its general appearance.

The single species representing the genus *Myrmecobius* (*M. fasciatus*) appears to be more plentiful in the Swan River Settlement than elsewhere; it nevertheless occurs in the Murray Scrub and other parts of South Australia, and from thence to the western coast it probably inhabits every locality suited to its habits and mode of life.

Like the *Myrmecobius*, the little honey-lapping *Tarsipes rostratus* stands quite alone—and a truly singular creature it is: to give the area over which it ranges is impossible, as we know far too little of these diminutive mammals to come to any positive conclusion on this point; at present, the neighbourhood of King George's Sound is one of the localities in which it has been seen in a state of nature.

Isolated in form and differing in the structure of its feet from every other known quadruped is the *Chœropus*, an animal which frequents the hard grounds of the interior, over which it is dispersed from New South Wales to Western Australia. The specific term of *ecaudatus*, first applied to this animal in consequence of the specimen characterized being destitute of the caudal appendage, must now sink into a synonym, that organ being as well developed in this as in any other of the smaller quadrupeds, the *Perameles* for instance, to which this singular animal is somewhat allied.

The root-feeding Dalgyte, or *Peragalea lagotis*, leads us still nearer to the genus *Perameles*: the fauna of Western Australia is greatly enriched by the addition of this beautiful species. I believe that South Australia may also lay claim to it; for I have seen a tail, said to have been obtained on the south coast, which greatly resembled that of the Swan River *Peragalea*; but it may have pertained to an allied animal with which we are not yet acquainted.

The members of the restricted genus *Perameles* are numerous in species, and universally dispersed over the whole of Australia and Van Diemen's Land; they also extend in a northerly direction to New Guinea and the adjacent islands. Of this genus there are two well-marked divisions: one distinguished by bands on their backs or crescentic markings across their rumps and by their diminutive tails, the other by a uniformity in their colouring. The species of the former division inhabit the hot stony ridges bordering the open plains; those of the latter the more humid forests, among grass and other dense vegetation. Figures of most of these Bandicoots, as they are called, and an account of the manners, habits, and economy of each, so far as known, will be found in their proper places in the body of the work.

The Phascogales, of which there are three, namely *P. penicillata*, *P. calura*, and *P. lanigera*, are all natives of the southern portions of Australia, from east to west; they are, however, rather denizens of the interior than of the provinces near the coast, but the *P. penicillata* is alike found in both. Their dentition indicates that they are sanguinary in their disposition,—a character which is confirmed by the *P. penicillata*, small as it comparatively is, being charged with killing fowls and other birds.

It might be thought that the *Phascogalæ* would naturally lead to the *Antechini*, but there is no real affinity between the two groups. I find it most difficult to arrange the Australian mammals in anything like a serial order; but the numerous species forming the genera *Antechinus* and *Podabrus* are, perhaps, as well placed here as elsewhere. Like the *Peramelides*, the members of those genera inhabit every part of Australia and the adjacent islands: the thick-tailed species, forming the genus *Podabrus*, frequent the interior rather than the coast; the *Antechini*, on the other hand, inhabit both districts; and wherever there are trees and shrubs, one or other of them may be found; some evince a partiality for the fallen boles lying on the ground, while others run over the branches of those that are still standing.

I now approach a better-defined section of the Australian Marsupiata than any of the preceding—the nocturnal Phalangers. These are divided into several genera—*Phascolarctos, Petaurista, Belideus, Phalangista, Cuscus, Acrobates,* and *Dromicia*. The extraordinary Koala is only found in the brushes of New South Wales. It stands quite alone—the solitary species of its genus, and it is well worth while to turn to my figures and description of this anomalous Sloth among the Marsupials. The *Petauristæ* are strictly brush-loving animals, and are almost entirely confined to New South Wales; some one or other of the *Belidei*, on the other hand, is found in all other parts of the Australian continent (except perhaps its western portion), wherever there are *Eucalypti* of sufficient magnitude for their branches to become hollow spouts wherein these nocturnes may sleep during the day. This form also occurs among the animals of the New Guinea group of islands. The little Opossum Mouse, *Acrobates pygmæus*, is a general favourite with the colonists; and well it may be so, for in its disposition it is as amiable as its form is elegant and its fur soft and beautiful: what the Dormouse is to the English boy, this little animal is to the juveniles of Australia. I have seen it kept as a pet, and its usual retreat in the day, while it sleeps, was a pill-box; as night approaches it becomes active, and then displays much elegance in its motions. The true *Phalangistæ* comprise many species; and are found in every colony, in Port Essington on the north, Swan River on the west, New South Wales and Queensland on the east, and Victoria and Van Diemen's Land on the south. They lead to the genus *Cuscus*, a form better represented in New Guinea and its islands than in Australia, where only one species has been discovered, in the neighbourhood of Cape York. Of the two fairy-like *Dromiciæ*, which live upon the stamens of flowers and the nectar of their corollas, one is found in Van Diemen's Land, the other in Western Australia. The description of a third species of this form has just been transmitted to the Zoological Society by Mr. Krefft, who states that it was taken from an example discovered by himself in New South Wales, and proposes to call it *D. unicolor*.

An equally remarkable and distinct division or group is composed of the Dasyures, to which the extraordinary *Sarcophilus ursinus* of Van Diemen's Land bears precisely the same degree of relationship that the Koala does to the Phalangers. Like the *Thylacinus*, the *Sarcophilus* is confined to Van Diemen's Land. And I would ask, why are these strange and comparatively large animals now restricted to so limited an area? for it can scarcely be supposed that they have not, at some time or other, inhabited the continent of Australia also. Had not Tasmania as well as the mainland been peopled for a long time by the human race, it might have been supposed that their extirpation from the continent had been effected by these children of nature. Whatever the cause may have been, it cannot now be ascertained, and we must be content to treat of the creatures that still exist. Of the true Dasyures, four very distinct species are dispersed over Australia from Van Diemen's Land to the shores of Torres' Straits. Tasmania is frequented by two (*Dasyurus maculatus* and *D. viverrinus*), the southern parts of the mainland by the same two species with the addition of a third (*D. Geoffroyi*), while the *D. hallucatus* inhabits the north. The animals of this genus are very viverrine both in their appearance and in their sanguinary disposition, and are probably the true representatives in Australia of that group of quadrupeds. The term 'sanguinary' is rightly applied to some of these animals, yet there is not one which a child might not conquer. The boldest of them are more troublesome than dangerous, and a robbery of the hen-roost is the utmost of the depredations their nature prompts them to commit.

I now come to the most bloodthirsty of the Australian mammals—the Wolf of the Marsupials—the *Thylacinus* of Tasmania's forest-clad country—the only member of its Order which gives trouble to the shepherd or uneasiness to the stockholder. Van Diemen's Land is the true and only home of this somewhat formidable beast, which occasionally deals out destruction among the flocks of the settler, to which it evinces a decided preference over the Brush Kangaroos, its more ancient food. To man, however, it is not an object of alarm; for the shepherd, aided by his dog, and stick in hand, does not for a moment hesitate about attacking and killing it. The large life-sized head and the reduced figures given in the body of the work well represent the *Thylacinus*, and all that is known of its habits will be found in the accompanying letter-press.

Until lately, only one species of *Phascolomys* or Wombat was clearly defined; but we now know that there are three, if not four, very distinct kinds; and in all probability others may yet be discovered, and prove that this form has a much more extended range than is at present supposed. The *P. Wombat* is still abundant in Van Diemen's Land and on some of the islands in Bass's Straits; and two or three species burrow in the plains of the southern countries of Australia generally. These huge, heavy, and short-legged animals, revelling in a state of obesity, feed most harmlessly on roots and other vegetable substances; they are the Rodents of their own Order, and the representatives of the Capybaras of South America. With this group I terminate the first volume; the next is devoted to the great family of the *Macropodidæ* or Kangaroos. This, the most important of all the Marsupial groups, both as to diversity of form and the number of species, is so widely and so universally dispersed over the Australian continent and its islands, that its members may be said to exist in every part of those countries. They are found in great abundance in the southern and comparatively cold island of Tasmania, while three species, at least, tenant that little-explored country, New Guinea, and some of the adjacent islands. Varied as the physical condition of Australia really is, forms of Kangaroos are there to be found peculiarly adapted for each of these conditions. The open grassy plains, sometimes verdant, at others parched up and sterile, offer an asylum to several of

e

the true *Macropi*; the hard and stony ridges and rocky crowns of the mountains are frequented by the great Osphranters; precipitous rocks are the home of the Petrogales; the mangrove-swamps and dense humid brushes are congenial to the various *Halmaturi*; in the more spiny brigaloe-scrubs the *Onychogaleæ* form their runs, and fly before the shouting of the natives when a hunt is the order of the day; among the grassy beds which here and there clothe the districts between the open plains and the mountain-ranges—the park-like districts of the country—the *Lagorchestes* sit in their " forms," like the Hare in England; and the *Bettongiæ* and *Hypsiprymni* shroud themselves from the prying eye of man and the eagle in their dome-shaped grassy nests, which are constructed on any part of the plains, the stony ridges, and occasionally in the open glades among the brushes. The species inhabiting New Guinea (the *Dendrolagus ursinus* and *D. inustus*) resort to the trees, and, monkey-like, ascend and live among the branches. Of the Filander of the same country we know little or nothing. How wonderfully are all these forms adapted to a separate and special end and purpose—an end and a purpose which cannot be seen to advantage in any but a compara-tively undisturbed country like Australia—a part of the world's surface still in maiden dress, but the charms of which will ere long be ruffled and their true character no longer seen! Those charms will not long survive the intrusion of the stockholder, the farmer, and the miner, each vying with the other to obliterate that which is so pleasing to every naturalist; and fortunate do I consider the circumstances which induced me to visit the country while so much of it remained in its primitive state.

I must revert to the Kangaroos; for it will be necessary to point out the situations affected by the various genera. In the body of the work three species of true *Macropi* are figured, and others are described, but not represented. These are all inhabitants of the southern districts of Australia and Van Diemen's Land. To say that no true *Macropus*, as the genus is now restricted, will be found in Northern Australia would be somewhat unwarrantable; at the same time, I have never seen an example from thence. The genus *Osphranter*, on the other hand, the members of which, as has been before stated, are always found in rocky situations, have their representatives in the north as well as in the south, but they are not found in Van Diemen's Land. The splendid *O. rufus* is an animal of the interior, and frequents the plains more than any other species of its genus. At present, the back settlements of New South Wales, Queensland, Victoria, and South Australia are the only countries whence I have seen specimens. The Great Black Wallaroo (*O. robustus*) forms its numerous runs among the rocks, and on the summits of mountains bordering the rivers Mokai and Gwydyr. The *O. Parryi* ranges over the rocky districts of the headwaters of the Clarence and adjacent rivers, while the *O. antilopinus* is as yet only known in the Cobourg Peninsula.

The smaller *Petrogalæ* differ from all the other Kangaroos, both in the form of their feet and the structure of their brushy dangling tails. With the exception of Tasmania, these rock-lovers dwell every-where, from north to south, and from east to west. The *P. penicillata* inhabits New South Wales; the *P. xanthopus*, South Australia; the *P. lateralis*, Western Australia; the *P. concinna* and *P. brachyotis*, the north-west coast; and the *P. inornata*, the opposite rocky shores of the east.

The true Wallabies, or *Halmaturi*, are all brush animals, and are more universally dispersed than any of the other members of the entire family. Tasmania is inhabited by two species, New South Wales by at least five, South Australia by two or three, and Western Australia by the same number; while the genus is represented on the north coast by the *H. agilis*. It will be clear, then, that the arboreal districts of the south, with their thick and impenetrable brushes, are better adapted for the members of this genus than the hotter country of the north.

The *Onychogaleæ* are, *par excellence*, the most elegantly formed and the most beautifully marked members of the whole family, and they are, moreover, as graceful in their actions as in their colouring they are pleasing to the eye.   One species, the *O. frænata*, inhabits the brigaloe-scrubs of the interior of New South Wales and Queensland, and probably South Australia.   The *O. lunata* plays the same part, and affects very similar situations, in Western Australia; while the *O. unguifera*, as far as we yet know, is confined to the north-eastern part of the continent.

The *Lagorchestes* are a group of small hare-like Kangaroos, which dwell in every part of the interior of the southern portion of the mainland, from Swan River on the west to Queensland on the east; one species has, however, been found in the northern districts—the *L. Leichardti*, as it has been named, in honour of its discoverer, the late intrepid and unfortunate explorer, Dr. Leichardt.   They are the greatest leapers and the swiftest runners among small animals I have ever seen; they sleep in forms, or seats, like the Common Hare (*Lepus timidus*) of Europe, and mostly affect the open grassy ridges, particularly those that are of a stony character.   The beautiful *L. fasciatus* of Swan River is one of the oldest known; the *L. Leichardti* the latest yet discovered.

The *Bettongiæ*, with their singular prehensile tails, also enjoy a wide range, the various species composing the genus being found in Tasmania, New South Wales, Southern and Western Australia, but, so far as we yet know, not in the north.   For a more detailed account of the localities favoured with the presence of these animals, and the manner in which their prehensile tails are employed in carrying the grass for their nest, I must refer to the history of the respective species, and particularly to the plate of *Bettongia cuniculus*.

The *Hypsiprymni* are the least and, perhaps, the most aberrant group of this extensive family.   They inhabit the southern and most humid parts of the country, and are to be found everywhere, from Tasmania to the 15th degree of latitude on the continent in one direction, and from the scrubs of Swan River and King George's Sound to the dense brushes of Moreton Bay in the other; like most other Kangaroos, they are nocturnal in their habits, grub the ground for roots, and live somewhat after the manner of the *Peramelides*, with which, however, they have no relationship.

To render my history of this group of animals the more complete, I have included in the work the three species inhabiting New Guinea: two of these belong to the genus *Dendrolagus*, and, as their name implies, dwell among the branches of trees, and rarely resort to the ground: the third forms the genus *Dorcopsis*, of which a single species only is known; it has doubtless some peculiar habits, but these must be left for a future historian to describe; at present they are unknown.

The great family of the Kangaroos, of which what I have here written must only be regarded as a slight sketch, is well worthy the study of every mammalogist.   It forms by far the most conspicuous feature in the history of Australian quadrupeds; and, numerous as are the species now known, I doubt not that others will yet be discovered when the north and north-western provinces of the country have been more diligently explored.

The third and concluding volume is devoted to the Rodents, Seals, and Bats, and ends with the *Canis Dingo*. These are the only Placental animals inhabiting the land of Australia, and, contrary to what was formerly supposed, the Rodents form no inconspicuous feature among the quadrupeds of that country. They are very numerous in species, and almost multitudinous in individuals. Every traveller who has visited the interior can testify to this fact. If exploration has been his object, the numerous runs and tracks of these little animals must have been frequently presented to his notice,—every grassy bed being tenanted by its own species of *Mus*, while all the sand-hills are run over by the same or other species, interspersed with the Jerboa-like *Hapalotides*. The sluggish river-reaches and water-holes of nearly every part, from Tasmania through all the southern portions of the continent, have their muddy banks traversed by the *Hydromys*, or Beaver-Rats, as they have been very appropriately called. Even New Zealand, a country which it was formerly supposed never had a more highly organized indigenous creature than a bird, has its Bats; it will not be surprising, therefore, that the sister country of Australia should be tenanted by numerous species of these Nocturnes; not only are they individually very plentiful, but many distinct forms or genera are there found. The brushes which abound in fruit-bearing fig-trees are frequented by Vampires or *Pteropi*—a form which appears to be mainly confined to the south-eastern and northern portions of the country, for I have not yet seen any examples from Tasmania, or Southern or Western Australia. The trees in this strange country which bear either fruit or berries are very few. Even the fruit of the stately parasitic Fig is a mere apology for that which we are accustomed to see, and hence but few species of these great frugivorous Bats occur in the fauna of Australia. At the same time, the paucity of species is amply compensated by the number of individuals; these, however, are confined to the brushes which stretch along the eastern coast. In these solitary forests they teem and hang about in thousands, frequently changing their *locale* when their food becomes scarce or has been entirely cleared off. The species I more particularly allude to is the *Pteropus poliocephalus*. The Cobourg Peninsula and other parts of the north coast are also inhabited by a species which, according to Gilbert and Leichardt, is very abundant. A third and very fine one frequents Fitzroy Island, lying off the eastern coast.

The extraordinary *Molossus australis* is a native of Victoria, and is the sole species of its genus yet discovered in Australia. The *Taphozoi* appear to be rock-loving Bats, and the single species as yet discovered is from the Peninsula of Cape York. The *Scotophili*, of which there are several species, are found in all parts of the country, from Van Diemen's Land to the most northern part of the continent.

The restricted genus *Vespertilio* is more feebly represented than the last-mentioned form, since only two species are known to exist in the country; these are very generally spread over the southern coast.

Of the leaf-nosed *Rhinolophi* I have figured three species—the *R. cervinus*, from Cape York, the *R. aurantius* (a very beautiful species) from North-western Australia, and the *R. megaphyllus* from New South Wales.

The *Nyctophili*, or Long-eared Bats, are well represented; four species, at least, frequenting every part of the continent from east to west, and also the island of Tasmania.

This, I am aware, is a very imperfect *résumé* of the *Cheiroptera* inhabiting Australia; could I have rendered it more complete, I would have done so; but it must be recollected that seventh-tenths of the country are yet unexplored.

A mere glance at the globe which stands in every school-room will show how greatly the sea preponderates over the land of this planet. Like the land, the ocean is tenanted by many remarkable animals, certain groups of which exist in one hemisphere and are not found in the other; and it is not often that even the great Cetaceans occur in both. Neither do the Seals: the equatorial region separates them most completely; that is, no species is common alike to the north and the south. I do not consider that either the Australian *Cetacea* or *Phocidæ* have been well made out, and this certainly is the part of the mammalian fauna of that country of which we know the least. I have omitted the former altogether, but it will be seen that I have figured two of the latter; these constitute two genera (*Stenorhynchus* and *Arctocephalus*); they both inhabit the shores and rocky islands of the southern portion of Australia, while the Dugong (*Halicore australis*) is, as far as I am aware, a native of the east coast only.

Whether the *Canis Dingo* be really indigenous, or has at some very remote period followed Man in his migrations, is a question on which naturalists are at variance. For my own part, I am inclined to the latter theory, as being the most philosophic mode of accounting for its presence there. That Man is the latest visitant to the soil of Australia there can be little doubt; the country is far too sparsely provided with fruits and other substances necessary for his existence to favour a contrary hypothesis.

In the following list of the Australian Mammals I shall refer to the volumes in which they are contained and to the plates on which they are respectively figured, and shall, moreover, give any additional information I may have acquired respecting them, together with an account of the new species which have been described by other writers, but which, from my not having been able to see examples, I have not figured.

Order MARSUPIATA.

Section MONOTREMATA.

Genus ORNITHORHYNCHUS, *Blumenb.*

1. Ornithorhynchus anatinus . . . . Vol. I. Pl. 1.
   Habitat. New South Wales and Tasmania. Victoria and South Australia?

Genus ECHIDNA, *Cuv.*

2. Echidna hystrix . . . . Vol. I. Pl. 2.
   Habitat. New South Wales, Victoria, the islands in Bass's Straits. Southern and Western Australia?

3. Echidna setosa, *Cuv.* . . . . Vol. I. Pl. 3.
   Habitat. Van Diemen's Land.

f

Genus MYRMECOBIUS, *Waterh.*

4. Myrmecobius fasciatus, *Waterh.* . . . . Vol. I. Pl. 4.
   Habitat. Western Australia, and parts of South Australia.

Genus TARSIPES, *Gerv. et Verr.*

5. Tarsipes rostratus, *Gerv. et Verr.* . . . . Vol. I. Pl. 5.
   Habitat. Western Australia.

Mr. Waterhouse is of opinion that this animal is most nearly allied to the *Dromiciæ*, yet he has not placed it near that form in his ' History of the Mammalia.'

Genus CHŒROPUS, *Ogilby.*

6. Chœropus castanotis, *Gray* . . . . Vol. I. Pl. 6.
   Habitat. Interior of New South Wales, South and Western Australia.

Genus PERAGALEA, *Gray.*

7. Peragalea lagotis . . . . Vol. I. Pl. 7.
   Habitat. Western Australia.

Genus PERAMELES, *Geoff.*

8. Perameles fasciata, *Gray* . . . . Vol. I. Pl. 8.
   Habitat. Interior of South Australia, Victoria, and New South Wales.

9. Perameles Gunnii, *Gray* . . . . Vol. I. Pl. 9.
   Habitat. Van Diemen's Land.

10. Perameles myosurus, *Wagn.* . . . . Vol. I. Pl. 10.
    Habitat. Western Australia.

11. Perameles nasuta, *Geoff.* . . . . Vol. I. Pl. 11.
    Habitat. New South Wales.

12. Perameles macroura, *Gould.*
    Perameles macroura, Gould in Proc. of Zool. Soc. part x. p. 4 ; Waterh. Nat. Hist. of Mamm. vol. i. p. 366.
    Perameles macrurus, Gray, List of Spec. of Mamm. in Coll. Brit. Mus. p. 96.

I have not figured this animal because, although twenty-one years have passed away since my description was published, I have never seen a second example ; still I have no doubt of its being a distinct species. It greatly resembles *P. obesula* and *P. nasuta,* but differs from both in its larger tail. I transcribe my original description from the ' Proceedings of the Zoological Society ' above referred to :—

" *Corpore supra nigro et flavescenti-albo penicillato, infra sordide albo, pilis rigidis obsito ; cauda pilis parvulis parce tecta, longitudine dimidio corporis æquante, supra nigra, infra fuscescenti-alba ; auris mediocribus.*

	unc.	lin.
" Longitudo ab apice rostri ad basin caudæ	16	3
———— caudæ	7	3
———— ab apice rostri ad basin auris	3	4
———— tarsi digitorumque	3	1
———— auris	1	2

" *Habitat.* Port Essington."

13. Perameles obesula, *Geoff.* . . . . Vol. I. Pl. 12.
    Habitat. South coasts of Australia and Tasmania generally.

14. Perameles Bougainvillei, *Quoy et Gaim.*

Perameles Bougainvillei, Quoy et Gaim. Zool. du Voy. de l'Uranie, p. 56, tab. 5, et Bull. des Sci. Nat. 1824, tom. i. p. 270 ; Waterh. Nat. Hist. of Mamm. vol. i. p. 385.

Habitat. Péron's Peninsula ; in Shark Bay, Western Australia.

Having never seen a specimen of this animal, I am unable to figure it, or to say if it be a good species.

Genus PHASCOLARCTOS, *De Blainv.*

15. Phascolarctos cinereus . . . . Vol. I. Pls. 13 & 14.
Habitat. New South Wales.

Genus PHALANGISTA, *Cuv.*

16. Phalangista fuliginosa, *Ogilby* . . . . Vol. I. Pl. 15.
Habitat. Van Diemen's Land. Victoria ?

In one of the letters from my son Charles, now engaged on a geological survey of Tasmania, the following passage having reference to this animal occurs :—

" I lay down, looking up at the moon and stars, thinking of home, and dreamily listening to the crackling of the fire, when a diabolical, chattering, grunting laugh overhead makes me start up, and discover that a Sooty Opossum is making an inspection of me, with comments, from the branch above ; his call is responded to by others, and a kind of concert commences, which is maintained at intervals throughout the night,—the smaller or Ring-tailed Opossums performing an active part in it also, and the ' More Pork ' (*Podargus Cuvieri*) lending a little lugubrious assistance occasionally."

17. Phalangista vulpina, *Desm.* . . . . Vol. I. Pl. 16.
Phalangista melanura, Wagn., Waterh. Nat. Hist. of Mamm. vol. i. p. 288.
————— *felina,* Wagn., Waterh. *ib.* p. 294.
Goö-mal, aborigines of Western Australia.
Habitat. Probably every part of Australia ; certainly all its southern portions.

18. Phalangista canina, *Ogilby* . . . . Vol. I. Pl. 17.
Habitat. New South Wales.

19. Phalangista Cookii, *Desm.* . . . . Vol. I. Pl. 18.
Ngö-ra, aborigines of Perth.
Ngork, aborigines of King George's Sound.

"This species," says Mr. Gilbert, " does not confine itself to the hollows of standing or growing trees, but is often found in holes in the ground, where the entrance is covered with a stump ; it is frequently hunted out of such places by the Kangaroo-dogs. It varies very much in the colour of the fur, from a very light grey to nearly a black ; in one instance I caught two, from the same hole, which exhibited the extremes of these colours."
Habitat. New South Wales.

20. Phalangista viverrina, *Ogilby* . . . . Vol. I. Pl. 19.
Habitat. Van Diemen's Land and Western Australia.

21. Phalangista laniginosa, *Gould* . . . . Vol. I. Pl. 20.
Habitat. New South Wales.

Genus CUSCUS, *Lacép.*

22. Cuscus brevicaudatus, *Gray* . . . . Vol. I. Pl. 21.
Habitat. The Cape York district.

Genus PETAURISTA, *Desm.*

23. Petaurista Taguanoïdes, *Desm.* . . . . Vol. I. Pl. 22.
    Habitat. New South Wales.

Genus BELIDEUS, *Waterh.*

24. Belideus flaviventer . . . . Vol. I. Pl. 23.
    Habitat. New South Wales.

25. Belideus sciureus . . . . Vol. I. Pl. 24.
    Habitat. New South Wales and Victoria.

26. Belideus breviceps, *Waterh.* . . . . Vol. I. Pl. 25.
    Habitat. New South Wales and Victoria.

27. Belideus notatus, *Peters* . . . . Vol. I. Pl. 26.
    Habitat. Victoria.

28. Belideus Ariel, *Gould* . . . . Vol. I. Pl. 27.
    Habitat. Cobourg Peninsula, on the north coast of Australia.

Genus ACROBATA, *Desm.*

29. Acrobata pygmæa, *Desm.* . . . . Vol. I. Pl. 28.
    Habitat. New South Wales and Victoria.

By some oversight the name of this species has been spelt on the plate and in the text *Acrobates pygmæus.*

Genus DROMICIA, *Gray.*

30. Dromicia gliriformis . . . . Vol. I. Pl. 29.
    Habitat. Van Diemen's Land.

31. Dromicia concinna, *Gould* . . . . Vol. I. Pl. 30.
    Dromicia Neillii, Waterh. Nat. Hist. of Mamm. vol. i. p. 315?
    Habitat. Western Australia.

32. Dromicia unicolor, *Krefft.*
    Dromicia unicolor, Krefft in Proc. Zool. Soc. Jan. 22, 1863.

" Fur of a uniform mouse-colour, lighter on the sides and beneath, with a blackish patch in front of the eye.

" All the hairs are slate-grey at the base, tipped with yellowish at the back and sides, and with grey beneath; longer black hairs, tipped with white, are interspersed, except on the under side of the body. Bristles black to within one-third of the tip, which is white; a few long bristly black hairs in front and behind the eye. Tail somewhat longer than the body, prehensile, thin, showing every joint; slightly enlarged at the base, and gradually tapering; covered with a mixture of light-coloured and black hairs; apical portion about $\frac{1}{4}''$ from the tip, wide beneath.

		inches.
" Length from tip to tip .	. . . .	$6\frac{1}{4}$
Tail . . .	. . . .	$3\frac{1}{4}$
Face to base of ear	. . . .	$\frac{7}{8}$
Ear . . .	. . . .	$\frac{1}{2}$
Arm and hands .	. . . .	$\frac{7}{8}$
Tarsi and toes .	. . . .	$\frac{4}{5}$

" This beautiful little creature was captured near St. Leonard's North Shore, Sydney, feeding upon the blossoms of the Banksias, and lived a few days in captivity. In its habits it is nocturnal. The tongue of this *Dromicia*

is well adapted for sucking the honey from the blossoms of the *Banksiæ* and *Eucalypti*, being furnished with a slight brush at the tip. This species differs from the *D. concinna* of Western Australia in being of a uniform dark colour, without the white belly, and having the base of the tail slightly enlarged; it is about the same size as *D. concinna*."
Habitat. New South Wales.

Genus PHASCOGALE, *Temm.*

33. Phascogale penicillata . . . . Vol. I. Pl. 31.
Bül-lard, aborigines of King George's Sound.
Habitat. New South Wales, Victoria, South Australia, and Swan River.

34. Phascogale calura, *Gould* . . . . Vol. I. Pl. 32.
King-goor, aborigines of Williams River.
Habitat. Interior of New South Wales and the colony of Victoria.

35. Phascogale lanigera, *Gould* . . . . Vol. I. Pl. 33.
Habitat. Interior of New South Wales.

Genus ANTECHINUS, *MacLeay.*

36. Antechinus Swainsoni . . . . Vol. I. Pl. 34.
Habitat. Van Diemen's Land.

37. Antechinus leucopus, *Gray* . . . . Vol. I. Pl. 35.
Habitat. Van Diemen's Land?

38. Antechinus ferruginifrons, *Gould* . . . . Vol. I. Pl. 36.
Habitat. New South Wales.

39. Antechinus unicolor, *Gould* . . . . Vol. I. Pl. 37.
Habitat. New South Wales.

40. Antechinus leucogaster, *Gray* . . . . Vol. I. Pl. 38.
Habitat. Western Australia.

41. Antechinus apicalis . . . . Vol. I. Pl. 39.
Habitat. Western and Southern Australia.

Mr. George French Angas having sent me a skin of this animal from South Australia, I am enabled to state that its range extends from Western Australia to that colony.

42. Antechinus flavipes . . . . Vol. I. Pl. 40.
Antechinus Stuartii, MacLeay in Ann. & Mag. Nat. Hist. vol. viii. p. 242; Waterh. Nat. Hist. of Mamm. vol. i. p. 416.
Mr. Waterhouse is of opinion that the animal described as *A. Stuartii* will prove to be identical with *A. flavipes.*
Dasyurus minimus, Geoff. Ann. du Mus. tom. iii. p. 362 ? ; Schreb. Säugeth. suppl. tab. 152 B. e?
Phascogale minima, Temm. Mon. de Mamm. tom. i. p. 59?
———— *affinis,* Grey, App. to Grey's Journ. of Two Exp. of Disc. in Australia, vol. ii. p. 406.
———— (*Antechinus*) *minima,* Waterh. Nat. Hist. of Mamm. vol. i. p. 419.
———— *affinis,* Waterh. *ib.* p. 421.
See Mr. Waterhouse's remarks on the animals indicated in the last five synonyms, Nat. Hist. of Mamm. vol. i. pp. 419, 421.
Habitat. New South Wales ; and Victoria ?

43. Antechinus fuliginosus, *Gould* . . . . Vol. I. Pl. 41.
Habitat. Western Australia.

g

44. Antechinus albipes . . . . Vol. I. Pl. 42.
    Habitat. Western Australia.

45. Antechinus murinus . . . . Vol. I. Pl. 43.
    Habitat. New South Wales.

46. Antechinus maculatus, *Gould* . . . . Vol. I. Pl. 44.
    Habitat. Queensland.

47. Antechinus minutissimus, *Gould* . . . . Vol. I. Pl. 45.
    Habitat. Queensland.

Genus PODABRUS, *Gould.*

48. Podabrus macrourus, *Gould* . . . . Vol. I. Pl. 46.
    Habitat. Darling Downs in Queensland.

49. Podabrus crassicaudatus, *Gould* . . . . Vol. I. Pl. 47.
    Habitat. Western and Southern Australia.

Genus SARCOPHILUS, *F. Cuv.*

50. Sarcophilus ursinus . . . . Vol. I. Pl. 48.
    Habitat. Van Diemen's Land.

Genus DASYURUS, *Geoff.*

51. Dasyurus maculatus . . . . Vol. I. Pl. 49.
    Habitat. Van Diemen's Land, New South Wales, and Victoria.

52. Dasyurus viverrinus . . . . Vol. I. Pl. 50.
    Habitat. Van Diemen's Land and Victoria.

53. Dasyurus Geoffroyi, *Gould* . . . . Vol. I. Pl. 51.
    Bur-jad-da, aborigines near Perth.
    Bar-ra-jit, aborigines of York and Toodyay districts.
    Ngoor-ja-na, aborigines of the Vasse district.
    Dju-tytch, aborigines of King George's Sound.
    Mr. Gilbert was informed that the stomach of this animal is frequently found to be filled with white ants.
    Habitat. South portions of the Australian continent generally.

54. Dasyurus hallucatus, *Gould* . . . . Vol. I. Pl. 52.
    Habitat. Northern Australia.

Genus THYLACINUS, *Temm.*

55. Thylacinus cynocephalus . . . . Vol. I. Pls. 53 & 54.
    Habitat. Van Diemen's Land.

Genus PHASCOLOMYS, *Geoff.*

56. Phascolomys Wombat, *Pér. et Les.* . . . . Vol. I. Pls. 55 & 56.
    Phascolomys platyrhinus, Owen, Cat. of Osteol. Ser. in Mus. Roy. Coll. Surg. Engl. p. 334 ?
    Habitat. Van Diemen's Land, and the islands in Bass's Straits.

57. Phascolomys latifrons, *Owen* . . . . Vol. I. Pls. 57 & 58.
    Habitat. Victoria and South Australia.

58. Phascolomys lasiorhinus, *Gould* . . . . Vol. I. Pls. 59 & 60.
Habitat. Victoria and South Australia.

59. Phascolomys niger, *Gould.*
Habitat. South Australia ?

Family MACROPODIDÆ.

Genus Macropus, *Shaw.*

60. Macropus major, *Shaw* . . . . Vol. II. Pls. 1 & 2.
Habitat. New South Wales, Victoria, and Van Diemen's Land.

61. Macropus ocydromus, *Gould* . . . . Vol. II. Pls. 3 & 4.
    Speaking of this animal, Mr. Gilbert states that, "if a female with a tolerably large one in the pouch be pursued, she will often by a sudden jerk throw the little creature out; but whether this be done for her own protection, or for the purpose of misleading the dogs, is a disputed point. I am induced to think the former is the case, for I have observed that the dogs pass on without noticing the young one, which generally crouches in a tuft of grass, or hides itself among the scrub, without attempting to run or make its escape; if the mother evades pursuit, she doubtless returns and picks it up.
    "Those inhabiting the forests are invariably much darker, and, if anything, have a thicker coat than those of the plains. The young are at first of a very light fawn-colour, but get darker until two years old, from which age they again become lighter, till in the old males they become very light grey. In summer their coat becomes light and hairy, while in winter it is of a more woolly character. It is a very common occurrence to find them with white marks or spots of white about the head, more particularly a white spot on the forehead between the eyes. A very curious one came under my notice, having the whole of the throat, cheeks, and upper part of the head spotted with yellowish white; and albinoes have been frequently seen by the hunters."
    Habitat. Western Australia.

62. Macropus fuliginosus . . . . Vol. II. Pl. 5.
Habitat. South Australia.

63. Macropus melanops, *Gould.*
    It will be seen that I have placed this name among the synonyms of *M. major*; but since my remarks on that species were written, I have seen other examples so closely accordant with the animal described by me under the above name in the 10th part of the 'Proceedings of the Zoological Society,' that I think there is a probability it will prove to be distinct, and therefore, for the present, I restore the animal to the rank of a species.
    Habitat. Southern and Western Australia.

Genus Osphranter, *Gould.*

Generic characters.
    Muffle broad and naked; *muzzle* broad and rather short; ears moderate, rounded at the apex; fore limbs comparatively long and stout, and the toes and claws very strong; hind limbs short and muscular; middle toe very large; lateral toes but little developed; two small inner toes, united in one common integument as in other Kangaroos, and terminating in a line with the small outer toe, or nearly so; under surface of the feet very rough, being covered with small horny tubercles.
    The above characters, especially the great expansion of the muzzle, the comparatively small development of the lateral toes of the hind feet, and the greater size of the middle toe, should, in my opinion, be regarded as generic or subgeneric rather than specific; and it was for these reasons that I proposed the new sectional title of *Osphranter.* See Proceedings of Zool. Soc. part ix. p. 80.

64. Osphranter rufus, *Gould* . . . . Vol. II. Pls. 6 & 7.
    Macropus (Osphranter) pictus, Gould in Proc. Zool. Soc. part xxviii. p. 373.
    Habitat. New South Wales, Victoria, and South Australia.

65. Osphranter Antilopinus, *Gould* . . . .   Vol. II. Pls. 8 & 9.
     Habitat. Cobourg Peninsula, Northern Australia.

66. Osphranter Isabellinus, *Gould.*
     Osphranter ? Isabellinus, Gould in Proc. Zool. Soc. part ix. p. 81.
     General colour bright fulvous or sandy red; fur rather short, and soft to the touch; hairs uniform in tint to the base; throat and under parts of the body white, faintly tinted with yellowish in parts; fur of the belly long and very soft; the white or whitish colouring of the under parts and the uniform fulvous colouring of the upper surface and sides of the body not blending gradually; tail similar in colour to the upper surface, but rather paler and uniform; hair of the fore feet and toes brown in front, yellowish on the sides.
     The above description was taken from an imperfect skin procured at Barrow Island, on the north-west coast of Australia, and transmitted to me by Captain Stokes of H.M.S. " Beagle," which, in my opinion, pertains to a species of which no other example has yet been sent to Europe. Under this impression I have bestowed upon it the above specific appellation.
     Habitat. Barrow Island, north-west coast of Australia.

67. Osphranter robustus, *Gould* . . . .   Vol. II. Pls. 10 & 11.
     Habitat. Mountain-ranges of the interior of New South Wales.

68. Osphranter ? Parryi . . . .   Vol. II. Pls. 12 & 13.
     Habitat. Rocky mountains of the east coast of Australia from Port Stephens to Wide Bay.

Genus HALMATURUS, *F. Cuv.*

69. Halmaturus ruficollis . . . .   Vol. II. Pls. 14 & 15.
     Habitat. New South Wales.

70. Halmaturus Bennettii . . . .   Vol. II. Pls. 16 & 17.
     Habitat. Van Diemen's Land.

71. Halmaturus Greyi, *Gray* . . . .   Vol. II. Pls. 18 & 19.
     Habitat. South Australia.

72. Halmaturus manicatus, *Gould* . . . .   Vol. II. Pls. 20 & 21.
     Habitat. Western Australia.

73. Halmaturus Ualabatus . . . .   Vol. II. Pls. 22 & 23.
     Habitat. New South Wales.

74. Halmaturus agilis, *Gould* . . . .   Vol. II. Pls. 24 & 25.
     Habitat. Northern Australia.

75. Halmaturus dorsalis, *Gray* . . . .   Vol. II. Pls. 26 & 27.
     Habitat. Interior of New South Wales.

76. Halmaturus Parma, *Gould* . . . .   Vol. II. Pl. 28.
     Habitat. Brushes of New South Wales.

77. Halmaturus Derbianus, *Gray* . . . .   Vol. II. Pls. 29 & 30.
     Thylogale Eugenii, Gray, Mag. Nat. Hist. vol. i. new ser. 1843, p. 583.
     Habitat. South Australia.

78. Halmaturus Houtmanni, *Gould.*
     Halmaturus Houtmanni, Gould in Proc. Zool. Soc. part xii. p. 31.
     " Of the whole of the islands forming Houtmann's Abrolhos," says Mr. Gilbert, " I found only two to be

inhabited by this species, viz. East and West Wallaby Islands. On both of these they are so numerous, and have been so little disturbed, that they will allow of a very near approach, and may in consequence be obtained in almost any number. The male weighs, on an average, about 12 lbs.; but several old bucks I obtained exceeded this, the heaviest weighing 15 lbs. A mature female weighs about 8 lbs. They appear to have no regular season for breeding, for *all* the females had young ones in the pouch, of very small size and quite naked; and none were seen or killed less than a year old, at which age their weight is about 5 lbs.

"The *Halmaturus Houtmanni* inhabits the dense scrub growing on almost every part of the two islands above mentioned; and its runs cross and recross almost every inch of them—even the sandy beaches close to the water's edge, and among the thick scrub and Mutton-bird holes; in these runs there are little sheltered spots, beneath which they lie during the heat of the midday sun, feeding for the most part during the night. On the approach of man it does not bound off at full speed as other Kangaroos do, but very leisurely takes two or three leaps, and then remaining stationary in an erect position, looks around with evident surprise, and is then easily shot. In fact, from having been so little disturbed, it will allow itself to be run down and caught. I was enabled to catch two in this way. Four or five of my men being on shore, I directed them to surround a bush into which I saw one of these Wallabies run, when the animal, seeing itself approached on all sides, became so bewildered that, instead of attempting to escape, it thrust its head into the thick scrub and allowed us to catch it by the tail.

"One I have alive has a habit of frequently crouching down like a Hare, with its tail brought forward between and before its fore feet."

Adult Male. Face dark grizzled grey, stained with rufous on the forehead; external surface of the ear and the space between the ears dark blackish grey; sides of the neck, shoulders, fore arms, flanks, and hind legs rufous, palest on the flanks; a line of obscure blackish brown passes down the back of the neck and spreads into the dark grizzled brown of the back; throat and chest buffy white; under surface of the body grey; tail grizzled grey, deepening into black on the upper side and the extremity. Fur somewhat short, coarse, and adpressed; the base bluish grey, succeeded by rufous, then white, and the extreme tip black.

Adult Female. Similar in colour to the male, but of a more uniform tint, in consequence of the rufous colouring of the shoulders and flanks being paler, and the grizzled appearance of the back not so bright.

Young. Dark grizzled grey approaching to black, particularly along the back.

	Adult Male. ft. in.	Female. ft. in.
Length from the nose to the tip of the tail	3 6	3 4
———— of tail	1 2¼	1 2
———— of tarsus and toes, including the nail	0 5¾	0 5⅜
———— of arm and hand, including the nails . . .	0 6	0 4
———— of face from the tip of the nose to the base of the ear .	0 4¼	0 4
———— of ear	0 2¼	0 2⅛

Notwithstanding Mr. Waterhouse's opinion that this animal is merely a variety of *H. Derbianus*, and what I have said in my account of that species tending to confirm his view of the subject, I have thought it best to append a copy of my original description taken from the examples sent home by Gilbert. Future research will determine whether it be identical with the *H. Derbianus* or distinct.

Habitat. Houtmann's Abrolhos, Western Australia.

79. Halmaturus Dama, *Gould.*

Halmaturus Dama, Gould in Proc. Zool. Soc. part xii. p. 32.
Dama, aborigines of Moore's River.

Mr. Gilbert states that this animal " is an inhabitant of the dense thickets of the interior, and is so exceedingly numerous that their tracks from thence to their feeding-grounds resemble well-worn footpaths. Its general habits and manners resemble those of the *Halmaturus Houtmanni.* Mr. Johnson Drummond informs me that it makes no nest, but merely squats in a clump of grass like a Hare; that it feeds in the night on the hills; and it is very difficult to procure specimens, as the places it frequents are so dense as to render shooting it almost impossible, nor can a dog even chase it. The only chance of obtaining it is by the aid of the natives, a number of whom

h

walking or, rather, pushing their way through and beating the bush as they go abreast, and loudly shouting ' *wow, wow, wow,*' drive the *Damas* before them, when, by waiting in a clear space, you get the chance of a shot.''

General colour of the fur grizzled brown, becoming of a reddish tint on the back of the neck, arms, and rump; face grey, washed with rufous on the forehead; outside of the ears and the space between them blackish grey; hinder legs light brown; tail grizzled grey; under surface of the body pale grey.

	ft.	in.
Length from the nose to the extremity of the tail	2	11
—— of tail	1	$2\frac{1}{2}$
—— of tarsus and toes, including the nail	0	$5\frac{3}{4}$
—— of arm and hand, including the nails	0	$4\frac{1}{4}$
—— of face from the tip of the nose to the base of the ear . . .	0	4
—— of ear	0	$2\frac{1}{3}$

This animal is closely allied to, and of nearly the same size as *H. Thetidis*, but has much larger ears, and a much more dense and lengthened fur, the base of which is bluish grey, to which succeeds reddish brown, then silvery white, the extreme tips being black.

The above is the description of a female; the male will doubtless prove to be of larger size.

Habitat. Houtmann's Abrolhos and Western Australia.

80. Halmaturus gracilis, *Gould.*

Macropus gracilis, Gould in Proc. Zool. Soc. part xii. p. 103.

Face and all the upper surface of the body grizzled grey and dark brown, the grizzled appearance being produced by each hair being greyish white near the tip; sides of the neck and outer side of the limbs washed with reddish brown; margin of the anterior edge and the base of the posterior edge of the ear buffy white; line from the angle of the mouth dark brown; line along the side of the face, chin, and throat buffy white; under surface buffy grey; tail clothed with short grizzled hairs similar to those of the upper surface of the body, and with a line of black on the upper side at the apex for about one-third of its length; fur somewhat soft to the touch, grey at the base, then brown, to which succeeds white, the points of the hairs being black; there are also numerous long black hairs dispersed over the surface of the body; feet grizzled grey and rufous.

	ft.	in.
Length from the tip of the nose to the tip of the tail	2	6
—— of tail	1	1
—— of tarsi and toes, including the nail	0	5
—— of arm and hand, including the nails	0	$3\frac{1}{4}$
—— of the face from the tip of the nose to the base of the ear . .	0	$3\frac{1}{4}$
—— of the ear	0	$2\frac{1}{4}$

This is a very elegantly-formed little animal. In size it is somewhat smaller than *H. Derbianus*, and has much slighter fore arms.

Gilbert, who had a good knowledge of the Kangaroos, believed this animal to be quite distinct from every other species; and, from a careful examination of the single specimen he sent me, I entertain the same opinion; but I have not figured it because the example alluded to is the only one I have seen.

Habitat. The scrubs of the interior of Western Australia.

81. Halmaturus Thetidis, *F. Cuv. et Geoff.* . . . . Vol. II. Pls. 31 & 32.
 Habitat. Brushes of New South Wales.

82. Halmaturus stigmaticus, *Gould* . . . . Vol. II. Pls. 33 & 34.
 Habitat. North-east coast of Australia.

83. Halmaturus Billardieri . . . . Vol. II. Pls. 35 & 36.
 Habitat. Van Diemen's Land.

84. Halmaturus brachyurus . . . . Vol. II. Pls. 37 & 38.
  Habitat. Western Australia.

Genus PETROGALE, *Gray.*

85. Petrogale penicillata, *Gray* . . . . Vol. II. Pls. 39 & 40.
  Heteropus albogularis, Jourd. Compt. Rend. Oct. 1837, p. 552, and Ann. des Sci. Nat. Dec. 1837, tom. viii. p. 368 ?
  Habitat. The rocky districts of New South Wales.

86. Petrogale lateralis, *Gould* . . . . Vol. II. Pls. 41 & 42.
  Habitat. Western Australia.

87. Petrogale xanthopus, *Gray* . . . . Vol. II. Pls. 43 & 44.
  Habitat. South Australia.

88. Petrogale inornata, *Gould* . . . . Vol. II. Pls. 45 & 46.
  Habitat. East coast of Australia.

89. Petrogale brachyotis, *Gould* . . . . Vol. II. Pl. 47.
  Habitat. North-western parts of Australia.

90. Petrogale concinna, *Gould* . . . . Vol. II. Pl. 48.
  Habitat. North-western Australia.

Genus DENDROLAGUS, *Müll.*

91. Dendrolagus ursinus, *Müll.* . . . . Vol. II. Pl. 49.
  Habitat. New Guinea.

92. Dendrolagus inustus, *Müll.* . . . . Vol. II. Pl. 50.
  Habitat. New Guinea.

Genus DORCOPSIS, *Müll.*

93. Dorcopsis Bruni . . . . Vol. II. Pl. 51.
  Habitat. New Guinea.

Genus ONYCHOGALEA, *Gray.*

94. Onychogalea unguifer, *Gould* . . . . Vol. II. Pls. 52 & 53.
  Habitat. North-eastern parts of Australia.

95. Onychogalea frænata, *Gould* . . . . Vol. II. Pl. 54.
  Habitat. Interior of New South Wales.

96. Onychogalea lunata, *Gould* . . . . Vol. II. Pl. 55.
  Habitat. Interior of Western Australia.

Genus LAGORCHESTES, *Gould.*

97. Lagorchestes fasciatus . . . . Vol. II. Pl. 56.
  Habitat. Western and Southern Australia.

98. Lagorchestes Leporoïdes, *Gould* . . . . Vol. II. Pl. 57.
  Habitat. South Australia.

99. Lagorchestes hirsutus, *Gould* . . . . Vol. II. Pl. 58.
  Habitat. Western Australia.

100. Lagorchestes conspicillatus, *Gould* . . . . Vol. II. Pl. 59.
  Habitat. Barrow Island, North-western Australia.

101. Lagorchestes Leichardti . . . . Vol. II. Pl. 60.
Habitat. The country bordering the Gulf of Carpentaria.

Mr. Blyth has described a species of this form under the name of *Lagorchestes gymnotus*, which he states is nearly allied to *L. conspicillatus*, and in all probability it is referable to one of the family figured in this work; but as the specimen is in the Museum of the Asiatic Society of Calcutta, it is impossible for me to determine this point. See "Report of Curator, Zoological Department, for May 1858," in Journ. Asiat. Soc. Bengal.

Genus BETTONGIA, *Gray*.

102. Bettongia penicillata, *Gray* . . . . Vol. II. Pl. 61.
Kangurus Gaimardi, Desm. Mamm. Supp. p. 542, sp. 842, 1822?
Hypsiprymnus Whitei, Quoy et Gaim. Voy. de l'Uranie, Zool. p. 62, pl. 10, 1824?
Kangurus lepturus, Quoy et Gaim. Bull. des Sci. Nat. Jan. 1824, tom. i. p. 271?
Hypsiprymnus Phillippi, Ogilb. in Proc. Zool. Soc. 1838, p. 62?
——————— *formosus*, Ogilb. *ib.* p. 62?
——————— *minor* (Potoroo), Cuv. Règ. Anim. p. 185?
——————— *Hunteri*, Skull in Roy. Coll. of Surg. of Engl.?
Habitat. New South Wales.

103. Bettongia Ogilbyi, *Gould* . . . . Vol. II. Pl. 62.
Wäl-ya, aborigines of Perth and the mountain districts.
Woile, aborigines of King George's Sound.
Habitat. Western Australia.

104. Bettongia cuniculus . . . . Vol. II. Pl. 63.
Habitat. Van Diemen's Land.

105. Bettongia Graii, *Gould* . . . . Vol. II. Pl. 64.
Habitat. Southern and Western Australia.

106. Bettongia rufescens, *Gray* . . . . Vol. II. Pl. 65.
Habitat. New South Wales.

107. Bettongia campestris, *Gould* . . . . Vol. II. Pl. 66.
Habitat. South Australia.

Genus HYPSIPRYMNUS, *Ill.*

108. Hypsiprymnus murinus . . . . Vol. II. Pl. 67.
Habitat. New South Wales.

109. Hypsiprymnus apicalis, *Gould* . . . . Vol. II. Pl. 68.
Habitat. Van Diemen's Land.

110. Hypsiprymnus Gilberti, *Gould* . . . . Vol. II. Pl. 69.
Habitat. Western Australia.

111. Hypsiprymnus platyops, *Gould* . . . . Vol. II. Pl. 70.
Habitat. Western Australia.

Order RODENTIA.

Genus Hapalotis, *Licht.*

112. Hapalotis albipes, *Licht.* . . . . Vol. III. Pl. 1.
Habitat. New South Wales, Victoria, and South Australia

113. Hapalotis apicalis, *Gould* . . . . Vol. III. Pl. 2.
Habitat. South Australia; and Van Diemen's Land?

114. Hapalotis hemileucura, *Gray* . . . . Vol. III. Pl. 3.
Habitat. Interior of the North-eastern portions of Australia.

115. Hapalotis hirsutus, *Gould* . . . . Vol. III. Pl. 4.
Habitat. Port Essington.

116. Hapalotis penicillata, *Gould* . . . . Vol. III. Pl. 5.
Habitat. Northern Australia.

117. Hapalotis conditor, *Gould* . . . . Vol. III. Pl. 6.
Habitat. Interior of New South Wales and Victoria.

118. Hapalotis murinus, *Gould* . . . . Vol. III. Pl. 7.
Habitat. Interior of New South Wales and South Australia.

119. Hapalotis longicaudata, *Gould* . . . . Vol. III. Pl. 8.
Habitat. Interior of Western Australia.

120. Hapalotis Mitchellii . . . . Vol. III. Pl. 9.
Habitat. Western and Southern Australia.

121. Hapalotis cervinus, *Gould* . . . . Vol. III. Pl. 10.
Habitat. The interior of South Australia.

I think it likely that the animal I have figured as *H. Mitchellii* may not be the *Dipus Mitchellii* of Ogilby, but that the true *H. Mitchellii* and my *H. cervinus* may be identical. If this should ultimately prove to be the case, *H. Gouldii* of Gray will be the correct designation of the animal I have called *H. Mitchellii*, to which the terms *H. macrotis* and *H. Richardsoni* of Gray, on the specimens in the British Museum, will also probably be referable.

122. Hapalotis arboricola, *MacLeay.*

This is another of the Australian mammals of which I have not had an opportunity of examining specimens.

Two coloured sketches, accompanied by the following notes, were kindly transmitted to me by Mr. Gerard Krefft:—

"The only example of this rarity which has yet been obtained has been presented to the Australian Museum by W. S. MacLeay, Esq. It was caught at Elizabeth Bay, where it inhabits the lofty *Eucalypti*, and builds a nest among the branches, with leaves and twigs, like that of a bird."

"Fur rather harsh to the touch, and of a slate-grey next the skin,—the longer hairs, or outer coat, being mingled ochreous and black; sides greyish, with an admixture of ochreous yellow, which becomes darker towards the back, and has the black hairs much longer than on any other part; outer surface of the ears clothed with very short white hairs; throat and abdomen white; tail thinly clothed with dark-brown hairs; toes of the hind and fore feet covered with short white hairs."

Genus Mus, *Linn.*

123. Mus fuscipes, *Waterh.* . . . . Vol. III. Pl. 11.
Habitat. The southern portions of Australia generally.

i

124. Mus vellerosus, *Gray* . . . . Vol. III. Pl. *12.*
Habitat. South Australia.

125. Mus longipilis, *Gould* . . . . Vol. III. Pl. *13.*
Habitat. Banks of the Victoria River.

126. Mus cervinipes, *Gould* . . . . Vol. III. Pl. *14.*
Habitat. Brushes of the eastern parts of New South Wales.

127. Mus assimilis, *Gould* . . . . Vol. III. Pl. *15.*
Habitat. New South Wales, and probably Western Australia.

128. Mus manicatus, *Gould* . . . . Vol. III. Pl. *16.*
Habitat. Port Essington.

129. Mus sordidus, *Gould* . . . . Vol. III. Pl. *17.*
Habitat. Darling Downs.

130. Mus lineolatus, *Gould* . . . . Vol. III. Pl. *18.*
Mus gracilicaudatus, Gould in Proc. Zool. Soc. part xiii. p. 77.
I now believe the animal I have thus named to be the same as *M. lineolatus.*
Habitat. Darling Downs.

131. Mus Gouldi, *Waterh.* . . . . Vol. III. Pl. *19.*
Habitat. The interior of New South Wales and Western Australia, and probably of the intermediate countries.

132. Mus nanus, *Gould* . . . . Vol. III. Pl. *20.*
Habitat. Interior of Western Australia.

133. Mus albocinereus, *Gould* . . . . Vol. III. Pl. *21.*
Habitat. Western Australia.

134. Mus Novæ-Hollandiæ, *Waterh.* . . . . Vol. III. Pl. *22.*
Habitat. New South Wales.

135. Mus delicatulus, *Gould* . . . . Vol. III. Pl. *23.*
Habitat. Port Essington.

Genus HYDROMYS, *Geoff.*

136. Hydromys chrysogaster, *Geoff.* . . . . Vol. III. Pl. *24.*
Habitat. Van Diemen's Land, New South Wales, Victoria, and South Australia.

137. Hydromys fulvolavatus, *Gould* . . . . Vol. III. Pl. *25.*
Habitat. The borders of the River Murray and Lake Albert in South Australia.

138. Hydromys leucogaster, *Geoff.* . . . . Vol. III. Pl. *26.*
Habitat. Banks of the Rivers Hunter and Clarence in New South Wales.

139. Hydromys fuliginosus, *Gould* . . . . Vol. III. Pl. *27.*
Habitat. King George's Sound, and the waters near Perth in Western Australia.

140. Hydromys Lutrilla, *MacLeay.*
I have never seen an example of the animal thus named by Mr. MacLeay, and of which two coloured sketches, one by Mr. G. French Angas, and the other by Mr. Gerard Krefft, were kindly sent to me by the latter gentleman; and without an inspection and comparison of it with the other species of *Hydromys*, it is quite impossible for me to say if it be really a species or not.

The following notes, by Mr. Krefft, accompanied the sketches :—

" The *Hydromys Lutrilla* was discovered by W. S. MacLeay, Esq., on the edge of the water in front of his beautiful seat, Elizabeth Bay. It is the only specimen yet seen, and Mr. MacLeay has presented it to the Australian Museum.

" Fur remarkably soft, and of a vinous or brownish grey next the skin, covered with dark brown and some sandy-coloured hairs on the flanks, and buffy hairs on the sides of the neck ; throat and abdomen white ; fore legs somewhat paler than the other parts of the body, with the exception of a brown patch on the upper surface of the feet, toes clothed with light-brown hairs ; nails white ; tarsi sepia-brown ; whiskers black and white intermixed, the upper and longer hairs being the dark-coloured ones ; tail about 7 inches long, five of which are covered with dark-brown coarse hair without any white at the tip.

	inches.
" Length from tip to tip	17
——— of tail	7
——— of face to base of ear	2
——— of tarsi and toes	2 "

Habitat. New South Wales.

Family CHEIROPTERA.

Genus PTEROPUS, *Briss.*

141. Pteropus poliocephalus, *Temm.* . . . . Vol. III. Pl. 28.
Habitat. Brushes of New South Wales.

142. Pteropus conspicillatus, *Gould* . . . . Vol. III. Pl. 29.
Habitat. Fitzroy Island, off the eastern coast of Australia.

143. Pteropus funereus, *Temm.* . . . . Vol. III. Pl. 30.
Habitat. The northern portions of Australia.

144. Pteropus scapulatus, *Peters.*
Pteropus scapulatus, Peters in Ann. and Mag. Nat. Hist. 3rd Series, vol. ii. p. 231.

A description of this species has been published by Dr. W. Peters of Berlin, in the number of the ' Annals and Magazine of Natural History ' for March 1863. As this description did not appear until after these pages were in type, I have had no opportunity of examining the specimen described, and must therefore content myself with transcribing Dr. Peters's remarks respecting it :—

" The present species nearly approaches *Pteropus medius* in size, and is very easily distinguished from all other species by two humeral spots " of ochreous-yellow, " and also by the golden-yellow colour of the abundant woolly hair on the ventral side of the wing-membranes, which appears near the lumbar region, on the humeral membrane, and near the fore arm almost to its end."

Habitat. Cape York, Northern Australia.

Genus MOLOSSUS, *Geoff.*

145. Molossus Australis, *Gray* . . . . Vol. III. Pl. 31.
Habitat. Victoria.

Genus TAPHOZOUS, *Geoff.*

146. Taphozous Australis, *Gould* . . . . Vol. III. Pl. 32.
Habitat. Northern coasts of Australia.

Genus RHINOLOPHUS, *Geoff.*

147. Rhinolophus megaphyllus, *Gray* . . . . Vol. III. Pl. 33.
Habitat. New South Wales.

148. Rhinolophus cervinus, *Gould* . . . . Vol. III. Pl. 34.
Habitat. Cape York and Albany Island, Northern Australia.

149. Rhinolophus aurantius, *Gould* . . . . Vol. III. Pl. 35.
Habitat. Port Essington.

Genus NYCTOPHILUS, *Leach.*

150. Nyctophilus Geoffroyi*, *Leach* . . . . Vol. III. Pl. 36.
Habitat. Western Australia.

151. Nyctophilus Gouldi, *Tomes.*
Nyctophilus Geoffroyi . . . . Vol. III. Pl. 37.
Habitat. New South Wales.

152. Nyctophilus unicolor, *Tomes* . . . . Vol. III. Pl. 38.
Habitat. Van Diemen's Land.

153. Nyctophilus Timoriensis . . . . Vol. III. Pl. 39.
Habitat. Western Australia.

154. Nyctophilus Australis, *Peters.*
Nyctophilus australis, Peters, in Abhandl. der Königl. Akad. der Wissenschaften zu Berlin, 1860, p. 135 and Tab.
See a valuable paper on the genus *Nyctophilus,* by Dr. Peters, in the above-mentioned Transactions of the Academy of Berlin.

Genus SCOTOPHILUS, *Leach.*

155. Scotophilus Gouldi, *Gray* . . . . Vol. III. Pl. 40.
Habitat. New South Wales and Victoria; and South Australia?

156. Scotophilus morio, *Gray* . . . . Vol. III. Pl. 41.
Habitat. New South Wales and Victoria; and Western Australia?

157. Scotophilus microdon, *Tomes* . . . . Vol. III. Pl. 42.
Vespertilio Muelleri, Beck. Trans. Phil. Inst. Victoria, vol. iv. part i. p. 41, with plate?
Habitat. Van Diemen's Land; and the south coast of Australia?

158. Scotophilus picatus, *Gould* . . . . Vol. III. Pl. 43.
Habitat. The interior of South Australia.

159. Scotophilus nigrogriseus, *Gould* . . . . Vol. III. Pl. 44.
Habitat. Queensland.

160. Scotophilus Greyi, *Gray* . . . . Vol. III. Pl. 45.
Habitat. Port Essington.

161. Scotophilus pumilus, *Gray* . . . . Vol. III. Pl. 46.
Habitat. New South Wales.

Genus VESPERTILIO, *Linn.*

162. Vespertilio macropus, *Gould* . . . . Vol. III. Pl. 47.
Habitat. South Australia.

163. Vespertilio Tasmaniensis .   .   .   .   Vol. III. Pl. 48.
Habitat. Tasmania.

Family PHOCIDÆ, *Gray.*

Genus ARCTOCEPHALUS, *F. Cuv.*

164. Arctocephalus lobatus .   .   .   .   Vol. III. Pl. 49.
Habitat. Southern coasts of New South Wales and Tasmania.

Genus STENORHYNCHUS, *F. Cuv.*

165. Stenorhynchus leptonyx .   .   .   .   Vol. III. Pl. 50.
Habitat. The coasts of Tasmania and the southern portions of Australia generally.

Family CANIDÆ.

Genus CANIS, *Linn.*

166. Canis Dingo, *Blumenb.* .   .   .   .   Vol. III. Pls. 51 & 52.
Dwer-da, aborigines of Western Australia.
Habitat. Australia generally.

Although I have omitted the Whales and Dugong, I cannot, in justice to Mr. Wm. Sheridan Wall, omit to call attention to his 'History and Description of the Skeleton of a New Sperm-Whale lately set up in the Australian Museum; together with some account of a new genus of Sperm-Whales called *Euphysetes,*' published by W. R. Piddington, Sydney, 1851. In like manner, I cannot leave unpublished the following interesting letter respecting the Dugong, which has been forwarded to me by my brother-in-law, Charles Coxen, Esq., of Brisbane, Queensland:—

"The Dugong (*Halicore australis,* Owen) occurs in considerable numbers in Moreton Bay, but, I am led to believe, is not found further south. To the north it is plentiful in all the bays, such as Wide Bay, Port Curtis, Keppel Bay, &c., and along the east and north coasts, in every situation suitable to its habits. In size it varies from six to nine feet in length, the latter being the size of a large 'bull'; the weight also varies from 600 to 1000 lbs.; the girth at the largest part, just behind the flippers, is about six-eighths of the length; near the root of the tail it is very taper and small. The head is very peculiar: the eyes and ears are small; the nostrils small and oblique; the fleshy upper lip, which depends some three or four inches from the jaw, is peculiarly truncate in form, and covered with short stout bristles; the lower lip is globular, pendulous, and attached by a small neck to the jaw. The name given to the Dugong by the aborigines is *Young-un.* The flesh is greedily eaten and much sought for by them; and when they have been successful in procuring one or two, which occasionally happens, they gorge themselves in a most unseemly manner, and grease themselves all over with the fat and oil until they glisten in the sun like a roll of butter in the dog-days.

"The female, or 'cow,' exhibits much tenderness in the care of her offspring, and when injured utters a low, plaintive, snuffling sound, which appears to be understood by the calf.

"In the spring or calving-time they frequent the smaller bays and inlets of Moreton Bay, and are found feeding, in the more tranquil spots, on the *Algæ* and other marine vegetable productions growing on the shoals near the mainland and the islands. During the winter months they are more frequently met with at sea, or outside the large bays. Their feeding-grounds vary from four to ten feet at high water.

"Harpooning is at present the only mode of procuring the Dugong. The aborigines are very expert in the use of the instrument, and the quickness of their sight renders them superior to Europeans for such service; but the loss of time, and consequent expense, owing to the unsettled habits of the natives, and at times the ruffled state of the water, have prevented its capture being entered upon as a business. A few years ago a party com-

k

menced setting nets on the shoals frequented by the Dugong, and for a time they answered the purpose; but the men engaged got careless, the nets were torn and destroyed by sharks and porpesses, and the affair fell to the ground.

"The oil, owing to its medicinal qualities, is in considerable demand, and very many persons have derived considerable benefit from its use; it is preferred to cod-liver oil, as being less disagreeable to the palate and more easily retained in the stomach. It is white and almost tasteless, and is occasionally used for frying fish. The quantity varies, according to the condition of the animal, from three to ten gallons. The meat is very good, is in flavour between beef and pork, and when salted is much like bacon.

"The head, back, sides, and tail are dark broccoli-brown; the belly and under part of the flippers light broccoli-brown, according to Werner's Nomenclature of Colours."

GENERAL INDEX.

	Vol.	Page
ACROBATA pygmæa. Vol. I. p. xxvi.		
Acrobates pygmæus	I.	85
Pigmy	I.	85
Al-wo-re	III.	30
Amblotis fossor	I.	63
Ant-eater, Aculeated	I.	5
Antechinus albipes. Vol. I. p. xxviii.	I.	49
apicalis. Vol. I. p. xxvii	I.	46
Dusky	I.	44
ferruginifrons. Vol. I. p. xxvii	I.	43
flavipes. Vol. I. p. xxvii	I.	47
Freckled	I.	46
fuliginosus. Vol. I. p. xxvii	I.	48
leucogaster. Vol. I. p. xxvii	I.	45
leucopus. Vol. I. p. xxvii	I.	42
maculatus. Vol. I. p. xxviii	I.	51
Minute	I.	52
minutissimus. Vol. I. p. xxviii	I.	52
Murine	I.	50
murinus. Vol. I. p. xxviii	I.	50
Rusty-footed	I.	47
Rusty-fronted	I.	43
Sooty	I.	48
Spotted	I.	51
Stuartii. Vol. I. p. xxvii.		
Swainsoni. Vol. I. p. xxvii	I.	41
Swainson's	I.	41
unicolor. Vol. I. p. xxvii	I.	44
White-bellied	I.	45
White-footed	I.	49
Arctocephalus lobatus. Vol. I. p. xxxix. III.	49	
Badger	I.	63
Bal-a-ga	I.	38
Balantia Cookii	I.	24
Bâl-lard. Vol. I. p. xxvii.		
Bal-lă-wa-ra	I.	38
Băm-ba	III.	89
Bangap	II.	36
Baŭ-gup	II.	44
Barbastellus Pacificus	III.	37
Bar-ra-jit. Vol. I. p. xxviii.		
Bar-roo	III.	1
Bat, Chocolate	III.	41
Fawn-coloured	III.	34
Gould's	III.	40
Great-footed	III.	47
Great-leaved Horse-shoe	III.	38
Little	III.	46
Little Black	III.	43
Orange Horse-shoe	III.	35
Small-toothed	III.	42
Tasmanian	III.	48
Bear, Native	I.	19
Sea	III.	50
Beaver-Rat, Fulvous	III.	25
Golden-bellied	III.	24
Sooty	III.	27

	Vol.	Page
Beaver-Rat, White-bellied	III.	26
Belidea Ariel	I.	34
Belideus, Ariel. Vol. I. p. xxvi	I.	84
Ariel	I.	84
breviceps. Vol. I. p. xxvi	I.	32
flaviventer. Vol. I. p. xxvi	I.	30
Long-tailed	I.	30
notatus. Vol. I. p. xxvi	I.	33
sciureus. Vol. I. p. xxvi	I.	31
Short-headed	I.	32
Squirrel-like	I.	81
Stripe-tailed	I.	83
Bettong	II.	77
Bettongia campestris. Vol. I. p. xxxiv.	II.	76
cuniculus. Vol. I. p. xxxiv	II.	73
fasciata	II.	65
Gouldi	II.	72
Graii. Vol. I. p. xxxiv	II.	74
Grayii	II.	74
Ogilbii	II.	72
Ogilbyi. Vol. I. p. xxxiv	II.	72
penicillata. Vol. I. p. xxxiv	II.	71
rufescens. Vol. I. p. xxxiv	II.	75
setosa	II.	73
Boomer	II.	2
Booŭ-dee	II.	74
Buĭ-loo-wa	I.	38
Bundary	II.	2
Buŭ-da	I.	10
Bur-jad-da. Vol. I. p. xxviii.		
Canis Dingo. Vol. I. p. xxxix	III.	51,52
familiaris, var. Australasiæ	III.	52
Chæropus castanotis. Vol. I. p. xxiv.	I.	10
Chestnut-eared	I.	10
ecaudatus	I.	10
Chrysæus australiæ	III.	52
Conilurus constructor	III.	1
Cuscus brevicaudatus. Vol. I. p. xxv.	I.	28
Short-tailed	I.	28
Dal-gyte	I.	11
Dama. Vol. I. p. xxxi.		
Dasyurus cynocephalus	I.	61
Geoffroyi. Vol. I. p. xxviii	I.	58
Geoffroy's	I.	58
hallucatus. Vol. I. p. xxviii	I.	59
macrourus	I.	56
maculatus. Vol. I. p. xxviii	I.	56
Maugei	I.	57
minimus. Vol. I. p. xxvii.		
North-Australian	I.	59
penicillatus	I.	38
Spotted-tailed	I.	56
Tafa	I.	38
ursinus	I.	55
Variable	I.	57
viverrinus. Vol. I. p. xxviii	I.	57

	Vol.	Page
Dasyurus (Sarcophilus) ursinus.	I.	55
Dendrolagus inustus. Vol. I. p. xxxiii.	II.	57
ursinus. Vol. I. p. xxxiii	II.	56
Devil	I.	55
Native	I.	55
Diabolus ursinus	I.	55
Diŭ-bler	I.	46
Didelphis Asiatica	II.	58
Brunii	II.	58
cynocephala	I.	61
gigantea	II.	2
guttatus	I.	57
Petaurus	I.	30
pygmæa	I.	85
ursina	I. 55, 63	
viverrina	I.	57
Didelphys lemurina	I.	22
macroura	I.	30
obesula	I.	16
penicillata	I.	38
sciurea	I.	31
volucella	I.	30
vulpina	I.	22
Dil-pea	III.	17
Dingo, The	III.	51, 52
Dipus Mitchelli	III.	9
Dju-tytch. Vol. I. p. xxviii.		
Djyr-dow-in	III.	9
Dol-goitch	I.	11
Dorcopsis Bruni. Vol. I. p. xxxiii	II.	58
Brunii	II.	58
Dromicia, Beautiful	I.	37
concinna. Vol. I. p. xxvi	I.	37
gliriformis. Vol. I. p. xxvi	I.	36
Neillii. Vol. I. p. xxvi.		
Thick-tailed	I.	36
unicolor. Vol. I. p. xxvi.		
Dun-ung-eŭ-de	I.	5
Dwer-da. Vol. I. p. xxxix.		
Echidna aculeata	I.	5
Australiensis	I.	5
breviaculeata	I.	5
Hairy	I.	7
Hystrix. Vol. I. p. xxiii	I.	5
longiaculeata	I.	5
setosa. Vol. I. p. xxiii	I.	7
Spiny	I.	5
Filander	II.	58
Forester	II.	2
Goŏl-a-wa	III.	8
Goo-mal. Vol. I. p. xxv.		
Goŏrh-a	II.	27
Gwĕn-dee	I.	16
Halicore australis. Vol. I. p. xxxix.		

a

	Vol.	Page
Halmaturus agilis. Vol. I. p. xxx .	II.	30, 31
Asiaticus	II.	58
Bennetti	II.	21, 22
Bennettii. Vol. I. p. xxx . .	II.	22
Billardieri. Vol. I. p. xxxii .	II.	41, 42
Binoë	II.	31
brachytarsus	II.	42
brachyurus. Vol. I. p. xxxiii .	II.	43, 44
Brunii	II.	58
Dama. Vol. I. p. xxxi.		
Derbianus. Vol. I. p. xxx .	II.	35, 36
dorsalis. Vol. I. p. xxx . .	II.	32, 33
elegans	II.	65
Emiliæ	II.	36
Eugenii	II.	36
gracilis. Vol. I. p. xxxii.		
Greyi. Vol. I. p. xxx . .	II.	24, 25
Greyii	II.	25
griseo-fuscus	II.	2
griseo-rufus	II.	20
Houtmanni. Vol. I. p. xxx .	II.	36
leptonyx	II.	22
Lessonii	II.	29
manicatus. Vol. I. p. xxx .	II.	26, 27
nemoralis	II.	29
nuchalis	II.	38
Parma. Vol. I. p. xxx . .	II.	34
Parryii	II.	18
ruficollis. Vol. I. p. xxx . .	II.	19, 20
stigmaticus. Vol. I. p. xxxii .	II.	39, 40
Thetidis. Vol. I. p. xxxii .	II.	37, 38
Ualabatus. Vol. I. p. xxx .	II.	28, 29
(Thylogale) brevicaudatus . .	II.	44
(——) Tasmanei	II.	42
Hapalotis albipes. Vol. I. p. xxxv .	III.	1
apicalis. Vol. I. p. xxxv . .	III.	2
arboricola. Vol. I. p. xxxv.		
Building	III.	6
cervinus. Vol. I. p. xxxv .	III.	10
conditor. Vol. I. p. xxxv .	III.	6
Elsey's	III.	3
Fawn-coloured	III.	10
Gouldii	III.	9
hemileucura. Vol. I. p. xxxv .	III.	3
hirsutus. Vol. I. p. xxxv . .	III.	4
Long-haired	III.	4
longicaudata. Vol. I. p. xxxv.	III.	8
Long-tailed	III.	8
melanura	III.	5
Mitchellii. Vol. I. p. xxxv .	III.	9
Mitchell's	III.	9
Murine	III.	7
murinus. Vol. I. p. xxxv .	III.	7
Pencil-tailed	III.	5
penicillata. Vol. I. p. xxxv .	III.	5
White-footed	III.	1
White-tipped	III.	2
Hepoona Cookii	I.	24
Heteropus albogularis. Vol. I. p. xxxiii	II.	46
Hyæna	I.	61
Hydromys chrysogaster. Vol. I. p. xxxvi.	III.	24
fuliginosus. Vol. I. p. xxxvi .	III.	27
fulvolavatus. Vol. I. p. xxxvi .	III.	25
leucogaster. Vol. I. p. xxxvi .	III.	26
Latrilla. Vol. I. p. xxxvi.		
Hypsiprymnus apicalis. Vol. I. p. xxxiv	II.	78
Bruni	II.	58
cuniculus	II.	73
formosus. Vol. I. p. xxxiv.		
Gilberti. Vol. I. p. xxxiv .	II.	79
Graii	II.	74
Hunteri. Vol. I. p. xxxiv.		

	Vol.	Page
Hypsiprymnus Lesueuri	II.	74
melanotis	II.	75
micropus	II.	79
minor. Vol. I. p. xxxiv.		
murinus. Vol. I. p. xxxiv .	II.	71, 77
myosurus	II.	77
Ogilbyi	II.	72
Peron	II.	77
Phillippi. Vol. I. p. xxxiv.		
platyops. Vol. I. p. xxxiv .	II.	80
rufescens	II.	75
setosus	II.	71, 77
Whitei. Vol. I. p. xxxiv.		
(Bettongia) campestris . . .	II.	76
(——) cuniculus	II.	73
(——) Graii	II.	74
(——) Ogilbyi	II.	72
(——) penicillatus	II.	71
(Potorous) Gilbertii	II.	79
(——) murinus	II.	77
(——) platyops	II.	80
Isoödon obesula	I.	16
Jeĕ-pin	I.	9
Jib-beetch	III.	20
Jŭp-pert	III.	21
Kangaroo à cou roux	II.	20
Branded Hare-	II.	65
Black Tree-	II.	56
Blue	II.	27
Bridled Nail-tailed	II.	62
Broad-faced Rat-	II.	80
Brown Tree-	II.	57
Brush-	II.	22, 27
Gilbert's Rat-	II.	79
Gray's Jerboa-	II.	74
Great Grey	II.	1, 2
Great Red-	II.	9, 10
Hare-	II.	67
Jerboa-	II.	71
Leichardt's Hare-	II.	70
Lunated Nail-tailed	II.	64
Nail-tailed	II.	60, 61
New South Wales Rat- . . .	II.	77
Rufous Hare-	II.	68
Rufous Jerboa-	II.	75
Ogilby's Jerboa-	II.	72
Old Man	II.	2
Plain-loving Jerboa-	II.	76
Rat-	II.	77
Sooty	II.	8
Spectacled Hare-	II.	69
Tasmanian Jerboa-	II.	73
Tasmanian Rat-	II.	78
West-Australian Great- . . .	II.	6, 7
Kangurus fasciatus	II.	65
Kangaroo	II.	2
Kangourou géant	II.	8
Kangurus Billardierii	II.	42
brachyurus	II.	44
Brunii	II.	29
fuliginosus	II.	8
Gaimardi. Vol. I. p. xxxiv.		
labiatus	II.	2
laniger	II.	10
lepturus. Vol. I. p. xxxiv.		
pencillatus	II.	46
penicillatus	II.	46
Rat, Forest	II.	73
ruficollis	II.	20

	Vol.	Page
Kangurus rufo-griseus	II.	20
rufus	II.	10
Ualabatus	II.	29
King-goor. Vol. I. p. xxvii.		
Koala	I.	18, 19
Koala, ou Coulak	I.	19
Koala Wombat	I.	19
Kor-tung	III.	8
Kurn-dyne	III.	19
Lagorchestes albipilis	II.	65
conspicillata	II.	69
conspicillatus. Vol. I. p. xxxiii .	II.	69
fasciatus. Vol. I. p. xxxiii .	II.	65
gymnotus. Vol. I. p. xxxiv.		
hirsutus. Vol. I. p. xxxiii .	II.	68
Leichardti. Vol. I. p. xxxiv .	II.	70
Leporoïdes. Vol. I. p. xxxiii .	II.	67
Leopard, Sea	III.	50
Lipurus cinereus	I.	19
Macropus Bennettii	II.	22
brachiotis	II.	54
Brunii	II.	58
elegans	II.	18
frænatus	II.	62
fuliginosus. Vol. I. p. xxix .	II.	2, 8
giganteus	II.	2
gracilis. Vol. I. p. xxxii.		
laniger	II.	10
lanigerus	II.	10
Leporides	II.	67
lunatus	II.	64
major. Vol. I. p. xxix . .	II.	1, 2
melanops. Vol. I. p. xxix .	II.	2
minor	II.	2
ocydromus. Vol. I. p. xxix .	II.	2, 6, 7
Parryi	II.	18
penicillatus	II.	46
robustus	II.	15
ruficollis	II.	20
ruficollis, var. Bennettii . .	II.	22
rufo-griseus	II.	20
Ualabatus	II.	29
unguifer	II.	61
veterum	II.	58
(Halmaturus) agilis	II.	31
(——) Antilopinus	II.	13
(——) Billardieri	II.	42
(——) brachyurus	II.	44
(——) Derbianus	II.	36
(——) dorsalis	II.	33
(——) fructicus	II.	22
(——) Greyi	II.	25
(——) Irma	II.	27
(——) manicatus	II.	27
(——) Parma	II.	34
(——) Parryi	II.	18
(——) robustus	II.	15
(——) ruficollis	II.	20
(——) rufiventer	II.	42
(——) rufus	II.	10
(——) Thetidis	II.	38
(——) Ualabatus	II.	29
(Heteropus) brachiotis . . .	II.	54
(——) concinnus	II.	55
(——) inornatus	II.	53
(——) lateralis	II.	49
(——) penicillatus	II.	46
(Lagorchestes) conspicillatus .	II.	69
(——) fasciatus	II.	65
(——) hirsutus	II.	68

	Vol.	Page
Macropus (Lagorchestes) Leporoïdes	II.	67
(Osphranter) pictus. Vol. I. p. xxix.		
(Petrogale) brachyotis	II.	54
(——) robustus	II.	15
Mäl-a	I.	14
Man-dui̇-da	I.	37
Marri̇-dera	I.	46
Mar-ra-a-woke	II.	13
Martin, Spotted	I.	56
Mät-tee-getch	III.	9
Molossus, Australian	III.	31
Australis. Vol. I. p. xxxvii	III.	31
Mo-lyne-be	III.	23
Moŏr-deet	III.	15
Moŏ-roo-rong	II.	49
Mouse, Delicate-coloured	III.	23
Greyish-white	III.	21
New Holland Field-	III.	22
White-footed	III.	19
Mus alboeinereus. Vol. I. p. xxxvi	III.	21
assimilis. Vol. I. p. xxxvi	III.	15
cervinipes. Vol. I. p. xxxvi	III.	14
conditor	III.	6
delicatulus. Vol. I. p. xxxvi	III.	23
fuscipes. Vol. I. p. xxxv	III.	11
Gouldi. Vol. I. p. xxxvi	III.	19
Gouldii	III.	19
gracilicaudatus. Vol. I. p. xxxvi.		
Greyii	III.	19
hirsutus	III.	4
lineolatus. Vol. I. p. xxxvi	III.	18
longipilis. Vol. I. p. xxxvi	III.	13
lutreola	III.	11
manicatus. Vol. I. p. xxxvi	III.	16
nanus. Vol. I. p. xxxvi	III.	20
Novæ-Hollandiæ. Vol. I. p. xxxvi	III.	22
penicillatus	III.	5
sordidus. Vol. I. p. xxxvi	III.	17
vellerosus. Vol. I. p. xxxvi	III.	12
Mustela Novæ-Hollandiæ	I.	56
Myrmecobius, Banded	I.	8
fasciatus. Vol. I. p. xxxiv	I.	8
Myrmecophaga aculeata	I.	5
Nagoor-ja-na. Vol. I. p. xxviii.		
Ngil-gyte	II.	79
Ngool-boon-goor	I.	9
Ngoŏr-joo	III.	27
Ngŏ-ra. Vol. I. p. xxv.		
Ngork. Vol. I. p. xxv.		
Ngwi̇r-ri-gin	III.	27
Noctulinia Tasmaniensis	III.	48
Noŏ-jee	III.	31
Noom-bat	I.	8
Nyctinomus	III.	37
Nyctophilus australis. Vol. I. p. xxxviii.		
Geoffroyi. Vol. I. p. xxxviii	III.	37
Geoffroyi*. Vol. I. p. xxxviii	III.	36
Geoffroy's	III.	36, 37
Gouldi. Vol. I. p. xxxviii.		
Tasmanian	III.	38
Timoriensis. Vol. I. p. xxxviii	III.	39
unicolor. Vol. I. p. xxxviii	III.	38
Western	III.	39
Nyĕm-mel	I.	14
Nyoong-arn	I.	5
Onychogalea frænata. Vol. I. p. xxxiii	II.	62
lunata. Vol. I. p. xxxiii	II.	64
unguifer. Vol. I. p. xxxiii	II.	61
unguifera	II.	60
Opossum, Dog-headed	I.	61

	Vol.	Page
Opossum hirsutum	I.	63
Javan	II.	58
New Holland	I.	24
Ring-tailed	I.	24
Spotted	I.	57
Vulpine	I.	22
White-tailed	I.	24
Zebra	I.	61
Ornithorhynchus	I.	1
anatinus. Vol. I. p. xxiii.	I.	1
brevirostris	I.	1
crispus	I.	1
fuscus	I.	1
Hystrix	I.	5
lævis	I.	1
paradoxus	I.	1
rufus	I.	1
Osphranter Antilopinus. Vol.I.p.xxx	II.	12, 13
? Isabellinus. Vol. I. p. xxx.		
? Parryi. Vol. I. p. xxx	II.	17, 18
robustus. Vol. I. p. xxx	II.	14, 15
rufus. Vol. I. p. xxix	II.	9, 10
Otăm-in	I.	49
Otaria cinerea	III.	49
jubata	III.	49
Lemairii	III.	49
Stelleri	III.	49
Pademelon	II.	38
Paragalia lagotis	I.	11
Peracyon cynocephalus	I.	61
Peragalea lagotis. Vol. I. p. xxiv	I.	11
Large-eared	I.	11
Perameles affinis	I.	16
arenaria	I.	14
aurita	I.	15
Banded	I.	12
Bougainvillei. Vol. I. p. xxv.		
fasciata. Vol. I. p. xxiv	I.	12
fossor	I.	63
fusciventer	I.	16
Gunnii. Vol. I. p. xxiv	I.	18
Gunn's	I.	18
lagotis	I.	11
Lawsoni	I.	15
Long-nosed	I.	15
macroura. Vol. I. p. xxiv.		
macrurus. Vol. I. p. xxiv.		
myosurus. Vol. I. p. xxiv	I.	14
nasuta. Vol. I. p. xxiv	I.	15
obesula. Vol. I. p. xxiv	I.	16
Saddle-backed	I.	14
Short-nosed	I.	16
Petaurista flaviventer	I.	30
Peronii	I.	29
Taguanoïdes. Vol. I. p. xxvi	I.	29
(Acrobata) pygmæa	I.	35
Petarus Ariel	I.	34
australis	I.	30
breviceps	I.	32
Cunninghami	I.	30
leucogaster	I.	29
macrourus	I.	30
Peronii	I. 29,	32
pygmæus	I.	35
sciureus	I.	31
Taguanoïdes	I.	29
(Acrobata) pygmæus	I.	35
(Belideus) Ariel	I.	34
(——) breviceps	I.	32
(——) flaviventer	I.	30
(——) notatus	I.	33

	Vol.	Page
Petaurus (Belideus) sciureus	I.	31
(Petaurista) taguanoïdes	I.	29
Petrogale brachyotis. Vol. I. p. xxxiii	II.	54
coneinna. Vol. I. p. xxxiii	II.	55
inornata. Vol. I. p. xxxiii	II.	52, 53
lateralis. Vol. I. p. xxxiii	II.	48, 49
penicillata. Vol. I. p. xxxiii	II.	45, 46
penicillatus	II.	46
xanthopus. Vol. I. p. xxxiii	II.	50, 51
Phalanger de Bougainville	I.	24
de Cook	I.	22, 24
Great Flying	I.	29
Woolly	I.	27
Phalangista Banksii	I.	24
Bougainvillii	I.	24
canina. Vol. I. p. xxv	I.	23
Cooki	I. 24,	25
Cookii. Vol. I. p. xxv	I.	24
Cook's	I.	24
Cuvieri	I.	21
felina. Vol. I. p. xxv	I.	21
fuliginosa. Vol. I. p. xxv	I.	21
gliriformis	I.	36
laniginosa. Vol. I. p. xxv	I.	27
melanura. Vol. I. p. xxv	I.	22
Short-eared	I.	23
Sooty	I.	21
viverrina. Vol. I. p. xxv	I.	25
Viverrine	I.	25
vulpina. Vol. I. p. xxv	I.	22
Vulpine	I.	22
xanthopus	I.	22
(Pseudochirus) Cookii	I.	24
(——) nudicaudata	I.	28
(Trichosurus) vulpina	I.	22
Phascogale affinis. Vol. I. p. xxvii.		
albipes	I.	49
apicalis	I.	46
Brush-tailed	I.	38
calura. Vol. I. p. xxvii	I.	39
crassicaudata	I.	45
flavipes	I.	47
Handsome-tailed	I.	39
lanigera. Vol. I. p. xxvii	I.	40
leucogaster	I.	45
leucopus	I.	42
minima. Vol. I. p. xxvii.		
murina	I.	50
penicillata. Vol. I. p. xxvii	I.	38
rufogaster	I.	47
Swainsonii	I.	41
Woolly	I.	40
(Antechinus) albipes	I.	49
(——) flavipes	I.	47
(——) leucogaster	I.	45
(——) leucopus	I.	42
(——) macroura	I.	53
(——) minima. Vol. I. p. xxvii.		
(——) murina	I.	50
(——) Swainsonii	I.	41
Phascolarctos cinereus. Vol. I. p. xxv.	I. 18,	19
Flindersi	I.	19
fuscus	I.	19
Phascolarctus cinereus	I.	19
Phascolomys Bassii	I.	63
fossor	I.	63
fusca	I.	63
lasiorhinus. Vol. I. p. xxix	I. 67,	68
latifrons. Vol. I. p. xxviii	I. 65,	66
niger. Vol. I. p. xxix.		
platyrhinus. Vol. I. p. xxviii.		
ursinus	I.	63

GENERAL INDEX.

	Vol.	Page
Phascolomys Wombat. Vol. I. p. xxviii	I.	62, 63
wombatus	I.	63
Phoca Homei	III.	50
leptonyx	III.	50
lobata	III.	49
ursina	III.	50
Phoque quatrième	III.	50
Platypus anatinus	I.	1
Plecotus Timoriensis	III.	39
Podabrus crassicaudatus. Vol. I. p. xxviii	I.	54
Large-tailed	I.	53
macrourus. Vol. I. p. xxviii	I.	53
Thick-tailed	I.	54
Poto-Roo	II.	77
Potoroüs murinus	II.	77
Pseudocheirus nudicaudata	I.	28
Pteropus conspicillatus. Vol. I. p. xxxvii	III.	29
funereus. Vol. I. p. xxxvii	III.	30
poliocephalus. Vol. I. p. xxxvii	III.	28
scapulatus. Vol. I. p. xxxvii.		
Quåk-a	II.	44
Quår-ra	II.	27
Quoint	I.	16
Rabbit	I.	11
Rabbit-Rat	III.	1
Rat, Allied	III.	15
Buff-footed	III.	14
Dusky-footed	III.	11
Fulvous Beaver-	III.	25
Golden-bellied Beaver-	III.	24
Little	III.	20
Long-haired	III.	13
Plain	III.	18
Sooty Beaver-	III.	27
Sordid	III.	17
Rabbit-	III.	1
Tawny	III.	12
White-bellied Beaver-	III.	26
White-footed	III.	16
Rhinolophus aurantius. Vol. I. p. xxxviii	III.	35
? cervinus. Vol. I. p. xxxviii	III.	34
megaphyllus. Vol. I. p. xxxviii	III.	33
Sarcophilus, Ursine	I.	55

	Vol.	Page
Sarcophilus Ursinus. Vol. I. p. xxviii.	I.	55
Sciurus Novæ-Hollandiæ	I.	30
Scotophilus, Blackish-grey	III.	44
Gouldi. Vol. I. p. xxxviii	III.	40
Gouldii	III.	40
Greyi. Vol. I. p. xxxviii	III.	45
Greyii	III.	45
Grey's	III.	45
microdon. Vol. I. p. xxxviii	III.	42
morio. Vol. I. p. xxxviii	III.	41
nigrogriseus. Vol. I. p. xxxviii	III.	44
picatus. Vol. I. p. xxxviii	III.	43
Pied	III.	43
pumilus. Vol. I. p. xxxviii	III.	46
Seal, Cowled	III.	49
from New Georgia	III.	50
Small-nailed	III.	50
Sloth, Native	I.	19
New Holland	I.	19
Squirrel, Norfolk Island Flying	I.	31
Sugar	I.	31
Stenorhynchus leptonyx. Vol. I. p. xxxix	III.	50
aux petits ongles	III.	50
Tachyglossus aculeatus	I.	5
setosus	I.	7
Taphozous, Australian	III.	32
Australis. Vol. I. p. xxxvii	III.	32
Tapoa-tafa	I.	38, 57
Tarsipes, Long-nosed	I.	9
rostratus. Vol. I. p. xxiv	I.	9
Spenseræ	I.	9
Thylacinus	I.	60, 61
cynocephalus. Vol. I. p. xxviii	I.	60, 61
Harrisii	I.	61
Thylogale Eugenii. Vol. I. p. xxx.		
Tiger	I.	61
Twoor-dong	I.	48
Vampire, Funereal	III.	30
Grey-headed	III.	28
Spectacled	III.	29
Vespertilio	III.	43
macropus. Vol. I. p. xxxviii	III.	47
Muelleri. Vol. I. p. xxxviii		
Tasmaniensis. Vol. I. p. xxxix.	III.	48

	Vol.	Page
Vespertilio Timoriensis	III.	39
Viverra maculata	I.	56
Wai-haw	I.	8
Wallaby	II.	42
Agile	II.	30, 31
Bennett's	II.	21, 22
Black	II.	28, 29
Black-gloved	II.	26, 27
Black-striped	II.	32, 33
Branded	II.	39, 40
Brush-tailed Rock	II.	45, 46
Derby's	II.	35, 36
Grey's	II.	24, 25
Little Rock	II.	55
Pademelon	II.	37, 38
Parma	II.	34
Rufous-necked	II.	19, 20
Short-eared Rock	II.	54
Short-tailed	II.	43, 44
Stripe-sided Rock	II.	48, 49
Tasmanian	II.	41, 42
Unadorned Rock	II.	52, 53
Yellow-footed Rock	II.	50, 51
Wallaroo, Black	II.	14, 15
Parry's	II.	17, 18
Red	II.	12, 13
Wål-ya. Vol. I. p. xxxiv.		
Warroon	II.	20
Wha Tapoa Roo	I.	22
Woile. Vol. I. p. xxxiv.		
Wolf, Zebra-	I.	61
Womback	I.	63
Wombat	I.	62, 63
Broad-fronted	I.	65, 66
Hairy-nosed	I.	67, 68
Koala	I.	19
of Flinders	I.	19
Wombatus fossor	I.	63
Wor-gi	I.	34
Work	II.	7
Wot-da	I.	10
Wÿ-a-lung	I.	46
Yerbua gigantea	II.	2
Yoõn-gur	II.	7

LIST OF PLATES.

VOL. I.

Ornithorhynchus anatinus	Ornithorhynchus	1
Echidna hystrix	Spiny Echidna	2
——— setosa	Hairy Echidna	3
Myrmecobius fasciatus	Banded Myrmecobius	4
Tarsipes rostratus	Long-nosed Tarsipes	5
Chœropus castanotis	Chestnut-eared Chœropus	6
Peragalea lagotis	Long-eared Peragalea	7
Perameles fasciata	Banded Perameles	8
——— Gunnii	Gunn's Perameles	9
——— myosurus	Saddle-backed Perameles	10
——— nasuta	Long-nosed Perameles	11
——— obesula	Short-nosed Perameles	12
Phascolarctos cinereus	Koala	13, 14
Phalangista fuliginosa	Sooty Phalangista	15
——— vulpina	Vulpine Phalangista	16
——— canina	Short-eared Phalangista	17
——— Cookii	Cook's Phalangista	18
——— Viverrina	Viverrine Phalangista	19
——— lanuginosa	Woolly Phalanger	20
Cuscus brevicaudatus	Short-tailed Cuscus	21
Petaurista Taguanoïdes	Great Flying Phalanger	22
Belideus flaviventer	Long-tailed Belideus	23
——— Sciureus	Squirrel-like Belideus	24
——— breviceps	Short-headed Belideus	25
——— notatus	Stripe-tailed Belideus	26
——— Ariel	Ariel Belideus	27
Acrobates pygmæus	Pigmy Acrobates	28
Dromicia gliriformis	Thick-tailed Dromicia	29
——— concinna	Beautiful Dromicia	30
Phascogale penicillata	Brush-tailed Phascogale	31
——— calura	Handsome-tailed Phascogale	32
——— lanigera	Woolly Phascogale	33
Antechinus Swainsoni	Swainson's Antechinus	34
——— leucopus	White-footed Antechinus	35
——— ferruginifrons	Rusty-fronted Antechinus	36
——— unicolor	Dusky Antechinus	37
——— leucogaster	White-bellied Antechinus	38
——— apicalis	Freckled Antechinus	39
——— flavipes	Rusty-footed Antechinus	40
——— fuliginosus	Sooty Antechinus	41
——— albipes	White-footed Antechinus	42
——— murinus	Murine Antechinus	43
——— maculatus	Spotted Antechinus	44
——— minutissimus	Minute Antechinus	45
Podabrus macrourus	Large-tailed Podabrus	46
——— crassicaudatus	Thick-tailed Podabrus	47
Sarcophilus ursinus	Ursine Sarcophilus	48
Dasyurus maculatus	Spotted-tailed Dasyurus	49
——— Viverrinus	Variable Dasyurus	50
——— Geoffroyi	Geoffroy's Dasyurus	51
——— hallucatus	North Australian Dasyurus	52
Thylacinus cynocephalus	Thylacinus	53, 54
Phascolomys Wombat	Wombat	55, 56
——— latifrons	Broad-fronted Wombat	57, 58
——— lasiorhinus	Hairy-nosed Wombat	59, 60

ORNITHORHYNCHUS ANATINUS.

ORNITHORHYNCHUS ANATINUS.

Ornithorhynchus.

Platypus Anatinus, Shaw, Nat. Misc., vol. x. pl. 385.—Ib. Gen. Zool., vol. i. part i. p. 229. pls. 66 & 67.—Gray. List of Mamm. in Coll. Brit. Mus., p. 191.

Ornithorhynchus paradoxus, Blumenbach in Voigt's Magaz., tom. ii. p. 305. pl. 41.—Home in Phil. Trans. 1800, p. 432, and 1802, p. 67.—Cuv. Règn. Anim. Edit. 1829, tom. i. p. 235.—Meckel, Ornith. paradoxi desc. anatom., Lips. 1826, fol.—Owen in Trans. Zool. Soc., vol. i. p. 221.—Bennett in Trans. Zool. Soc., vol. i. p. 229.

——————— *fuscus* et *rufus*, Peron, Voy. de Découv., tom. i. pl. 34. figs. 1 & 2.—Leach, Zool. Misc., vol. ii. p. 136. pl. 3.—Desm. Mamm., part ii. p. 380.

——————— *brevirostris*, Ogilby in Proc. of Comm. of Sci. and Corr. of Zool. Soc., part i. p. 150.

——————— *crispus* et *lævis*, MacGill. in Mem. of the Wernerian Soc. 1832, p. 127.

——————— *Anatinus*, Waterh. Nat. Hist. of Mamm., vol. i. p. 25.

On commencing a history of the *Ornithorhynchus*, the mind naturally reverts to the period of its first discovery; a period so recent, that the animal was unknown to Linnæus and the older authors. It was in 1799 that a description of this singular quadruped first appeared in the "Naturalists' Miscellany" of Dr. Shaw; about this time also, the *Koala, Wombat, Kangaroo, Emu, Menura, Cereopsis,* and *Black Swan* were made known. These important discoveries gave an extraordinary impulse to the study of natural history, and set the whole scientific community wondering at the paradoxical creations of the distant country known by the name of Australia. Unquestionably the most singular and anomalous of all these animals was the *Ornithorhynchus*, with the habits and economy of which, as well as the mode of its reproduction, we are even now, after an interval of fifty-five years, but imperfectly acquainted. It is true that Professor Owen has given an elaborate paper on its anatomy and physiology in the "Transactions of the Zoological Society of London," and that the same work contains Mr. Bennett's interesting account of his observations of the animal in a state of nature and in captivity; still I am persuaded that much more remains to be ascertained and made known respecting this extraordinary type among quadrupeds. Although the ornithology of Australia almost exclusively engrossed my attention during my interesting visit to that country, I did not fail to notice the mammals which crossed my path and by which I was always surrounded. The *Ornitho-rhynchus* especially attracted my attention, as I frequently met with it both while ascending the rivers in Van Diemen's Land and while encamped beside the quiet pools of New South Wales. I endeavoured to determine the centre of its area and to trace the extent of its range, but was not entirely successful, nor have they yet been accurately ascertained: Van Diemen's Land, and the south-eastern part of the continent from Moreton Bay to Port Philip, are the only portions of that great country whence I have received specimens, or where I have heard of it existing. In New South Wales it is common in the streams and rivers flowing from the mountain ranges to the sea, as well as in those descending towards the interior. It is equally numerous in all the tributaries which feed the great rivers Darling and Murray; and if it be not now plentiful in the Hawkesbury, Hunter, &c., the diminution in its numbers is solely due to the wholesale destruction dealt out to it by the settlers, which, if not restrained, will ere long lead to the utter extirpation of this harmless and inoffensive animal, a circumstance which would be much to be regretted; it is in fact often killed from mere wantonness, or at most for no more useful purpose than to make slippers of its skin. Some zoologists have entertained the opinion that there are more than one species of this form, and that the animal inhabiting Van Diemen's Land, with stiff wiry hairs, particularly on the tail, where they, moreover, nearly cross each other at right angles, is specifically different from that found on the continent, which is generally of a smaller size, and of which the hairy covering is more sleek and glossy; I believe, however, that no tangible specific differences will be found, and that the variations in question are due to localization alone; much variety is also found in the colouring of the under surface, but as this occurs both in island and continental specimens, it cannot be regarded as a matter of importance.

In many of its habits and actions, and in much of its economy, the *Ornithorhynchus* assimilates very closely to the Common Water Vole of this country (*Arvicola amphibius*, Desm.); frequenting as it does similar situations, climbing stumps of trees and snags which lie prostrate in the beds of rivers, and burrowing in the bank side in an upward direction, a retreat to which it resorts during the day or on the approach of danger. If it be not strictly nocturnal, it is in the early morning and evening and in lowery weather only that it is to be seen during the daytime. It swims with great ease, and frequents alike the rushy banks of the great rivers near the sea, and the silent, tranquil pools of the interior. Its mode of swimming is very singular and not always alike; sometimes the body of the animal, beaver-like, is partly raised above the surface, while at others, particularly in the still pools, every part is submerged except the upper surface of the bill and nostrils, and these being but sufficiently elevated above the water to enable the animal to breathe, it is only by the little rings which this operation creates upon the glassy surface that its presence can be detected. I have frequently come suddenly upon it while ascending the reedy sides of the Derwent in a boat, when it instantly dived, with an audible splash, caused apparently by the hasty flap of its broad tail. I could say much more respecting the habits of this curious quadruped. In the volume of the " Transactions of the Zoological Society" above referred to, Mr. Bennett states that:—

" The *Ornithorhynchus* is known to the colonists by the name of *Water-Mole*, from some resemblance which it is supposed to bear to the common European Mole, *Talpa Europæa*, Linn.: by the native tribes at Bathurst and Goulburn Plains, and in the Yas, Murrumbidgee, and Tumat countries, I found it designated by the name of *Mullangong* or *Tambreet*; but the latter is more in use among them than the former. It is very abundant in the river Yas, particularly in the tranquil parts of the stream called 'ponds,' the surface of which is covered with various aquatic plants. On perceiving it, the spectator must remain perfectly stationary, as the slightest noise or movement would cause its instant disappearance, so acute is it in sight or hearing, or perhaps both; and it seldom reappears when it has been frightened. By remaining perfectly quiet when the animal is 'up,' the spectator is enabled to obtain an excellent view of its movements; it seldom, however, remains longer than one or two minutes, playing and paddling on the surface, soon diving again and reappearing a short distance above or below, generally according to the direction in which it dives, which it does head foremost.

" The various contradictory accounts that have been given, on the authority of the aborigines, as to the animal laying eggs and hatching them, induced me to take some pains to find out the cause of error, and being perfectly satisfied, from an internal examination of a female, that *ova* were produced in the *uteri*, I could the more readily determine the accuracy or inaccuracy of the accounts which I might receive from the natives.

" The Yas natives at first asserted that the animal lays eggs, but shortly afterwards contradicted themselves. In the Tumat country the answers were readily and satisfactorily given—'No egg tumble down; pickaninny make tumble down'—which accorded with my own observations.

" On the 7th of October, I accompanied an aborigine, called Daraga, to the banks of the Yas, to see the burrow of an *Ornithorhynchus*, from which, he told me, the young had been taken last summer. I asked him, 'What for you dig up *Mullangong*?' 'Murry budgeree patta' (Very good to eat), was his reply. On arriving at the spot, situated on a steep bank close to the river, about which long grass and various other herbaceous plants abounded, my guide, putting aside the long grass, displayed the entrance to the burrow, distant rather more than a foot from the water's edge. In digging up this retreat the natives had not laid it entirely open, but had delved holes at certain distances, and introduced a stick to ascertain its direction previously to again digging down upon it. By this method they were enabled to explore the whole extent with less labour than by laying it open from end to end. The termination of the burrow was broader than any other part, nearly oval in form, and strewed at the bottom with dry river weeds, &c., a quantity of which still remained. The whole of the interior was smooth, extending about twenty feet in a serpentine direction up the bank. It had one entrance near the water's edge, and another under the water, communicating with the interior by an opening just within the upper entrance. It is no doubt by the latter that the animal seeks refuge when it is seen to dive and not to rise again to the surface.

" On examining the cheek-pouches or the stomachs of these animals, I always observed the food to consist of river insects, very small shell-fish, &c., comminuted and mingled with mud or gravel: this latter might be required to aid digestion, as I never observed the food unmingled with it. The natives say that they also feed on river weeds; but I never found remains of that description of food in their pouches. Mr. George MacLeay informed me that he had shot some, in a part of the Wollondilly River, having river weeds in their pouches; but he observed that in that part of the river aquatic insects were very scarce. The young are suckled at first, and afterwards fed with insects, &c., mingled with mud.

" Having captured one alive, I placed it in a cask, with grass, mud, water, and everything necessary to make it comfortable. It ran round its place of confinement, scratching and making great efforts to get out; but finding them useless, became quite tranquil, contracted itself into a small compass, and soon fell asleep. At night it became very restless, and diligently sought to escape, going round the cask with the fore paws raised against the sides and the webs thrown back, and scratching violently with the claws of the fore feet, as if to burrow its way out. In the morning I found it fast asleep, with the tail turned inwards, the head and beak under the breast, and the body contracted into a very small compass; subsequently, however, I observed it sleeping with the tail turned inwards, the body contracted, and the beak protruded. When disturbed from its sleep, it uttered a noise something like the growl of a puppy, but in a softer and more harmonious key. Although quiet for the greater part of the day, it constantly made efforts to escape, and uttered a growling noise during the night."

Shortly after this, Mr. Bennett started for Sydney, taking with him his interesting captive. "*En route*," he availed himself of the vicinity of some ponds, inhabited by these animals, to give it a little recreation; " and accordingly tied a long cord to its leg and roused it from its sleep; when placed on the bank it soon found its way to the water, and travelled up the stream, apparently delighting in those places which most abounded with aquatic weeds. Although it dived in deep water, it appeared to prefer keeping close to the bank, occasionally thrusting its beak (with a motion similar to that of a *Duck* when it feeds) among the mud, and at the roots of the weeds lining the margin of the ponds, which we may readily suppose to be the resort of insects. After it had wandered some distance, it crawled up the bank, and lying down on the grass, enjoyed the luxury of scratching itself and rolling about. In this process of cleaning itself, the hind claws were alone brought into use; first the claws of one hind leg, then those of the other. The body being so capable of contraction was readily brought within reach of the hind feet, and the head also was brought so close as to have its share in the cleaning process. The animal remained for upwards of an hour thus engaged, after which it had a more sleek and glossy appearance. It permitted me to smooth it gently over the back, but disliked being handled.

"On the 28th of December I visited a very beautiful part of the Wollondilly River, which has the native name of Koroa, and explored a burrow, the termination of which was thirty-five feet from the entrance. Extensive as this may appear, burrows have been found of even fifty feet in length. On arriving at the termination a growling was distinctly heard, which upon further search was found to proceed from two full-furred young ones, a male and a female, coiled up asleep, and which growled exceedingly at being exposed to the light of day. They measured ten inches from the extremity of the beak to that of the tail; had a most beautifully sleek and delicate appearance, and seemed never to have left the burrow. When awakened and placed on the ground, they moved about, but did not make such wild attempts to escape as the old ones do when caught. Shortly afterwards a female was captured, which was no doubt the mother; she was in a ragged and wretchedly poor condition; her fur was rubbed in several places and she seemed in a very weak state. The eyes of the natives glistened and their mouths watered when they saw the fine condition of the young *mullangongs*, and they frequently and earnestly exclaimed, 'Cobbong fat' (large, or very fat), and 'Murry budgeree patta' (very good to eat). They said they were more than eight moons old; if so, they must have been the young of the previous season.

"The young animals sleep in various postures; sometimes in an extended position, and often rolled up like a hedgehog. One lies curled up like a dog, keeping its beak warm with the flattened tail brought over it; while another lies stretched on its back, the head resting by way of pillow upon the body of the old one, lying on its side, the delicate beak and smooth clean fur of the young contrasting with the rough and dirtier appearance of the mother. The favourite posture appears to be that of lying rolled up like a ball: this is effected by the fore paws being placed under the beak, with the head and mandibles bent down towards the tail, the hind paws crossed over the mandibles, and the tail turned up, thus completing the rotundity of the figure.

"Although furnished with a good thick coat of fur, they seemed particular about being kept warm. They would allow me to smooth the fur, but if their mandibles were touched they darted away immediately, those parts appearing to be remarkably sensitive. I could permit the young to run about as they pleased, but the old one was so restless, and damaged the walls of the room so much by attempts at burrowing, that I was obliged to keep her close prisoner. The little animals appeared often to dream of swimming, as I have frequently seen their fore paws in movement as if in that act. If placed on the ground in the daytime, they sought some dark corner for repose; but when put in a dark corner or in a box, they huddled themselves up as soon as they became reconciled to the place, and went to sleep. They would sleep on a table, sofa, or indeed in any place; but, if permitted, would always resort to that to which they had been accustomed. Still, although for days together they would sleep in the place made up for them, yet on a sudden they would repose behind a box or in some dark corner in preference.

"When running they are exceedingly animated, their little eyes glisten, and the orifices of the ears contract and dilate with rapidity; if then taken into the hands for examination, they struggle violently to escape, and their loose integuments render it difficult to retain them. Their eyes being placed high in the head, they do not see objects well in a straight line, and consequently run against everything in their perambulations, spreading confusion among all light and readily overturnable articles. Occasionally they elevate the head, as if to observe objects above or around them. Sometimes I have been able to enter into play with them by scratching and tickling them with my finger; they seemed to enjoy it exceedingly, opening their mandibles, biting playfully at the finger, and moving about like puppies indulged with similar treatment. Besides combing their fur to clean or dry it when wet, I have also seen them peck it with the beak, as a *Duck* would clean its feathers, by both which processes their coats acquire a clean and glossy appearance.

"I was often surprised to find them on the summit of a book-case or some other elevated piece of furniture, and equally at a loss to imagine how they came there, until I at length discovered that it was effected by the animal placing its back against the wall and its feet against the book-case, and by means of the strong cutaneous muscles of the back and the claws of the feet, contriving to reach the top very expeditiously."

The number of young produced at one time has not been satisfactorily ascertained; it has been stated that they are from two to four in number, but I believe that they rarely exceed two. When first born they are naked, and the beak does not resemble that of the adult, but is short, broad and thick, and fitted to embrace the mammary areola concealed by the hairs of the mother; "the tongue too," says Professor Owen, "which in the adult is lodged far back in the mouth, advances in the young animal close to the end of the lower mandibles; and its disproportionate breadth is plainly indicative of the importance of the organ to the young animal, both in receiving and swallowing its food; the thin fold of integument also, which surrounds the base of the mandibles, and extends the angle of the mouth from the base of the lower jaw to equal the breadth of the base of the upper one, must increase the facility for receiving the milk ejected from the mammary areola of the mother." "While sucking," says M. Verreaux, "the young continually rub or triturate the mother's belly with the fore feet, and occasionally with the hinder ones. At the end of fifteen to twenty days the new born are covered with a silky hair, and are able to swim." M. Verreaux also describes another mode by which the young obtain the lacteal fluid:—"Having a considerable number of adults and young at my disposition, I saw the latter accompany their mothers, with which they played, especially when they were too far from the bank to take their nourishment. I observed that when they wished to procure it, they profited by the moment when the mother was amongst the aquatic plants near the land, where there is no current. The female having her back exposed, by the exercise of a strong pressure the milk floats to a little distance, and the young may suck it up with facility; and thus they do, turning about so as to lose as little as possible. I cannot, perhaps, better compare the appearance of the greasy milk, under these circumstances, than to the iridescent colours produced by the solar rays upon stagnant water. I have witnessed this fact repeatedly, both daily and nightly. I have also remarked that the young, when fatigued, climbed upon the back of the mother, who brought it to land, where it caressed her.

"The body of this singular animal is covered with a fine, long and thick hair, underneath which is a finer, short, very soft fur, resembling the two distinct kinds of fur found in the *Seal* and *Otter*; on the abdomen, breast and throat, the fur and hair are of a much finer quality and of a more silky nature than on the other parts of the body; while on the upper surface of the tail the hair is longer and coarser. The general colour of the upper surface is a light black; the under short fur is greyish; the whole of the under surface is ferruginous; immediately below the inner angle of the eye is a small spot of a light or pale yellow; the legs are short, pentadactyle and webbed; on the fore feet (which seem to have the greatest muscular power, and are in principal use for burrowing and swimming) the webs extend a short distance beyond the claws, are loose, and fall back when the animal burrows; the claws are strong, blunt, and well adapted for burrowing; the hind feet are short, narrow, turned backwards, and when the animal is at rest, have, like those of the *Seal*, some resemblance to a fin; their action is backwards and outwards; the nails are all curved backwards, and are longer and sharper than those of the fore feet; the web does not extend further than the base of the claws. The head is rather flat, from which project two flat lips or mandibles, resembling the beak of a *Shoveller Duck*, the lower of which is shorter and narrower than the upper, and has its internal edges channeled with numerous *striæ*, resembling in some degree those seen in the bill of a *Duck*. The colour of the superior mandible is of a dull dirty greyish-black, covered with innumerable minute dots; the under part of the upper mandible is of a pale pink or flesh-colour, as is the internal or upper surface of the lower mandible, the under surface of which is either perfectly white or mottled,—in young specimens usually the former, in old ones the latter; at the base of both mandibles is a transverse loose fold or flap of integument, always similar in colour to the skin covering the mandibles, that is, dull greyish-black above, and white or mottled below. In the upper mandible this is continued to the eyes, and may perhaps afford protection to those organs when the animal is burrowing or seeking food in the mud; the upper fold or flap is continuous with another portion arising from the lower mandible also at its base; the eyes are very small, but brilliant, and of a light brown.

"In young specimens, the under surface of the tail, as well as the hind and fore legs near the feet, are covered by fine hair of a beautiful silvery-white appearance; this is lost, however, in the adult, in which the under surface of the tail is almost entirely destitute of hair. Whether this proceeds from its trailing along the ground, I know not; but the prevailing opinion among the colonists, for which, however, I could not discover any foundation, is that it is occasioned by the animal using the tail as a trowel in the construction of its dwelling.

"The only external difference in the sexes is the presence in the male of a spur, situated on the internal part of the leg, some distance above the claws; this spur, which is moveable and turned backwards and inwards, was considered to be poisonous, but some experiments" (instituted by Mr. Bennett) "prove that it is innocuous: it is entirely wanting in the females.

"The size of the *Ornithorhynchus* varies, but the males are usually found to be slightly larger than the opposite sex; the average length is from 18 to 20 inches."

In conclusion, I must not omit to call attention to the very valuable details respecting the anatomy of this animal, given by Professor Owen and Mr. Bennett, in the "Transactions of the Zoological Society" above referred to. There will also be found in the "Revue Zoologique" for 1848 some very interesting particulars respecting the reproduction and other points in the economy of this animal, by M. Jules Verreaux, acquired by personal observation in Van Diemen's Land. Professor Owen's remarks on M. Verreaux's observations, published in the "Annals and Magazine of Natural History" for 1848, may also be consulted with advantage.

The Plate represents the two sexes about three-fourths of the natural size.

ECHIDNA HYSTRIX, *Cuv.*

ECHIDNA HYSTRIX, *Cuv.*

Spiny Echidna.

Myrmecophaga aculeata, Shaw, Nat. Misc., vol. iii. pl. 109.
Aculeated Ant-eater, Shaw, Gen. Zool., vol. i. pt. 1. p. 175.
Ornithorhynchus Hystrix, Home, Phil. Trans. 1802, p. 348.
Echidna Hystrix, Cuv. Règ. Anim.—Leach, Zool. Misc., vol. ii. t. 90.—List of Mamm. in Brit. Mus., p. 192.
Tachyglossus aculeatus, (Illiger) Schreb. Saugth., t. lxiii. B.
Echidna longiaculeata, Tiedem. Zool., tom. i. p. 592.
———— *Australiensis*, Lesson.
———— *aculeata*, Waterh. Nat. Hist. Mamm., vol. i. p. 41.
Dun-ung-er-de, Aborigines of the Toodyay and Guildford Districts of Western Australia.
Nyoong-arn, Aborigines of the York district.

THE sandy and sterile districts which so frequently occur over the whole of the southern portions of the Australian continent constitute the native habitat of the *Echidna Hystrix*, but although so very generally dispersed, it is nowhere abundant; I have also met with it in the islands in Bass's Straits, and Mr. Gilbert obtained a single example in Western Australia, which had been taken on a farm situate on the upper part of the eastern branch of the river Avon; he subsequently learnt from the natives that it had been seen in the Toodyay district and in the vicinity of Guildford. No instance of its occurring to the northward of the colonies has yet been recorded, and in all probability, like the Ornithorhynchus, it is strictly confined to the southern part of the country.

As I had but little opportunity of acquiring a knowledge of the habits of this animal in a state of nature, and my friend George Bennett, Esq., has been more fortunate in this respect, I cannot perhaps do better than transcribe the account published by him in his 'Wanderings in New South Wales, &c.'

" Among other extraordinary animals furnished to the naturalist in this interesting country is the *Echidna*, or 'native Porcupine,' the Nickobejan and Jannocumbine of the natives. It inhabits mountain ranges, burrowing with extraordinary facility, and producing its young in December.

" At Goulburn Plains the natives brought me a young living specimen of this animal which they had just caught upon the ranges: they called it Jannocumbine, and fed it upon ants and ants' eggs. It was often taken to an ant-hill to provide itself with food: from being so young it had an unsteady walk, and was covered with short sharp spines projecting above the fur. On expressing a fear to the natives of not being able to keep it alive, they replied that 'it would not now die, as it had prickles on;' meaning, I suppose, that it could feed and provide for itself, not requiring the fostering care of its parents. On asking whether it was a male or female, they examined the hind feet for the spurs, and, seeing them, declared it to be a male. It sleeps during the day, running about and feeding at night. Its movements are tardy, the principal exertions being made when burrowing. When touched upon the under surface, or uncovered part s of its body, or when attacked by dogs, it rolls, like the hedgehog, into a spherical form, the prickly coat forming a good defence against the canine race, who have a decided aversion to have their noses pricked. When attacked, it has been known to burrow to a great depth in a surprising short period of time.

" The Echidna is eaten by the natives, who declare it to be 'very good, and, like pig, very fat.' Europeans who have eaten of them confirm this opinion, and observe that they taste similar to a sucking-pig. This animal, when scratching, or rather cleaning itself, uses only the hind claws, lying in different positions, so as to enable it to reach the part of the body to be operated upon. The power of erecting the spines and rolling itself into a spherical form makes an excellent defence against many of its enemies.

" I consider that there are two species of this genus existing : first, *E. Hystrix*, or Spiny Echidna, which is found on the mountain ridges in the colony of New South Wales ; and the second, *E. setosa*, or Bristly Echidna, which is found more common in Van Diemen's Land. The first species attains a large size ; it is stated in our works of natural history as being the size of a hedgehog ; my young specimen was fully that. At 'Newington,' the residence of John Blaxland, Esq., I had an opportunity of seeing a specimen full fourteen inches long and of proportionate circumference ; it fed upon milk and eggs, the eggs boiled hard and chopped up small, with rice ; its motion was heavy and slow ; it was of a perfectly harmless disposition. When disturbed from its place of retreat it would feed during the day, but was difficult to remove from the cask in which it was placed, on account of its firmly fixing itself at the bottom; it feeds by thrusting out the tongue, to which organ the food is attached, and then withdrawing it. Mine moved about, and drank milk at night, taking little other food. After keeping it for nearly seven months, I found it one morning dead."

In a state of nature the food consists of ants, of which a never-failing supply can at all times be procured, since this tribe of insects is probably more numerous in Australia than in any other part of the world ; they are procured by means of its protractile, lengthened, slender and flexible tongue, which is constantly kept lubricated with a viscous matter, to which the ants adhere. " To supply this secretion," says Mr. Water-house in the work above quoted, " the Echidna is provided with two enormous submaxillary glands, which extend from behind the ear to the fore-part of the chest. There are no teeth to the jaws, but the palatal portion of the mouth is armed with several rows of strong horny spines, the points of which are directed backwards ; and on the upper surface of the tongue are numerous small horny warts, between which and the palatal spines the prey of the animal is, no doubt, crushed before passing into the stomach." Lieut. Breton states that " occasionally the tongue is curved laterally, and the food as it were swept into the mouth."

The muzzle is covered with a naked purplish black skin ; the eyes are small and black ; the rather short and stout body is covered with a thick skin, particularly on the back, where it has to support the strong spines ; these are of a dirty white colour, more or less broadly tipped with black, sharply pointed, and about one inch and three-quarters in length ; they commence on the back part of the head, and extend over the whole upper surface of the body ; their points are directed backwards, and on the back inwards, so that they cross each other in the mesial line ; near the root of the tail they form a large tuft, radiating from two approximating centres, and hide the small rudimentary tail ; the head, with the exception of the hinder half of the upper surface, and the lower half of the sides of the body, as well as the whole of the under surface and limbs, are covered with coarse brownish black hairs ; the legs are short and strong ; the fore feet short and broad, and armed with large, solid and nearly straight nails, that of the middle toe being about an inch in length and a quarter of an inch in width ; the shortest, that of the inner toe, is four or five lines in length ; all are rounded at the extremity ; the lined feet are narrower and less powerful than the others, and have the inner toe very short, apparently slightly opposable, and with a short and broad nail rounded at the extremity ; the toe next the inner one is the longest, and is armed with an enormous claw, measuring sometimes an inch and a half in length ; it is curved and nearly cylindrical, but concave beneath ; the claws of the other toes are progressively shorter. The hind foot, when in its natural position, rests on its inner side, and perhaps in a great measure upon the thumb or great toe, by which arrangement the claws are protected from wear when the animal is walking, and have the concave surface presented outwards ; the use of these claws, it would appear, is to cast away the earth which is loosened by the stronger fore-feet and claws. Like the Ornithorhynchus, the heel in the male sex is armed with a strong spur, which is moveable, perforated, and supplied with a gland and muscles capable of ejecting the secretion of the gland through the canal of the spur. Messrs. Quoy and Gaimard tried, by irritating the animal, to induce it to inflict a wound upon themselves, in order to ascertain whether this apparatus was poisonous, but were unsuccessful ; and after repeated inquiries could not learn that any accident had ever happened from a wound of the spur.

The figures are of the natural size.

ECHIDNA SETOSA: *Cuv.*

ECHIDNA SETOSA, *Cuv.*

Hairy Echidna.

Echidna setosa, Cuv. Règne Anim., Edit. 1817, tom. i. p. 226 ; Nouv. Edit., tom. i. p. 235.—Waterh. Nat. Hist.
of Mamm., vol. i. p. 47.—Geoff. Bull. Soc. Phil., tom. iii t. 15.—Gray, List of Mamm. in Brit. Mus.,
p. 192.
Echidna breviaculeata, Tiedemann, Zoologie, tom. i. p. 592.
Tachyglossus setosus, Ill. Schreb. t. 63.

WHETHER there be one or two species of the present genus is a question on which the opinions of zoologists
are divided, but in either case it becomes necessary that animals exhibiting so great a difference as do the
Echidnas from New South Wales and Van Diemen's Land, should each be figured in a work on the
Mammals of Australia. No instance has come under my notice of the Hairy Echidna or the animal here
figured having occurred on the continent of Australia, while in Van Diemen's Land it is very common. I
am aware that the hairy covering has been considered indicative of youth, and also as due to the colder
climate of Van Diemen's Land ; nevertheless I have not failed to remark, that not only is the animal generally speaking of larger size, but the spines are shorter and more slender ; it is however, I admit, a matter
still wrapped in uncertainty, and one which I would recommend to the attention of zoologists resident in
Australia, since it is by their observations that the doubt is most likely to be cleared up.

The *Echidna setosa* is universally dispersed over the sandy districts of Van Diemen's Land, and so common is it in the neighbourhood of Hobart Town, that living specimens are frequently brought in and exhibited for sale, the usual price being half-a-crown. Several examples kept by me for some time during my
residence there, ran about the room in which I was engaged without exhibiting any signs of alarm ; at the
same time they appeared impatient of restraint, and made many attempts to escape ; but that it might to
a certain degree be domesticated or trained to bear captivity, is proved by several examples having lately
been brought alive to this country, which unfortunately did not long survive.

Like the other species, it feeds upon ants and other insects, which it procures by protruding and
retracting the tongue, covered with a thick glutinous fluid ; in captivity sopped bread and milk forms an
agreeable substitute for its natural food.

General colour brown ; all the upper surface of the body thickly beset with pale yellowish spines tipped
with black ; the fur on the back dark brown, and so dense and lengthened as nearly to hide the spines ;
eye brown ; snout slate-colour ; tongue and soles of the feet pink ; claws blackish brown.

"The *E. setosa*," says Mr. Waterhouse, " is subject to some slight variation in tint, as well as in texture
of the fur ; the spines also vary slightly, being rather longer in some specimens than others ; yet the
differences observable in individuals are not such as to render it difficult to distinguish the *E. setosa* from the
E. hystrix ;" and he adds, that he suspects the more hairy clothing of this animal may be due to the comparatively humid climate of Van Diemen's Land, which may have had the effect of causing the fur to become
longer and more dense ; and if so, the increased development of the fur would in all probability affect the
growth of the spines, by robbing them of their nutriment.

The figures are of the natural size.

MYRMECOBIUS FASCIATUS: *Waterh.*

J. Gould and H.C. Richter, del. et lith.

Hullmandel & Walton Imp.

MYRMECOBIUS FASCIATUS, *Waterh.*

Banded Myrmecobius.

Myrmecobius fasciatus, Waterh. in Proc. of Zool. Soc., Part IV. pp. 69 and 131.—Ib. Trans. of Zool. Soc., vol. ii.
p. 149. pl. 27.—Ib. Nat. Lib. Mamm., vol. ix. (Marsupialia), p. 145. pl. xi.—List of Mamm. in Brit.
Mus. Coll., p. 100.
Noom-bat, Aborigines of the York and Toodyay districts of Western Australia.
Wai-haw, Aborigines of King George's Sound.

THE beautiful animal forming the subject of the present Plate is a native of Western Australia, where it is very generally dispersed over the interior of the Swan River Settlement, from King George's Sound on the south to the neighbourhood of Moore's River on the north, and as far westward as civilized man has yet been able to penetrate. Although it must have been known to the settlers from the foundation of the colony, yet it is only within the last ten years that specimens have been sent to Europe, and brought under the notice of the scientific world. For the first description of this elegant marsupial we are indebted to Mr. Waterhouse, who, from the scanty materials of a single skin, formed a just view of its affinities and assigned it to the order—the Marsupialia—to which it naturally belongs. Sterile sandy districts thinly studded with moderately sized trees appear to be congenial to its habits and mode of life. As the form of its teeth would indicate, insects constitute a great part of its food; but I believe that it also feeds upon honey and a species of manna which exudes from the leaves of the *Eucalypti*. Wherever the Myrmecobius takes up its abode, there ants are found to be very abundant, and in all probability, for I have no direct evidence that such is the case, it is upon this insect or its larvæ that it mainly subsists.

As regards the ornamental appearance of this animal, I need only call the attention of my readers to the accompanying figures, where it is represented of the natural size. When running on the ground with its beautiful tail spread out to the full extent, it offers a great resemblance to the Squirrels. On the slightest appearance of danger it secretes itself in a hollow tree, from which it is not easily driven. Much diversity exists in the markings of different individuals, and these variations are common to both sexes. In animals of the same age the male considerably exceeds the female in size. The young from their earliest youth are marked with fasciæ like the adult, so that the latter are to be distinguished only by size, or ascertained by dissection.

The following remarks, which I give in the words of the respective writers, and in the order they have reached me, may not prove uninteresting :—

"Two of these animals," says Mr. Dale (from one of which Mr. Waterhouse took his description), "were discovered about ninety miles south-east of Swan River, and within a few miles of each other. They were first observed on the ground, and on being pursued, both directed their flight to some hollow trees which were near. The country in which they were found abounded in decayed trees and ant-hills." (Waterhouse's Marsupialia, p. 147.)

"You may place great reliance," writes His Excellency G. Grey, Esq., Governor of South Australia, "on the following description of the habits of Myrmecobius; it is partly derived from the natives, and partly from the observations of Mrs. Grey, who has seen several in a state of captivity. It cannot run very fast. Its tongue is about as thick as a common tobacco-pipe and gradually tapers; it is extensile and can be protruded from the mouth for several inches, and when in this state the animal moves it about with great rapidity. In the daytime it lives in decayed trees; at night it runs about and climbs the trees like an opossum. One that was kept in confinement was fed on sugar and milk, in which it dipped its tongue."

In a letter lately received from Mr. Gilbert he states, "I have seen a good deal of this beautiful little animal. It appears very much like a squirrel when running on the ground, which it does in successive leaps, with its tail a little elevated; every now and then raising its body and resting on its hind-feet. When alarmed it generally takes to a dead tree lying on the ground, and before entering the hollow invariably raises itself on its hind-feet to ascertain the reality of approaching danger. In this kind of retreat it is easily captured, and when caught is so harmless and tame as scarcely to make any resistance, and never to attempt to bite. When it has no chance of escaping from its place of refuge it utters a sort of half-smothered grunt, apparently produced by a succession of hard breathings. I have heard of this animal being frequently kept in confinement and fed for several weeks together upon no other food than bran.

"The female is said to bring forth her young in a hole in the ground or in a fallen tree, and to produce from five to nine in a litter. I have not myself observed more than seven young attached to the nipples. Like the members of the genus *Antechinus*, this animal has no pouch for protection or envelopment of the young." The only protection afforded their delicate offspring is the long hairs which clothe the under surface of the mother.

The hair of the Myrmecobius is harsh and bristly to the touch. A black stripe passes from the nose through the eye to the neck; shoulders and upper part of the back bright rusty red, which gradually fades into rusty brown on the crown of the head, face and ears; back distinctly banded with lines of buffy white and blackish brown, the number of bands varying in different individuals; chin, throat and all the under surface yellowish white; upper part of both fore- and hind-feet rusty yellow; tail bushy for its whole length and parti-coloured, some of the hairs being black, while others are rusty red or yellowish white; in some instances the hairs of the tail are black at the base, then yellowish white, and terminate in rusty red.

The Plate represents an adult male and female of the size of life.

TARSIPES ROSTRATUS. *Gerv. et Verr.*

J. Gould and W. Hart del. et lith.   Hullmandel & Walton Imp.

TARSIPES ROSTRATUS, *Gerv. et Verr.*

Long-nosed Tarsipes.

Tarsipes rostratus, Gerv. and Verr. in Proc. of Zool. Soc., Part X. p. 1.—Ib. in Guerin's Mag. de Zool., 1842,
  Mamm., pls. 35, 36, 37.
——— *Spenseræ,* Gray in Ann. and Mag. of Nat. Hist., vol. ix. p. 40.—List of Mamm. in Brit. Mus. Coll., p. 87.
Jeë-pin, Aborigines around Perth.
Ngool-boon-goor, Aborigines of King George's Sound.

THIS highly curious little animal was first brought before the notice of the scientific world by M. Paul
Gervais, who in his own name and that of M. Jules Verreaux read a lengthened memoir, illustrated by
drawings, respecting its structure and affinities, at the scientific meeting of the Zoological Society of
London, held on the 11th of January, 1842; immediately after which period specimens were sent to
this country from King George's Sound by His Excellency Governor Grey; and Mr. J. E. Gray, conceiving
the differences they exhibited from M. Gervais' animal to be of specific importance, applied to the animal
in question the term *Spenseræ,* from the maiden name of His Excellency's amiable lady. As soon as I
became aware of the existence of so interesting an animal in Western Australia, I wrote to Mr. Gilbert,
and directed him to pay particular attention to the subject; and he has since transmitted to me several
examples both from the neighbourhood of Swan River and from King George's Sound, a careful examination of which with those above-mentioned has fully satisfied me of their identity.

The following notes accompanied the specimens :—

" The Tarsipes is generally found in all situations suited to its existence from Swan River to King
George's Sound, but from its rarity and the difficulty with which it is procured, notwithstanding the high
rewards I offered, the natives only brought me four specimens; one of these, a female, I kept alive for
several months, and it soon became so tame as to allow itself to be caressed in the hand without evincing
any fear or making any attempt to escape. It is strictly nocturnal, sleeping during the greater part of
the day and becoming exceedingly active at night : when intent upon catching flies it would sit quietly in
one corner of its cage, eagerly watching their movements, as, attracted by the sugar, they flew around ; and
when a fly was fairly within its reach it bounded as quick as lightning and seized it with unerring aim, then
retired to the bottom of the cage and devoured it at leisure, sitting tolerably erect and holding the fly
between its fore-paws, and always rejecting the head, wings and legs. The artificial food given it was
sopped bread made very sweet with sugar, into which it inserted its long tongue precisely in the way
in which the Honey-eaters among birds do theirs into the flower-cups for honey ; every morning the sop
was completely honey-combed, as it were, from the moisture having been drained from it by the repeated
insertion of the tongue ; a little moistened sugar on the end of the finger would attract it from one part of
the cage to the other ; and by this means an opportunity may be readily obtained for observing the beautiful
prehensile structure of the tongue, which I have frequently seen protruded for nearly an inch beyond the
nose ; the edges of the tongue near the tip are slightly serrated. The tail is prehensile, and is used when
the animal is climbing precisely like that of the *Hepoona.* The eyes, although small, are exceedingly
prominent and are placed very near each other ; the ears are generally carried quite erect. When sleeping
the animal rests upon the lower part of the back, with its long nose bent down between its fore-feet and its
tail brought over all and turned down the back. Mr. Johnson Drummond shot a pair in the act of sucking
the honey from the blossoms of the *Melaleuca* ; he watched them closely, and distinctly saw them insert
their long tongues into the flower precisely after the manner of the birds above-mentioned."

The figures on the accompanying Plate are of the natural size, and being carefully coloured after nature,
renders a minute description unnecessary. The sexes are similarly marked, and may be thus briefly described :—

All the upper surface grey with a dorsal stripe of black, on either side of which is a broader one of
reddish brown. The under surface and feet are buffy white, the buff tints becoming of a deeper hue on the
flanks, the forehead inclining to rufous, and the space round the eye buffy white.

The singular plant upon which the three figures are placed is a species of *Petrophila,* the specific name
of which I am unacquainted with : like many others of the Western Australian plants, it is probably
undescribed.

CHÆROPUS CASTANOTIS.

CHŒROPUS CASTANOTIS.

Chestnut-eared Chœropus.

Chœropus castanotis, Gray, Ann. and Mag. of Nat. Hist., vol. ix. p. 42.—List of Mamm. in Brit. Mus. Coll., p. 96.
Chœropus ecaudatus, Ogilby in Proc. of Zool. Soc., Part VI. p. 26.
Chœropus ecaudatus, Mitch. Trav. in Australia, vol. ii. p. 132. pl. 27.—Waterh. Nat. Lib. Mamm., vol. ix. (Marsupialia), p. 163.
Buř-da, Aborigines of the Walzemara district.
Wot-da, Aborigines of the interior from York, Western Australia.

For our first knowledge of this very singular animal we are indebted to the researches of Major Sir Thomas L. Mitchell, who during one of his expeditions into the interior of South-eastern Australia procured a specimen on the left bank of the Murray, and of which he gave a figure in the second volume of his " Travels." The specimen itself is deposited in the museum at Sydney, but a drawing by Sir Thomas Mitchell having been submitted to Mr. Ogilby's inspection, he at once perceived that it differed from every other known group of animals, and consequently made it the type of a new genus, assigning to it, from the presumed absence of any tail, the specific appellation of *ecaudatus*. Since that period an example from nearly the same locality has been sent to this country by His Excellency George Grey, Esq., Governor of South Australia, and two others by Mr. Gilbert from Western Australia. All these specimens are furnished with a well-developed tail, and the want of that organ in the Major's animal was doubtless the result of accident ; hence it became necessary that the specific term applied to it by Mr. Ogilby should be exchanged for one more appropriate, and Mr. Gray has therefore assigned to it that of *castanotis*, from the deep chestnut colouring of the ears.

" That the *Chœropus*," remarks Mr. Gilbert, " should occasionally lose its tail is not singular, for I have frequently found examples of the *Mala* (*Perameles myosurus*) with their tail shortened or entirely lost, apparently by some accident."

The specimen in the Sydney museum and that from South Australia above mentioned, and which is now in the British Museum, differ considerably, both in colour and in the length of the hair that covers the body, from those from Western Australia, so much so in fact as almost to induce a belief of their being distinct ; but until further information has been obtained respecting this very curious form, I prefer considering them as identical, and figuring them as such under Mr. Gray's name of *castanotis* ; but should future research prove the Western Australian animal to be distinct, the specific term *occidentalis* might be applied to it.

In Western Australia the *Chœropus* is confined to the interior ; it makes a nest precisely similar to that of *Perameles myosurus*, except that it is more abundantly supplied with leaves. It is sometimes found in the densest scrub, where from the thickness of the vegetation it is extremely difficult to be procured.

As its dentition would indicate, its food consists of insects and their larvæ, and vegetables of some kind, probably the bark of trees, bulbous and tuberous roots.

One of the two specimens received from Western Australia is in the collection of the British Museum, the other in that of the Earl of Derby.

The two front figures in the Plate represent the animals from Western Australia, and the central one that from South Australia. All the figures are of the natural size, and coloured so accurately as to render a description unnecessary.

PERAGALEA LAGOTIS.

J. Gould and H.C. Richter del. et lith.

Hullmandel & Walton Imp.

PERAGALEA LAGOTIS.

Large-eared Peragalea.

Perameles lagotis, Reid in Proc. of Zool. Soc., Part IV. p. 129.—Waterh. Nat. Lib. Mamm., vol. ix. (Marsupialia),
    p. 153. pl. xii.
Paragalia lagotis, Gray, App. to Gray's Trav., vol. ii. p. 401.
Peragalea lagotis, List of Mamm. in Brit. Mus. Coll., p. 96.
Dol-goitch or *Dal-gyte*, Aborigines of Western Australia.
Rabbit of the Colonists.

THE western portion of Australia is the only locality in which this fine animal has yet been discovered, evidencing with our comparatively recent acquisitions *Myrmecobius* and *Tarsipes*, that the mammalogy of that part of the continent is fully as interesting, both for novelty and singularity of form, as that of the eastern coast, which is inhabited by the Ornithorhynchus, Koala, &c.

The first notice of this animal on record is that published in the "Proceedings of the Zoological Society of London" for 1836, Mr. Reid, a member of the Society, having described it, from a skin exhibited at the scientific meeting of the 13th of December in that year, under the name of *Perameles lagotis*: he was in error, however, when he stated it to inhabit Van Diemen's Land. At the end of his paper, wherein the external characters and the dentition of the animal are minutely described, Mr. Reid gave it as his opinion that the distinctions between it and the other members of the genus *Perameles* were so marked, that it might be considered the type of a distinct genus, for which the term *Macrotis* would be an appropriate designation; he did not, however, publish any generic characters, and as the term *Macrotis* is objectionable from its similarity to the specific name, I am induced to adopt the generic designation proposed for it by Mr. Gray.

Were any attempts to be made at introducing the indigenous animals of Australia into Europe for ornamental purposes, or as additions to our articles of food, the present would be one of those with which it would be most desirable to make the trial. That it bears confinement well and contentedly, is proved by the fact of one having lived in the Gardens of the Zoological Society of London for some time; its death was doubtless attributable to the want of a suitable substitute for its natural food.

That its flesh is sweet and delicate, I have abundant testimony. When boiled it resembles that of the rabbit; prejudice would therefore be the only obstacle to its general adoption as an article of food, and this surely might easily be combated. I trust from what I have here said, that a sufficient hint has been thrown out to induce those who have the opportunity to import it into Europe.

The *Peragalea lagotis* is tolerably abundant over the whole extent of the grassy districts of the interior of the Swan River colony, where it lives for the most part in pairs, usually selecting spots where, the soil being loose, its powerful claws enable it to excavate the earth and form burrows with amazing rapidity. Into these holes it always retreats for safety; and as these subterraneous runs are both deep and long, it frequently eludes the pursuit of the natives, who hunt it for the sake of its flesh.

Its food consists of insects, their larvæ, and the roots of trees and plants; a favourite article is a large grub, the larva of a species of *Cerambyx*? which is deposited in the roots of the *Acaciæ*, and which is equally in request with the natives, who never fail to cut it out from an exposed root whenever they observe the *Dal-gyte* has been unsuccessful.

The number of young brought forth at a time has not yet been satisfactorily ascertained, but we may fairly presume that they are at least three or four.

The sexes present no difference whatever in their colouring, but the female is smaller than the male.

General tint of the upper parts of the head and body ashy grey; sides of the head, shoulders, and the sides of the body very pale vinous rust-colour; under parts of the head and body and the inner side of the limbs white; fore-legs and feet white, with a dark greyish patch on the outside of the former; tarsi white above, the hairs covering the under surface of a smoky brown colour; forepart of the outer legs white, outer and hinder part blackish grey; a whitish line extends backwards on the sides of the rump; soft long hair, coloured like that of the body, clothes about an inch and a half of the base of the tail; beyond this, for about three inches and a half, the tail is covered with black and somewhat harsh hairs; on the under side of the tail they are scarcely half an inch in length, but on the upper side most of them are upwards of one inch in length; the remainder of the tail is covered with white hairs, which increase in length on the upper side to the tip, where they are about two inches in length; on the under side they are short, and decrease in length towards the apex of the tail, the extreme point of which is naked; moustaches moderately long and black; ears almost naked, the margins fringed with whitish hairs; externally on the forepart they are covered with minute brown hairs.

The figure is of the natural size.

PERAMELES FASCIATA: Gray.

PERAMELES FASCIATA, *Gray.*

Banded Perameles.

Perameles fasciata, Gray in App. to Capt. Grey's Journ. of Two Exp. of Discovery in N.W. and W. Australia,
    vol. ii. p. 407.—Waterh. Nat. Hist. of Mamm., vol. i. p. 379.—Gray, List of Mamm. in Brit. Mus.,
    p. 95.

THIS elegant species of *Perameles* enjoys a wide range over the eastern and southern portions of Australia, but is more frequently met with in the country within the ranges, or what is commonly called in the colony " the interior," than in the districts lying between the mountains and the sea. In New South Wales, the stony ridges which branch off from the ranges towards the rivers Darling and Namoi, are localities in which it may always be found; in South Australia I hunted it myself on the stony ranges and spurs which run down towards the great bend of the river Murray. On reference to my notes, I find the following entry:—"July 1, 1839. Killed for the first time the Striped-backed Bandicoot, on the ranges bordering the great scrub on the road to the Murray. I started the animal from the crest of one of the stony ridges, and after a sharp chase of about a hundred yards it took shelter under a stone, and was easily captured; it passed over the ground with considerable rapidity, and with a motion precisely similar to the galloping of a pig, to which animal it also assimilates in the tenacity with which its skin adheres to the flesh; on dissection its stomach was found to contain the remains of caterpillars and other insects, a few seeds and fibrous roots; the flesh on being roasted proved delicate and excellent food; as is also that of most, if not all, the other members of the genus." His Excellency Governor Grey transmitted examples to this country during his residence at Adelaide, accompanied by the following note: " This animal is found in the vast open plains near the head of St. Vincent's Gulf, and where no other species is to be met with."

The sexes assimilate in colour, the female being as conspicuously marked as the male, but of a smaller size; the markings of the back are also as apparent in the young animal as in the adult.

The *Perameles fasciata* is very nearly allied to *P. Gunnii*, but is of a much smaller size, has the ears proportionately rather longer and broader at the base, the tail longer and dusky along the whole upper surface, instead of for a small space at the base; the feet and muzzle are also more slender.

Fur moderately long and harsh to the touch; upper surface pencilled with black and yellow in about equal proportions; on the sides of the body the yellow, and on the hinder part of the back the black prevails as a ground colour, but here are three broad yellowish-white bands, the foremost of which crosses the back, the other two run obliquely downwards and backwards from the mesial line; the posterior of these two is almost longitudinal, and the one in front of this joins the foremost band; these bands are interrupted in the middle of the back; under surface of the body and the feet white; the tail is also white, but along the whole upper surface the hairs are partly black and partly yellow.

The Plate represents the two sexes of the size of life.

PERAMELES GUNNII, *Gray.*

PERAMELES GUNNII, *Gray*.

Gunn's Perameles.

Perameles Gunnii, Gray in Proc. of Zool. Soc., part vi. p. 1.—Ib. Ann. of Nat. Hist., vol. i. p. 108.—Waterh. in
Jard. Nat. Lib. *Marsupialia*, p. 156, pl. 15.—Gray, List of Spec. of Mamm. in Coll. Brit. Mus., p. 95.
—Waterh. Nat. Hist. of Mamm., vol. i. p. 376.—Gunn in Proc. of Roy. Soc. of Van Diemen's Land,
vol. ii. p. 83.

It is well that the name of Mr. Ronald C. Gunn, a gentleman devotedly attached to natural history, and long
resident in Tasmania, should be perpetuated in a work of science, as the author would fain have the present
considered to be, since he has not only paid considerable attention to the Botany, but also to the Zoology
and even to the physical features of his adopted country; and hence I have great pleasure in figuring so
conspicuous an animal as the present under his name.

The *Perameles Gunnii* is an inhabitant of Tasmania, and appears to be more common in the northern than in the southern parts of that country; it also, if I mistake not, inhabits the islands in Bass's
Straits, and even the southern portions of the continent of Australia. I say this, however, somewhat doubt-
fully, because I possess no certain evidence that the animal has been killed at Port Phillip, though I have
received it direct from thence; but it is just possible that it had been taken there before its transhipment to
this country; I incline, however, to consider it a native of those parts as well as of Tasmania. In size this
species ranges next to *Perameles nasuta*, being in fact intermediate between that animal and *P. myosurus*;
but from these, as well as all others, it differs in its short and white tail; in the banding of its back it
approaches *P. fasciata*—but these marks are not so dark or so well defined. Mr. Gunn has given us an
interesting account of the destructive habits of this animal when gardens come within its range. This
account, moreover, indicates the kind of food upon which it naturally subsists, a point well worthy of the
attention of those who have the charge of menageries, and who would wish to be successful in their mode
of treating and preserving this tribe of animals.

" It has sometimes been doubted," says Mr. Gunn, " whether the *Perameles* feed upon roots. For several
years past my garden at Launceston has suffered severely from the attacks of *P. Gunnii*. Two beds of *Ixia
maculata*, var. *viridis*, were entirely eaten, so as to eradicate the species. Some other *Ixiæ* and *Babianæ*
were afterwards attacked; but many *genera* of Cape bulbs close to them were left untouched. The *Crocus*
seems an especial favourite, as wherever they occurred they were diligently sought out, rooted up, and
eaten, and that too at a season when no leaves appeared above ground to indicate their position. Tulips
seem to be less relished, although they are occasionally eaten. In the bush I lately discovered a new species
of tuberous fungus partly eaten, at the bottom of a hole about nine inches deep, which I believe had been
the work of a *Perameles*; my impression is that the *Perameles* live a good deal, if not principally, upon roots
and fungi."

The fur is moderate in length and not so hard to the touch as that of *P. obesula* or *P. nasuta*; the hairs of
the upper surface are grey at the root, the visible portion of each being pencilled with black and ochreous
yellow; on the sides the general hue is somewhat paler, the hairs having a smaller amount of the black
pencilling and a delicate vinous tint; on the hind quarters the ground-colour is blackish brown, and on this
part are three broad light-coloured bands, the first of which crosses the back slightly in front of the thigh,
the second is nearly transverse, and the third longitudinal; under surface pure white; the feet and tail are
also white, with the exception of a dusky patch on the base of the latter, and on the sides of the heel of the
hind foot the sides are dusky; ears internally clothed with very small pale yellow hairs, but on the hinder
part they are nearly white; a broad dusky mark crosses the outer surface of the ear, commencing about the
middle of the anterior margin, and running obliquely backwards as it descends to the base.

The Plate represents the male, female, and young, of the size of life.

PERAMELES MYOSURUS: *Wagn.*

PERAMELES MYOSURUS, *Wagn.*

Saddle-backed Perameles.

Perameles myosurus, Wagn. in Wiegm. Archiv, 7th Year, p. 289 ; and Schreb. Saugth., pl. 155 A.d., Part 111-112,
   Nov. 1842.
Perameles arenaria, Gould in Proc. of Zool. Soc., Part XII. p. 104.
Màl-a, Aborigines of the York and Toodyay districts.
Nyèm-mel, Aborigines of King George's Sound.

HAVING lately had an opportunity of consulting Schreber's " Saugthiere," I find therein the figure of an animal so nearly resembling my *Perameles arenaria,* that I am induced to believe it to be the same species ; I have consequently, in justice to the first describer, Dr. Wagner, placed my own name as a synonym to his.

Dr. Wagner, whose labours display great care and no ordinary extent of information, has very accurately pointed out the distinctions between it and the *P. fasciata,* the most nearly allied species yet discovered ; but as he has not mentioned the habitat of the animal he has described, I am unable to come to a positive conclusion on the subject : if it be from Western Australia, it is doubtless identical with the one here figured.

The present animal inhabits the whole line of coast of the Swan River colony, but, so far as I can learn, is not found to the westward of the Darling range of hills. It resides in the densest scrub, thickets of the seedling *Casuarinæ* being its favourite resort. It makes a compact nest in a hollow on the ground, of grasses and other materials, which assimilate so closely in colour and appearance to the surrounding herbage, that it is very difficult of detection, the difficulty being much increased by there being no visible opening for the ingress and egress of the animals. The nest is generally inhabited by pairs. The young are either three or four in number.

Its food consists of insects, seeds and grain. It excavates holes in the earth with rapidity and ease, and to these and the hollow trunks of fallen trees it flies for shelter when pursued by its natural enemies.

Mr. Gilbert remarks that this species is, without exception, the most difficult to skin of all the marsupials with which he is acquainted ; the skin in fact is so tender, that the weight of one of the limbs, if left hanging by the skin, is sufficient to separate it from the body ; and living specimens are often met with minus a portion or the whole of the tail.

The sexes are alike in colour, but when adult the female is smaller than the male. Examples are frequently seen of all sizes, which appears to be solely occasioned by a difference of age.

The fur is harsh to the touch, and of a greyish brown hue, interspersed with numerous long black hairs, which form a broad indistinct band across the flanks immediately above the hind-legs, and a kind of saddle-like mark on the centre of the back ; ears of three colours, rusty red near the base, then dark brown, and the apex of a light greyish brown ; sides of the muzzle and all the under surface buffy white ; line along the upper surface of the tail dark brown, the remainder buffy white ; outside of the fore-legs brownish grey ; feet and claws buffy white.

The figures represent the two sexes of the natural size. The flowering plant is a species of *Melaleuca,* probably undescribed.

PERAMELES NASUTA, *Geoff.*

PERAMELES NASUTA, *Geoff.*

Long-nosed Perameles.

Perameles nasuta, Geoff. Ann. du Muséum, tom. iv. p. 62. pl. 44.—Waterh. in Jard. Nat. Lib. Mamm., vol. xi.
(Marsupialia) p. 155. pl. 13.—Gray, List of Mamm. in Coll. Brit. Mus., p. 96.—Waterh. Nat. Hist.
of Mamm., vol. i. p. 374.

———— *Lawsoni*, Quoy et Gaim. Voy. de l'Uranie, Zoologie, pp. 57 & 711.

———— *nasuta* et *aurita* of the Paris Museum.

ALTHOUGH this animal inhabits the portion of Australia which has been longest known to us, it is remarkable
how little is the information that has been obtained respecting it; I procured many specimens during my
sojourn in the country, and ascertained that it is sparingly dispersed over the districts lying between the
mountain ranges and the sea. It frequents stony and sterile localities, and in all parts of this character,
even in the neighbourhood of Sydney, it occurs as frequently as elsewhere. It is perhaps the largest
species of the genus yet discovered, and is distinguished from every other by the great length of its snout,
which circumstance has obtained it the specific appellation of *nasuta*. I have never met with this species
in collections from any other part of Australia than New South Wales: I mention this because Dr. Gray
considers the *Perameles Bougainvillii* of MM. Quoy and Gaimard, which inhabits Western Australia, to be
identical with it; but, in my opinion, such is not the case. Independently of the genus *Paragalea*, there
are two other very distinct sections of the *Peramelinæ*, one of them inhabiting low swampy grounds covered
with dense vegetation; the other, the stony ridges of the hotter and more exposed parts: the former is
represented by the *Perameles obesula* and its allies, the latter by the beautiful banded group comprising
P. fasciata, *P. Gunni*, *P. myosurus*, &c. To this latter section the present species, though destitute of the
dorsal markings, also belongs.

The food of this animal consists of bulbous and other roots, which it readily obtains by means of its
powerful fore feet and claws.

The sexes, as is usual with the other members of the family, do not differ in colour, but the female never
attains the size of the male.

The fur, which is almost entirely composed of harsh, flattened hairs with a scanty under-fur of finer hairs, is
of a pale grey on the upper surface of the body; the longer and coarser hairs of the back are pencilled with
pale brown and blackish; on the sides the black is nearly obsolete, and here, as well as on the sides of the head,
the general tint is pale vinous-red; the under surface of the body is white, the hairs being uniform to the
root; feet white; the fore leg is grey externally at the base, and the hind leg has a dusky patch immediately
above the heel; ears clothed with very small hairs, which are whitish on the inner side, dusky on the outer,
and pale brown near the anterior angle; the small stiff hairs of the tail are brownish on the upper surface
and dirty-white on the under.

The front figure is of the natural size.

PERAMELES OBESULA, Geoff.

PERAMELES OBESULA, *Geoff.*

Short-nosed Perameles.

Didelphys obesula, Shaw, Nat. Misc., vol. viii. pl. 298.—Ib. Gen. Hist., vol. i. p. 490.
Perameles obesula, Geoff. Ann. du Mus., tom. iv. p. 64. pl. 45.—Waterh. Nat. Hist. of Mamm., vol. i. p. 368.—
    Gunn in Proc. Roy. Soc. of Van Diem. Land, vol. ii. p. 82.
Isoodon obesula, Desm. in Nouv. Dict. d'Hist. Nat., tom. xvi. p. 409.
Perameles fusciventer, Gray in App. to Grey's Journ., vol. ii. p. 407.
————— *affinis*, Gray, List of Mamm. in Coll. Brit. Mus., p. 96.
Gwën-dee, Aborigines of Perth, Western Australia.
Quoint, Aborigines of King George's Sound.

HAVING had many opportunities of observing this animal in a state of nature, both in Van Diemen's Land and New South Wales, I am enabled personally to state, that it does not, like some of the other species of the genus, such as *Perameles Gunni*, *P. fasciata* and *P. myosura*, dwell among the stony ridges of the open country, but evinces a preference for the low, damp, swampy places, overgrown with dense green herbage, which occur on the borders, and even within the great forests. In Van Diemen's Land it is more frequently met with on the southern side of the River Derwent than elsewhere. This great river, indeed, forms the line of demarcation to many species both of quadrupeds and birds; its southern side being clothed with vast forests of *Eucalypti*, growing on a stiff clayey soil, while the opposite bank is of a light sandy character, suitable to the growth of *Banksiæ* and *Acaciæ*; the former is the kind of country preferred by the animal under consideration, and, as might be supposed, it is found in all parts of Van Diemen's Land wherever similar localities occur; it is also to be found in like situations on the islands in Bass's Straits, in New South Wales, and in Southern and Western Australia. Specimens from all these countries are now before me, and although the range extends over an area of nearly three thousand miles, I am unable to detect any differences of sufficient value to warrant the establishment of a second species. The only perceptible difference between the examples from Western Australia and those from New South Wales and Van Diemen's Land, is a slightly darker tint in the colouring of the under surface of the former: very old males from each country attain to nearly a foot in length, exclusive of the tail, while the adult female is considerably smaller, and immature animals may be found of all sizes, according to age and sex. While engaged in my observations on the "Birds of Australia," I have very frequently trodden upon the almost invisible nest of this species and aroused the sleeping pair within, which would then dart away with the utmost rapidity, and seek safety in the dense scrub, beneath a stone, or in the hollow bole of a tree; that is, if their career were not stopped by a discharge from my gun, or by my dogs.

The following note is from the pen of the late Mr. Gilbert, and comprises his observations of the animal in Western Australia, which, although they do not quite agree with my own, I give in his own words:—

"This little animal is abundant in every part of the colony, and is found in every variety of situation; in thick scrubby places, among the high grass growing along the banks of rivers and swamps, and also among the dense underwood both on dry elevated land and in moist situations. It makes a nest of short pieces of dried sticks, coarse grasses, leaves, &c., sometimes mixed with earth, and so artfully contrived to resemble the surrounding ground, that only an experienced eye can detect it. When built in dry places, the top is flat, and on a level with the ground, but in moist situations the nest is often raised in the form of a heap, to the height of about twelve inches; the means of access and exit being most adroitly closed by the animal both on entering and emerging. The *P. obesula* is generally found in pairs: when driven from its nest, it takes to the first hollow log or hole in the ground that occurs. Although its usual food consists of insects, it occasionally feeds on grain, and I have several times seen it in great numbers in a wheat-stack. Specimens are sometimes met with of a very large size, which circumstance has induced a belief among the settlers that there are two species, but such is certainly not the case."

No one has more diligently endeavoured to unravel the confusion which has hitherto existed respecting the synonymy of this animal than Mr. Waterhouse; it will be but fair, therefore, to give his remarks on the subject:—

"The Short-nosed Perameles has an unusually wide range, being found in New South Wales, South Australia, King George's Sound, the Swan River district, and Van Diemen's Land. I have examined specimens from each of these localities, and taken much pains to satisfy myself of their specific identity. The males I have usually found larger than the females; their fore-feet are proportionately larger, and so are the canine teeth. The colouring varies somewhat in different individuals, and is darker than that of other species, if we except *P. macroura*."

The *Perameles affinis* of Dr. Gray "is founded upon a small animal from Van Diemen's Land, which appears to me to be a young individual of *P. obesula*: excepting in size, I can perceive no difference; its length from the tip of the nose to the root of the tail is 8 inches. When of this size, the young *P. obesula* has so much the general appearance of an adult animal in the character of the fur, &c., that I supposed, like Dr. Gray, there really existed a second species resembling *P. obesula*; but after examining the skulls, removed from two such specimens, I was convinced that their small size merely indicated immaturity."

Of the *Perameles fusciventer* of Dr. Gray, Mr. Waterhouse remarks, "Two specimens in the British Museum are labelled *Perameles fusciventer*; one agrees in every respect with the *P. obesula*, excepting that its head is rather shorter. . . . The other is considerably smaller than the adult *P. obesula*, and differs in being more strongly pencilled with black on the upper parts of the body, and in having the under parts of the body of a pale brownish-yellow, and the hairs on this part are slightly tinted with grey at the root. The head bears the same proportion to the body in length as in *P. obesula*. I question much whether the shortness of the head in the larger specimen does not arise from the mode in which the specimen has been stuffed; and with regard to the yellowish tint of the abdomen, I may observe, that in specimens which are undoubtedly the *P. obesula*, the under parts of the body are sometimes tinted with yellow, though less strongly than in the little animal above described. I cannot see any good grounds for regarding the specimens called *fusciventer* as specifically distinct from the *P. obesula*."

The animal here represented is one of the very commonest of the Australian mammals, and is, moreover, one of the oldest known, having been figured and described in some of the earliest works on that country.

The hairs composing the fur of this animal are of two kinds: all that are visible are harsh to the touch, flattened, pointed and glossy: upon dividing these coarse hairs, a soft, somewhat scanty fur becomes visible: on the upper parts of the body the coarse hairs are greyish-white at the root, black at the point, and broadly annulated in the middle with ochreous-brown, giving the whole the appearance of being pencilled in about equal proportions with black and ochreous-brown; the under-fur is grey; on the under parts of the body the hairs are yellowish-white at the tip and white at the base, and the under-fur is also white: towards the end of the muzzle the hairs are of a uniform dusky-brown; the lips, chin and throat are whitish; hairs clothing the inner side of the ears yellowish, becoming brownish on the margin; on the outer side dusky, becoming paler on the posterior part, and there is a faint indication of a pale spot at the base, near the anterior margin; fore-feet whitish; tarsi dirty-white, tinged with yellowish, and freckled with black on the upper surface; on the inner side they are delicate yellow; hairs of the base of the tail similar to those of the body; beyond this the upper surface is dusky, and of a dirty yellowish tint on the under surface.

The figures are somewhat less than the size of life.

PHASCOLARCTOS CINEREUS.

Gould and H.C.Richter, del. et lith.   Hullmandel & Walton, Imp.

PHASCOLARCTOS CINEREUS.

Koala.

HEAD AND FORE LEG, OF THE SIZE OF LIFE.

LIKE the *Ornithorhynchus*, this remarkable creature is only found in the south-eastern portion of the great land of the South. It is in the brushes which skirt the sea side of the mountain-ranges between the district of Illawarra and the River Clarence that it is most numerous; here, among the leafy branches of the great trees, the Koala remains sleeping during the daytime; but at nightfall this lethargy gives place to more active habits, and it then moves about with agility in search of its natural diet, which is said to be the tender buds and shoots of the *Eucalypti*.

Like too many others of the larger Australian mammals, this species is certain to become gradually more scarce, and to be ultimately extirpated; I have not hesitated, therefore, to give a life-sized head, as well as reduced figures, which, with a full account of the economy of the animal, will be found to follow the present page.

PHASCOLARCTOS CINEREUS.

J. Gould and H.C. Richter del. et lith.

Hullmandel & Walton, Imp.

PHASCOLARCTOS CINEREUS.

Koala.

Lipurus cinereus, Goldf. in Oken's Isis, 1819, p. 271.
Phascolarctos fuscus, Desm. Mammalogie, p. 276.—Ib. Dict. des Sci. Nat., tom. xxxix. p. 448.—Wallich in Jard.
    Nat. Lib., Marsupialia, p. 295.
———— *Flindersi*, Less. Man. de Mamm., p. 221.
————*fuscus* et *cinereus*, Fisch. Syn. Mamm., p. 285.—Wagn. Schreb. Saugth., 111-112 Heft, p. 92.
———— *cinereus*, List of Mamm. in Coll. Brit. Mus., p. 87.
Koala Wombat, Home, Phil. Trans. 1808, p. 304.
Le Koala ou Colak, Desm. Nouv. Dict. d'Hist. Nat., tom. xvii. p. 110. tab. E. 22. fig. 4.
Wombat of Flinders, Knox in Edinb. New Phil. Journ. 1826, p. 111.
Phascolarctus cinereus, Waterh. Nat. Hist. of Mamm., vol. i. p. 259.—Gray, Ann. Phil. 1821.
New Holland Sloth, Perry, Arcana, t. .
Native Bear and *Native Sloth* of the Colonists.

DURING my two years' ramble in Australia, a portion of my time and attention was directed to the fauna of the dense and luxuriant brushes which stretch along the south-eastern coast, from Illawarra to Moreton Bay. I also spent some time among the cedar brushes of the mountain ranges of the interior, particularly those bordering the well-known Liverpool Plains. In all these localities the Koala is to be found, and although nowhere very abundant, a pair, with sometimes the addition of a single young one, may, if diligently sought for, be procured in every forest. It is very recluse in its habits, and, without the aid of the natives, its presence among the thick foliage of the great *Eucalypti* can rarely be detected. During the daytime it is so slothful that it is very difficult to arouse and make it quit its resting-place. Those that fell to my own gun were most tenacious of life, clinging to the branches until the last spark had fled. However difficult it may be for the European to discover them in their shady retreats, the quick and practised eye of the aborigine readily detects them, and they speedily fall victims to the heavy and powerful clubs which are hurled at them with the utmost precision. These children of nature eat its flesh, after cooking it in the same manner as they do that of the Opossum and the other brush animals.

I believe the Koala to be extremely local in its habitat, as up to the present time the south-eastern portion of the continent of Australia is the only part in which it is known to exist.

No difference occurs in the external appearance of the sexes.

An excellent account of the habits of this animal was given in the "Philosophical Transactions" for 1808, by Colonel Patterson, formerly Governor of New South Wales. It was known to this gentleman as an inhabitant of the forests about fifty or sixty miles to the south-west of Port Jackson, whence, it is stated, the first specimens were brought. "The New Hollanders," says Colonel Patterson, "eat the flesh of this animal, and therefore readily join in the pursuit of it : they examine with wonderful rapidity and minuteness the branches of the loftiest gum-trees, and, upon discovering a Koala, they climb the tree with as much ease and expedition as a European would mount a tolerably high ladder. Having reached the branches, which are sometimes 40 or 50 feet from the ground, they follow the animal to the extremity of a bough, and either kill it with a tomahawk or take it alive. The Koala feeds upon the tender shoots of the blue gum-tree, being more particularly fond of this than of any other food ; it rests during the day on the tops of these trees, feeding at ease or sleeping. In the night it descends and prowls about, scratching up the ground in search of some particular roots ; it seems to creep rather than walk : when incensed or angry, it utters a long shrill yell, and assumes a fierce and menacing look. They are found in pairs, and the young is carried by the mother on her shoulders. This animal appears soon to form an attachment to the person who feeds it."

"It has been frequently compared to a bear in its movements and mode of climbing," observes Mr. Waterhouse, " and, indeed, in appearance the animal is not unlike a small bear."

Mr. Waterhouse has given so correct a description of this animal in his "Natural History of the Mammalia," that I cannot perhaps do better than transcribe it into these pages :—

"The Koala is usually about 2 feet in length, and when on all-fours stands 10 or 11 inches in height; the girth of the body is about 18 inches. Its limbs are of moderate length, and powerful; the hands and feet large, and admirably adapted by their structure to tree-climbing habits. The toes of the fore feet are so arranged, that the two innermost of the five are opposed to the other three ; and all the toes, both of the fore and hind feet (if we except the innermost one of the latter), are provided with large, curved, very deep, and compressed claws. The innermost toe of the hind foot is large, nail-less, assumes the form of a thumb, and is used as such, being opposed to the toes in grasping, as is the thumb of the human hand to the fingers. The head is rather large, the muzzle short and nearly naked both on the sides and on the upper surface, these parts being merely and rather sparingly clothed with small velvet-like hairs; the part thus sparingly clothed is most extended on the upper surface of the muzzle, here reaching back about $1\frac{1}{4}$ inch from the tip of the nose, while at the sides only $\frac{1}{4}$ an inch or rather more of the muzzle is destitute of the ordinary fur. The ears are of moderate size and pointed, and entirely hidden by the very long hairs with which they are clothed, these latter being for the most part about 2 inches in length; on the inner side of the ears the hairs are white, and on the outer side of the same grey hue as those of the head, excepting those which spring from the anterior margin of the ear, which are chiefly black. The eyes are rather large, and, like those of other Marsupial animals (with the exception of the Kangaroos), are not protected by eyelashes; there are, however, a few long bristly hairs springing from immediately above the eye; the hairs of the moustaches are small and scanty. The fur is tolerably long, dense, of a wool-like quality, and rather soft to the touch; its general hue ashy grey somewhat suffused with brown,—a tint produced by the hairs being brown before, and whitish at the point. The hinder part of the back is of a dirty yellowish white hue. The under parts of the head and body, as well as the inner side of the fore legs and the posterior part of the hind legs, are white, but not very pure; the hairs covering the feet have the visible portions whitish, but they are dusky brown at the root, and a slight pencilling of this darker hue is generally observable on the toes. The inner side of the hind legs is of a brownish rust-colour. The muffle is naked, and, like the naked soles of the feet, appears to have been black in the living animal.

"A very young Koala in the Museum of the Zoological Society presents some features worthy of notice. Instead of having the woolly fur of the adult, it is clothed with hairs which are moderately soft, short, and closely applied to the skin; on the mesial line of the back a little behind the shoulders, the hairs radiate, and running forwards over the neck meet those of the head having an opposite direction, and form a kind of crest at the line of junction; on the rump there is another of these centres from which the hairs radiate. The ears, which are much pointed and have the posterior edge emarginated, are clothed with hairs of about a quarter of an inch in length. Its colouring is the same as in the adult."

One of the accompanying Plates represents the head of the animal, of the size of life; the other, a reduced figure of a female and young.

PHALANGISTA FULIGINOSA. *Ogilby.*

PHALANGISTA FULIGINOSA, *Ogilby.*

Sooty Phalangista.

Phalangista fuliginosa, Ogilby in Proc. of Comm. of Sci. and Corr. of Zool. Soc., Part I. p. 135.—Gray, List of
    Mamm. in Brit. Mus., p. 85.—Waterh. Nat. Hist. of Mamm., vol. i. p. 288.
Phalangista Cuvieri, Gray.
———— *felina,* Wagn. ?

A QUESTION has been raised by Mr. Waterhouse, no mean authority as regards mammalia, whether the
Phalangista fuliginosa of Mr. Ogilby is really a distinct species from *Phalangista vulpina* : admitting that he
has some slight grounds for the suspicion alluded to, I am myself induced to consider them to be distinct ;
and I have come to this conclusion from having seen much of the two animals in a state of nature ; I have
taken them alive, fed upon their flesh, and their skins have served me for a covering in the country they
both inhabit. The true and, I believe, exclusive habitat of the animal to which Mr. Ogilby gave the name
of *fuliginosa* is Van Diemen's Land, while the continent of Australia is as exclusively the native country of
the *P. vulpina.* Two important points of difference between the two animals are found to exist : the island
species or *P. fuliginosa* far exceeds the other in size, and is subject to great variety in its colouring, varying
as it does from an almost jet-black to light grey, while many are characterized by a large admixture of red
of a greater or lesser degree of intensity : on the other hand, the continental species or *P. vulpina* are of a
uniform light grey,—at least that was the colouring of all those I saw while resident in the country ; the fur
of the Van Diemen's Land animal is also of a more dense and frizzly character. The skins of the island and
continental animals are both made into sleeping rugs, but the former are esteemed so much more highly,
that a rug formed of them is considered worth three times the price of one of the latter. I am aware that
climate has considerable influence over many animals, but it is not usual to find increase of size and depth
of colouring in the colder latitudes. The habits of both animals are as nearly alike as may be : strictly
nocturnal, they spend the entire day in sleep in the hollows of the boles and large limbs of the *Eucalypti*
and other trees of the forest ; on the approach of night they sally forth, and sometimes seek their food on
the ground, but more frequently among the branches ; the food consisting of the leaves and tender shoots,
and the flowers and honey-cups of the *Eucalypti.* They both form a considerable article of food for the
natives, who having discovered their retreat cut a hole in the branch, fearlessly insert their hand in the
hole, seize the animals by the tail, drag them forth, and despatch them by beating the head against the
tree ; when roasted the flesh is white and delicate, and not unlike that of a rabbit.

The animal in the British Museum which has been named *P. Cuvieri* by Mr. Gray, is considered by Mr.
Waterhouse to be identical with the present species, as the feet and incisor teeth, which are larger than
those of *P. vulpina,* precisely agree with those of *P. fuliginosa,* from which it only differs in being paler and
in having the tail less bushy ; and with respect to these differences, he remarks, that the animal had lived in
confinement for some time prior to its death, and when it died had shed the greater portion of the longer
and coarser hairs of the fur ; I may add that I have myself examined the animal, and believe that it is
referable to one or other of the above species, and I conclude from its greater size that it must be regarded
as synonymous with *P. fuliginosa.* I have had no opportunity of examining the *P. felina* of M. Wagner ;
Mr. Waterhouse remarks that it agrees in size with *P. fuliginosa,* with some of the varieties of which it
also agrees in colouring ; he does not believe it to be a distinct species, and as I know of no other it resembles, I regard it, at least for the present, as synonymous with the animal here represented.

Much variation in colour is found to exist in this animal, some having the general colour almost black,
especially on the back, with a rich brown hue on the sides and the throat ; chest and under parts of a rich
brownish fulvous hue, rather deeper on the abdomen than elsewhere ; the chin and muzzle, back of the ears,
feet and tail almost entirely black ; others are of a very deep rufous brown tint, much suffused with black
on the back ; others are of a rich rufous grey, and others again are entirely grey.

The figure in the accompanying Plate, which was drawn from life, represents a dark variety, somewhat
less than the natural size.

PHALANGISTA RULPINA, *Desm.*

PHALANGISTA VULPINA, *Desm.*

Vulpine Phalangista.

Didelphys vulpina, Shaw, Gen. Zool., vol. i. p. 503.
————— *lemurina*, Shaw, Gen. Zool., vol. i. p. 487. pl. 110.
Phalangista vulpina, Desm. in Nouv. Dict. d'Hist. Nat., tom. xxv. p. 475.—Ib. Ency. Méth. Mammalogie, part i.
    p. 267.—Temm. Monog. de Mamm., tom. i. p. 5.—Gray, List of Mamm. in Coll. Brit. Mus., p. 86.—
    Martin in Proc. of Zool. Soc., part 4. p. 2.
————— *melanura*, Wagn. in Schreb. Saug. Suppl. 111, 112. Heft, p. 81.
————— *xanthopus*, Ogilby, Proc. of Comm. of Sci. and Corr. of Zool. Soc., part i. p. 135.—Waterh. Nat. Hist.
    of Mamm., vol. i. p. 294.
Phalanger de Cook, F. Cuv. et Geoff. Mammifères, pl. 45.
Vulpine Opossum, Phillips's Voy. to Botany Bay, p. 150, and pl.
Wha Tapoa Roo, White's Journ. of a Voy. to New S. Wales, p. 278, and pl.
Phalangista (Trichosurus) vulpina, Waterh. Nat. Hist. of Mamm., vol. i. p. 284. pl. 9. fig. 1.

OF all the Opossums inhabiting Australia, the *Phalangista vulpina* is by far the commonest, and the one most widely distributed over the country, being found in all parts of New South Wales, Port Philip, and Southern and Western Australia. According to Mr. Waterhouse, it is also found in Northern Australia; but I observe that specimens from that country are larger than those obtained in the countries above mentioned, and a doubt exists in my mind as to their identity.

During my travels in Australia no living mammal was more frequently presented to my notice, and no one was more often brought by the natives to the camp-fire for the purpose of eating. All these examples were of a uniform grizzly-grey, and in no instance did I meet with the dark colouring of *Phalangista fuliginosa*, in my account of which species I stated, that I believed its true and exclusive habita to be Van Diemen's Land; that opinion I now find to be incorrect, as I have recently received examples from the dense brushes near the coast of the continent; its range therefore, independently of Van Diemen's Land, evidently extends over the brushes of New South Wales, and perhaps future research may prove that it enjoys a still wider range of habitat.

The *Phalangista vulpina*, like its congener, is strictly nocturnal in its habits, living in the hollow spouts and holes of the large gum-trees during the day, and ascending the branches during the night to feed upon the buds, leaves and fruit; sometimes descending to the ground, where it probably finds herbs to its taste. "While climbing," says Mr. Waterhouse, "its prehensile tail assists it to maintain a firm hold of the branches: in captivity I have noticed, that in descending from one perch of its cage to another, or to the floor, the tail invariably encircled the perch it was quitting until the animal was again securely lodged. Numerous specimens have from time to time formed part of the Zoological Society's living collection, and, from my own observations, they appeared to be by no means intelligent animals. During the day-time they were usually asleep, but towards evening they became active, and on the alert for their food, consisting of bread and milk, and various vegetable substances, including fruits. Whatever eatable was given to them, was taken by and held between the hands, in the manner a squirrel holds a nut. Occasionally a dead bird was given to them, when they evinced an evident fondness for such food, and more particularly for the brain, which was the part first consumed."

This animal constitutes a considerable part of the food of the natives, who diligently search for it, and having discovered a tree in which it is secreted, ascend it with surprising agility; the position of the animal being ascertained, a hole is cut with their little axes sufficiently large to admit the naked arm; it is then seized by the tail, the chopping and jarring of the tree not inducing it to leave its retreat, and before it has time to bite, or use its powerful claws, it is deprived of life by a blow against the side of the tree, and thrown to the ground; its captor proceeding to his encampment with a dinner in perspective. I have frequently eaten its flesh myself, and found it far from disagreeable. Mantles and sleeping rugs are made of its skin, but, as I have mentioned in my account of *P. fuliginosa*, are not considered so valuable as those made from the skin of that animal.

A very elaborate account, by Mr. Martin, of the internal anatomy of a female of this species will be found in the " Proceedings of the Zoological Society " above quoted.

Fur long, loose, and moderately soft ; general colour grey, the visible portions of the hairs being partly black and partly white ; fur of the back of a somewhat deeper hue than on the sides, owing to a plentiful interspersion of long black hairs ; muzzle and chin blackish, the former pale near the tip, and the naked muffle of a whitish flesh-colour ; eyes encircled with blackish hairs ; skin of the inner surface of the ears brownish-pink, with a few scattered pale-coloured hairs ; outer surface of the ear, excepting near the point and a narrow space along the anterior margin, clothed with a dense and moderately long fur, which is white at the posterior angle and towards the apex, but black elsewhere ; the hairs of the moustaches are long, numerous and black, and there are a few long bristly black hairs springing from above the eyes ; throat, under surface of the body, and inner side of the limbs pale buffy-yellow, with a large oblong patch of deep rust-coloured hairs along the chest ; feet yellowish-white, suffused with brown on the toes ; naked soles flesh-coloured ; claws dusky ; tail clothed at the root like the body ; beyond, the fur is more bushy, of a harsher character, and black, the last inch or so being in some instances white ; the extreme point of the tail, and the apical half of the under surface are naked.

The figure represents the animal rather under the size of life.

PHALANGISTA CANINA, Ogilby

PHALANGISTA CANINA, *Ogilby*.

Short-eared Phalangista.

Phalangista canina, Ogilby in Proc. of Zool. Soc., part iv. p. 191.—Gray, List of Mamm. in Coll. Brit. Mus., p. 85.—Waterh. Nat. Hist. of Mamm., vol. i. p. 296.

THIS is a powerful animal, fully equalling in size the *P. fuliginosa*. It is at once distinguished from that, and from every other known species of the genus, by the short and rounded form of its ears. It is much more restricted in its habitat, being, so far as my knowledge extends, exclusively confined to the brushes of New South Wales, particularly those in the neighbourhood of the Hunter, Clarence, and Richmond rivers, and the cedar brushes of the Liverpool range. Its habits and economy closely resemble those of its near ally the *P. vulpina*, but it is much more fierce in its disposition. Like the *P. fuliginosa*, it is subject to much variation in its colouring, some specimens being black, while others have a reddish tinge pervading the shoulders and flanks; the prevailing tint is a dark grizzly-grey, similar to that represented in the accompanying Plate.

The following is Mr. Waterhouse's description of this animal, taken from the typical specimen in the Museum of the Zoological Society of London:—

The fur is long, dense, and somewhat woolly; its general hue is grey, being finely pencilled with black and white; on the under surface of the body it is white, but each hair is indistinctly suffused with yellow externally, and is greyish next the skin; on the chest is a narrow rusty-brown mark; ears nearly naked internally; externally they are furnished at the base with fur of the same kind as that on the head, and of a blackish hue, but towards the exterior margin the hairs are whitish; muzzle dusky, and the eye surrounded by the same dark hue; feet blackish; tail very bushy, coloured at the base like the body, the thick bushy hairs on the remaining portion black; apical third of the under surface and the tip of the tail naked; moustaches black; claws pale horn-colour.

The figure is somewhat smaller than the natural size of the animal.

PHALANGISTA COOKI, *Desm.*

J. Gould and H.C. Richter, del. et lith.   Hullmandel & Walton, Imp.

PHALANGISTA COOKI, *Desm.*

Cook's Phalangista.

Phalangista Cookii, Desm. Nouv. Dict. d'Hist. Nat., tom. xxv. p. 478.—Temm. Mon. de Mamm., tom. i. p.   .—
      Gray, Ann. Nat. Hist. new ser. 1838, vol. i. p. 107.
——————— *Banksii*, Gray, Ann. Nat. Hist. new ser. 1838, vol. i. p. 107.
Phalanger de Cook, Cuv. Règn. Anim. ed. 1817, tom. i. p. 179 ; ed. 1829, tom. i. p. 183.
——————— *Bougainville*, Cuv. Règn. Anim. ed. 1829, tom. i. p. 183.
Phalangista Bougainvillii, Wagn. in Schreb. Saug. 111, 112. Heft, p. 82.
New Holland Opossum, Penn. Hist. of Quad., vol. ii. p. 301.
White-tailed Opossum, Shaw, Gen. Zool., vol. i. p. 504.
Phalangista (Pseudochirus) Cookii, Waterh. Nat. Hist. of Mamm., vol. i. p. 299.—Gunn in Proc. of Roy. Soc.
      Van Diem. Land, vol. ii. p. 84.
Balantia Cookii, Kuhl, Beitr. 63.
Hepoona Cookii, Gray, List of Mamm. in Coll. Brit. Mus., p. 84.
Ring-tailed Opossum of the Colonists.

———————————

A QUESTION has been raised by some modern mammalogists, whether the Ring-tailed Opossum of New South Wales, characterized by the rusty-red hue of its colouring, and the animal of the same form inhabiting Van Diemen's Land, which is principally of a sooty blackness, are distinct species, or merely varieties of one and the same animal. Mr. Waterhouse considers them to be identical : Mr. Ogilby, on the other hand, is of opinion that they are not, and has given the name of *viverrina* to the island or darker-coloured specimens, retaining that of *Cooki* for the animal from New South Wales : now, it may be asked, what is the opinion of one who has seen these animals in a state of nature ? In reply, I may say, that I have hunted them upon very many occasions in both countries, and that I invariably found the black specimens to be confined to Van Diemen's Land, and the red ones to New South Wales ; besides which, I observed that the island examples were of a larger size and were always dressed in a softer and longer fur.

My figure of *Phalangista Cooki* represents the animal as it generally appears in the brushes of New South Wales, while that of *P. viverrina* as correctly portrays the one killed in Van Diemen's Land. It will be seen that, at least for the present, I have regarded them as distinct.

The *Phalangista Cooki* is strictly nocturnal in its habits, sleeping in the hollow spouts and holes of the larger trees during the day and leaving its retreat on the approach of darkness, sometimes descending to the ground, but more frequently ascending to the smaller branches to feed upon the flowers and tender shoots of the *Eucalypti*.

Its flesh is delicate, juicy, and well-tasted, and is much prized by the aborigines.

I have spoken of the brushes of New South Wales as being the part of the country inhabited by this animal ; it is just possible that it may also be found on the plains and *Angophora* flats between the lower mountain ranges ; but it must not be confounded with a smaller and more woolly species which is there found, the *Phalangista laniginosa*, a figure and description of which will be found in its proper place in this work.

The fur is dense and somewhat harsh to the touch ; its colour on the upper surface of the body and tail in some specimens is dark brown, grizzled with grey ; in others a greyer hue prevails ; face, cheeks, sides of the neck and body, the outer surface of the limbs and the under surface of the basal portion of the tail, rich deep rust-red ; sides of the muzzle blackish ; eyes surrounded by a series of black hairs ; chin, under surface of the body and inner sides of the limbs tawny, increasing to rufous in some specimens ; in some examples the apical half of the tail is white, in others the apical third, and in others it is of nearly the same hue as the basal portion, but this latter state rarely occurs.

The Plate represents the sexes somewhat under the size of life.

PHALANGISTA VIVERRINA, *Ogilby*

Gould and H.C.Richter, del. et lith.

Hullmandel & Walton, Imp.

PHALANGISTA VIVERRINA, *Ogilby*.

Viverrine Phalangista.

Phalangista viverrina, Ogilby in Proc. of Zool. Soc., part v. p. 151.—Waterh. Nat. Hist. of Mamm., vol. i. p. 303.
——————— *Cooki*, Gunn, Ann. of Nat. Hist., vol. i. 1838, p. 102.

On reference to the description of the preceding species, *Phalangista Cooki*, will be found some general observations respecting the Ring-tailed Opossums of Van Diemen's Land and New South Wales. A lengthened discussion between Mr. Ogilby and Mr. Gray, as to whether the island or continental animals should be called *Cooki*, will be found in the "Annals of Natural History" for the year 1838, into the merits of which I shall not enter, but merely remark, that my observations of the animals in a state of nature lead me to coincide with Mr. Ogilby, and, for the present at least, to consider the Van Diemen's Land animal as distinct from that of New South Wales, and as the one for which his name of *viverrina* should be retained. I saw much of this animal during my sojourn in the island, and frequently hunted it in company with some of the settlers, and the servants who accompanied me. Like many other species, it evinced a great partiality for certain trees and localities, some districts being resorted to by great numbers, while in others it was almost entirely absent. It may extend its range to the continent, as I have lately seen a dark-coloured specimen which had been obtained in the brushes. The Ring-tailed Opossums procured by Mr. Gilbert in Western Australia, of which I have several examples, are fully equal in size, and even blacker in colour, than the Van Diemen's Land animals; but he did occasionally meet with greyer specimens in the neighbourhood of Perth. I mention this, that mammalogists may form opinions for themselves as to whether these animals constitute one or more species; whether they do or do not, I have considered it advisable to give correct representations of the very opposite colours they exhibit, and to state all I know respecting them.

I found this animal gave a decided preference to those districts of Van Diemen's Land that are of a sandy character, and where the large gum-trees were sparingly dispersed, such as the islands on the River Derwent and the plains on the northern side of that stream; but it was not to be found in the more dense and humid scrub of its opposite shore. Our usual mode of hunting this animal was to go out in a small party on moonlight nights, when, with the aid of one or two small cur dogs, it was soon discovered, either on the ground or among the branches of the trees, where, if looked for with the face towards the moon, it is not difficult to see, and when seen, much less difficult to shoot, as it never attempts to retreat.

Mr. Gunn states that this animal "is common near Launceston, and is there usually called Ring-tail Opossum as a specific name. All the opossums come out of the holes of the trees, in which they usually sleep all day, about twilight; and for about an hour or two after sunset they may be seen busily employed eating the leaves of the various species of *Eucalypti*; on the branches in moonlight nights they are usually shot, and opossum-shooting is sometimes fine sport where a few join together. Orchards in country places suffer sometimes from the opossums eating all the leaves and young branches."

Mr. Gilbert says this animal is called *Ngö-ra* by the aborigines of Perth, and *Ngork* by those of King George's Sound; and states that "it does not confine itself to the hollows of trees, but is often found in holes in the ground, where the entrance is covered with a stump, and from which it is often hunted out by the Kangaroo dogs. It varies very much in the colour of the fur, from a very light grey to nearly black. In one instance I caught a pair in the same hole exhibiting these extremes of colour."

At page 303 of Mr. Waterhouse's "Natural History of the Mammalia," he mentions that there are five specimens from Van Diemen's Land in the British Museum, in which the general hue of the fur is pale rufous-grey on the back, and bright rust-colour on the sides of the body and limbs; this statement induced me carefully to examine these specimens, and I feel confident that, by some accident, the labels they originally bore have been lost, probably during the process of mounting; and that they are from New South Wales, and not from Van Diemen's Land.

The following is the description of a specimen from Van Diemen's Land sent to me by R. C. Gunn, Esq. :—

Fur of the head, all the upper surface of the body, the outer side of the limbs, and basal half of the tail, sooty-grey, grizzled with whitish, and with numerous interspersed long black hairs; sides of the face and orbits dusky; posterior part of the exterior of the ear, and a patch behind and below it, white; throat, chest, abdomen, inner side of the limbs and apical half of the tail white; moustaches black.

Other examples from that island have the upper surface very much darker.

Specimens from Swan River are still darker, not only on the upper but on the under surface, which latter is of a brownish-grey only a trifle lighter than the hue of the upper parts. The colour of one example now before me is almost wholly black, somewhat paler beneath, and with a small patch of white on the chest, and an interrupted line of white down the throat and abdomen. The colouring, in fact, is much varied between grey and black, but never exhibits the rufous hue of the New South Wales animal, *Phalangista Cooki.*

The figures are somewhat less than the natural size.

PHALANGISTA LANIGINOSA; *Gould.*

J. Gould and H.C Richter, del. et lith.   Hullmandel & Walton, Imp.

PHALANGISTA LANIGINOSA, *Gould.*

Woolly Phalanger.

AT the period of my visit to Australia, this species was abundant on most of the *Angophora* or "Apple-tree" flats of the Upper Hunter, particularly those of the Dartbrook district, and it is doubtless to be found there still, and in all probability will be for ages to come. I mention this locality especially because there are two nearly allied *Phalangistæ* in New South Wales, which, when brought to this country and exposed in our museums, undergo so great a change in the colouring of their fur as to render it exceedingly difficult to distinguish them. These two nearly allied species are the *Phalangista Cooki* and the *P. laniginosa* figured on the accompanying Plate. I am the more certain of the specific distinctness of these two animals as those keen observers, the natives, particularly impressed upon my attention that the animal from the flats was different from the one frequenting the brushes which clothe the "corries" of the great Liverpool Chain. While in the country I had no difficulty in distinguishing them, and never had a doubt of their being distinct; but what was plain to me in Australia, I am unable to render so clear to the Mammalogists of Europe; I have no doubt, however, that when the great country of Australia has sons of her own interested in the subject, my views will be borne out and strictly verified, and it is for this reason that I have given so particularly the precise locality in which my specimens were obtained; doubtless all similar districts in Eastern Australia will also be favoured with the presence of this animal. I may remark that there is a greater difference between the *P. laniginosa* and *P. Cooki* than there is between *P. Cooki* and *P. fuliginosa*, which, indeed, may possibly be mere varieties of each other, although I have treated them as distinct.

My figure of *P. laniginosa* is taken from a fully adult male now before me. This animal is clothed in a thick, short, woolly kind of fur, of a greyish hue, with a wash of rufous on the outer side of the limbs; has the throat and all the under surface white, and the tail not so extensively tipped with white as in its near allies; it is also of smaller size.

The following is a more minute description of the animal:—Fur soft and yielding to the touch; general colour of the upper surface brownish grey, interspersed on the back with numerous greyish-white hairs; head and neck suffused with rufous, particularly round the eyes and on the outer surface of the ears; lower edge of the ear buff; whiskers black; outer side of the limbs rusty red; throat, under surface of the body and inner side of the limbs greyish white; basal fourth of the tail brownish grey, suffused with rufous; apical fourth white, the middle portion blackish brown.

The figures are fully the size of life.

CUSCUS BREVICAUDATUS.

CUSCUS BREVICAUDATUS, *Gray.*

Short-tailed Cuscus.

Phalangista (Pseudocheirus) nudicaudata, Gould in Proc. of Zool. Soc. 1849, p. 110.
Cuscus brevicaudatus, Gray in Proc. of Zool. Soc., part xxvi. p. 102.—Ib. Cat. of Mamm. and Birds of New Guinea
in Coll. Brit. Mus., p. 7.
Pseudocheirus nudicaudata, Macg. Voy. of H.M.S. Rattlesnake, vol. i. p. 129.

THAT a member of the genus *Cuscus* should be found in the extreme north-east of the Australian continent is not surprising, from the contiguity of New Guinea and the adjacent islands, where various members of the genus abound. It will be seen, by the synonyms given above, that when I described this animal, I regarded it as a *Phalangista,* and gave it the name of *Phalangista (Pseudocheirus) nudicaudata*; now, as all the *Cusci* have the extremity of the tail denuded of hairs, this specific appellation is certainly inappropriate, and I have therefore adopted that of *brevicaudatus,* proposed for it by Dr. Gray.

When speaking of the objects observed near Port Albany, Mr. Macgillivray says,—"The natives one day brought down to us a live Opossum, quite tame and very gentle. It turned out to be a new species, and has since been described by Mr. Gould under the name of *Pseudocheirus nudicaudatus*;" and this, unfortunately, is all we know of the natural history of this pretty animal.

Dr. Gray, in his "Observations on the Genus *Cuscus,*" published in the 'Proceedings of the Zoological Society of London' for the year 1858, says,—"This species" (the *Cuscus brevicaudatus*) "is very like the ashy variety of *Cuscus maculatus,* but the front lower cutting-teeth are much broader, and the tail is considerably shorter than any of the specimens of the *C. maculatus* contained in the British Museum collection.

"The light mark on the rump is common to all the species of *Cuscus,* and is probably produced by the habit of the animal sitting on that part of the body, rolled up into a ball, on the fork of the branches of a tree."

Head, all the upper surface, the sides of the body, and the outer sides of the limbs brownish grey; the tips of the hairs with a silky appearance; under surface of the neck and body and the inner sides of the limbs pale buff; the colouring of the upper and under surface distinctly defined on the sides of the body, but gradually blending on the limbs, the rump, and root of the tail, which is thickly clothed on its basal third, and naked for the remainder of its length; hands, feet, and naked portion of the tail pinky flesh-colour.

	inches.
Length from tip of nose to root of tail	12
„ of tail	8
„ of fore-feet, including the nails	3
„ of hind-feet, including the nails	3¼

The animal is figured in two positions, of the size of life; at the same time, it must be mentioned that it is believed to be immature, and not more than two-thirds of the size it would be when adult.

PETAURISTA TAGUANOÏDES.

Hullmandel & Walton, Imp.

PETAURISTA TAGUANOÏDES, *Desm.*

Great Flying Phalanger.

Petaurus Taguanoïdes, Desm. Nouv. Dict. d'Hist. Nat., tom. xxv. p. 400.—Waterh. in Jard. Nat. Lib. Mamm.,
   vol. xi. (Marsupialia) p. 283. pl. 27.
Petaurista Taguanoïdes, Desm. Mamm., pt. 1. p. 269.— Gray, List of Mamm. in Coll. Brit. Mus., p. 84.
Petaurista Peronii, Desm. Nouv. Dict. d'Hist. Nat., tom. xxv. p. 400.
Petaurus Peronii, Benn. Cat. of Australian Museum, Sydney, p. 3. no. 10.
Petaurus Leucogaster, Mitch. Three Exp. into Eastern Australia, vol. i. p. xvii. ?
Petaurus (Petaurista) taguanoïdes, Waterh. Nat. Hist. Mamm., vol. i. p. 322.

The Great Flying Phalanger is strictly an inhabitant of the extensive brushes which stretch along the south-eastern and eastern portions of New South Wales, the forests between the mountain ranges and the sea from Port Philip to Moreton Bay being in fact its native habitat.   Strictly nocturnal in its movements, this fine animal secretes itself during the day in hollow trees of the largest growth, and on the approach of evening emerges from its retreat in quest of the newly opened blossoms of the *Eucalypti*, in which, together with the tender buds and shoots of the same trees, it finds a description of food congenial to its well-being.   It passes along the branches with the utmost celerity, and, when necessitated to remove from one tree to another, effects its object by leaping from the higher branches, and floating through the air in easy and elegant sweeps, its progress being greatly aided by the parachute-like membrane at its sides. Although plentiful in the districts above mentioned, examples are not procured without difficulty, owing to the thickness of the brush or forest; the natives, however, readily detect its retreat by the presence of a few straggling hairs at the entrance of its hole, or by the impressions made by its sharp claws in the bark, and having found it speedily cut it out with their hatchets.   It is not a little surprising that this very singular animal should not have been captured alive and sent to this country, like the smaller members of the family; it would be by far the most interesting and attractive : its power of inflicting most severe lacerations with its sharp teeth and strong hooked claws may be one reason why this has not been done.

At present this is the only well-established species of the genus *Petaurista*, but I doubt not that others exist in the extensive forests which stretch along the eastern coast of Australia, and which have as yet been but imperfectly explored.   It is subject to very great variety in the colouring of its fur, some specimens being entirely blackish brown on the upper surface, while in others it is blackish brown suffused with grey; others are of a uniform cream colour, and others again quite white : these latter I have always regarded as mere varieties ; I am not, however, prepared to say that they had red eyes, like true albinos.

The sexes offer no external difference, except that the female is somewhat smaller than the male.

Fur very long, loose, and soft to the touch, of a brownish black hue on the upper surface and on the flank-membrane, and of a browner tint on the head and back of the neck; the flank-membrane is, moreover, pencilled with white; feet, muzzle and chin nearly black ; throat, chest, under side of the body and of the flank-membrane, and the inner side of the limbs pale buffy white ; the wrists and ankles are, however, black both on the inner as well as on the outer side; the long hairs near and at the posterior margin of the ear are whitish and project from the edge of the ear like a fringe ; tail black or brownish black, almost always paler at the root and along the under surface for a considerable distance from the base, sometimes of a yellowish brown, at others of a brownish white.

The drawing represents the animal rather more than two-thirds of the size of life.

BELIDEUS FLAVIVENTER.

BELIDEUS FLAVIVENTER.

Long-tailed Belideus.

Petaurista flaviventer, Desm. Mamm., p. 269.
Didelphis Petaurus, Shaw, Gen. Zool., vol. i. p. 496.
Petaurus (Belideus) flaviventer, Waterh. Nat. Lib. Mamm., vol. ix. (Marsupialia), p. 286.
———— *australis*, Shaw, Nat. Misc., vol. ii. pl. 60.—List of Mamm. in Brit. Mus. Coll., p. 83.
Didelphys volucella, Meyer, p. 26.
Petaurus Cunninghami, Gray, MSS. B.M.
Sciurus Novæ-Hollandiæ, Meyer.
Didelphys macroura, Shaw, Zool. of New Holl., p. 33. pl. xii. young.—Ib. Gen. Zool., vol. i. p. 500. pl. 113. young.
Petaurus macrourus, Waterh. Nat. Lib. Mamm., vol. ix. (Marsupialia), p. 288. young.

MR. WATERHOUSE, who has paid considerable attention to the Marsupialia, in speaking of this animal, says, "The Hepoona Roo of White's Journal, the original also of Shaw's *Didelphis Petaurus*, is still in existence in the Museum of the College of Surgeons; it proves to be the present species, and not the *P. Taguanoides*, as has always been supposed. This ought therefore to be regarded as the type of Shaw's genus *Petaurus*, if authors are right in attributing that genus to Shaw; but I do not perceive that he ever regarded the animal in question as constituting a genus, or that he applied the name in a generic sense."

This fine species is common in all the brushes of New South Wales, particularly those which stretch along the coast from Port Philip to Moreton Bay. In these vast forests, trees of one kind or another are perpetually flowering, and thus offer a never-failing supply of blossoms upon which the Long-tailed Belideus feeds; the flowers of the various kinds of gums, some of which are of great magnitude, are the principal favourites. Like the rest of the genus it is nocturnal in its habits, dwelling in holes and in the spouts of the larger branches during the day, and displaying the greatest activity at night while running over the small leafy branches, frequently even to their very extremities, in search of insects and the honey of the newly-opened blossoms. Its structure being ill adapted for terrestrial habits it seldom descends to the ground, except for the purpose of passing to a tree too distant to be attained by springing from the one it wishes to leave. The tops of the trees are traversed by this animal with as much ease as the most level ground is by such as are destined for terra firma. If chased or forced to flight, it ascends to the highest branch, and performs the most enormous leaps, sweeping from tree to tree with wonderful address; a slight elevation gives its body an impetus, which with the expansion of its membrane enables it to pass to a considerable distance, always ascending a little at the extremity of the leap; by this ascent the animal is prevented from receiving the shock which it would otherwise sustain.

It is now very generally believed that the *Petaurus macrourus* is only the young of *Petaurus flaviventer*; I have therefore placed the former name as a synonym.

General colour either greyish or yellowish brown; head clouded with black, particularly round the eyes, at the base of the ear, on the muzzle and chin; a black mark extends from the occiput along the middle of the back; the fore and hind legs and the side membrane blackish brown; edge of the membrane and under surface of the body buff; basal half of the tail yellowish brown, the remainder black.

The sexes are alike in colour.

The accompanying Plate represents the animal a trifle less than the natural size.

BELIDEUS SCIUREUS.

J. Gould and H.C. Richter del. et lith. Hullmandel & Walton imp.

BELIDEUS SCIUREUS.

Squirrel-like Belideus.

Didelphys sciurea, Shaw, Zool. of New Holl., pl. xi. p. 29.

Petaurus sciureus, Desm. Nouv. Dict. d'Hist. Nat., 2nd Edit., tom. xxv. p. 403.

Petaurus (Belideus) sciureus, Waterh. in Proc. of Zool. Soc., Part VI. p. 152.—Nat. Lib. Mamm., vol. ix. (Marsupialia), p. 289. pl. xxviii.—List of Mamm. in Brit. Mus. Coll., p. 83.

Norfolk Island Flying Squirrel, Phillip's Voy., pl. in p. 151.

Sugar Squirrel, Colonists of New South Wales.

THIS is not only one of the most elegant and beautiful species of the genus to which it belongs, but is also one of the commonest animals of the country ; being very generally dispersed over the whole of New South Wales, where, in common with other Opossums, it inhabits the large and magnificent gum-trees. Nocturnal in its habits, it conceals itself during the day in the hollows and spouts of the trees, where it easily falls a prey to the natives, who capture it both for the sake of its flesh and its skin, which in some parts of the colony they dispose of to the colonists, who occasionally apply it to the same purposes as those to which the fur of the Chinchilla and other animals is applied in Europe,—the trimming of dresses, boas, &c.

At night it becomes as active and agile in its motions as it is sluggish and torpid in the daytime. I observed that it prefers those forests which adorn the more open and grassy portions of the country to the thick brushes near the coast. By expanding the beautiful membrane attached to its sides it has the power of performing enormous leaps, and of passing from tree to tree without descending to the ground : like other animals provided with a similar means of transit, it slightly ascends at the extremity of its leap, and thereby avoids the shock which a direct contact with the branch upon which it alights would cause it to sustain.

It is of course marsupial, and I believe produces two young at a time, as I found two animals about half-grown in the same hole with the adults.

For a beautiful living example of this animal I am indebted to the kindness of my most estimable friend Mrs. Simpkinson, to whom it had been sent by her sister Lady Franklin, who procured it from Port Philip. It has become very tame, and its actions when permitted to run about the apartments are amusing and attractive in the extreme : the slightest projection affording it support, it passes over the cornices, picture-frames and hangings with the greatest ease ; it becomes exceedingly animated at night, leaping from side to side of its cage, spreading its membrane and tail, and repeatedly turning completely over, or performing several summersaults in succession. Its usual food is sopped bread and milk, upon which it thrives, and which appears to be a good substitute for its natural food, which consists of insects, the honey of flowers, and the tender buds and leaves of the *Eucalypti*.

Fur extremely soft and of moderate length ; general tint of the upper surface ashy grey ; a blackish brown line extends from the nose along the middle of the back nearly to the root of the tail ; the upper surface of the flank membrane, and the anterior and posterior portion of the fore and hind legs black or brownish black ; just below the ear a brownish black patch ; feet dusky grey ; chin, throat, inner side of the limbs and the under surface of the body white ; under side of the flank membrane dusky ; the margin fringed with white hairs ; tail smoke-grey for somewhat more than the basal half of its length, the remainder deep black ; ears nearly naked except at the base, where they are clothed with a black fur, and the posterior margin which is white ; eyes very full and black.

The figures represent fully adult animals of the natural size on a branch of one of the commonest of the *Eucalypti* of New South Wales.

BELIDEUS BREVICEPS: *Walsh.*

BELIDEUS BREVICEPS, *Waterh.*

Short-headed Belideus.

Petaurus (Belideus) breviceps, Waterh. in Proc. of Zool. Soc., Part VI. p. 152.—Ib. vol. xi. Nat. Lib. (Marsu-
pialia), p. 290. pl. 29.—Ib. Nat. Hist. of Mamm., vol. i. p. 334.
Petaurus Peronii, G. Benn. Cat. Aust. Mus., p. .? not of Desmarest.
Petaurus breviceps, Gray, List of Mamm. in Brit. Mus., p. 83.

THIS species of *Belideus* is not so widely dispersed over New South Wales as the *B. Sciureus*; it is in fact a much more local species; judging from the great number of specimens I have seen from Port Philip, I presume that district to be its great stronghold. I have two specimens in my collection, sent by Mr. Strange, one of which is labeled Wollongong, and the other Torrumbong; the former, as is well known, is the port of the rich district of Illawarra, and I presume the latter to be the name of an adjacent locality, as both bear the date of June 9. It is a somewhat singular circumstance, that, so far as we yet know, no example of this form has been found in Southern or Western Australia, nor in Van Diemen's Land.

In general appearance this animal closely resembles the *B. Sciureus*, but differs in being of a smaller size, and in having the tail more slender and cylindrical: the head is so much shorter, that the difference is readily perceptible in the living or recent animal, and conspicuously so in the denuded crania; it is from this character that Mr. Waterhouse assigned to it the specific name of *breviceps*.

In its habits and general economy there is no marked difference from those of *B. Sciureus*; like that species, it secretes itself in the hollows of trees, and sallies forth in search of food on the approach of evening, when it becomes exceedingly active, and readily transports itself from tree to tree by means of the expanding membrane attached to the sides and limbs.

The accompanying drawing was made from living examples in the possession of Mr. Harrington which had bred and reared two young ones, either in London or during their passage to this country.

In a state of nature its food consists of the tender buds of trees and flowers, honey, and insects; in captivity, bread and milk sweetened with sugar forms an excellent substitute for its natural food.

Fur soft; upper surface ashy grey; a dusky longitudinal line extends from between the eyes along the back until lost in the general hue of the rump; tail dusky grey, rather more than two inches of its apical portion black; flank-membrane blackish above, white on the edge, this white fringe extending along the hinder part of the arm to the tip of the little finger; upper surface of the arm sooty black; a dusky mark along the outer side of the legs; under surface white, greyish white or greyish buff; ears black at the base, white at the posterior angles.

The figures are of the natural size.

BELIDEUS NOTATUS, *Peters*

J. Gould and H.C. Richter del. et lith. R. Hanhart Chr. & Walters, Imp.

BELIDEUS NOTATUS, *Peters.*

Stripe-tailed Belideus.

Petaurus (Belideus) notatus, Peters in Monatsb. der Königl. Preuss. Akad. der Wissensch. zu Berlin, 1859, p. 14.

I HAVE been kindly favoured by Dr. W. Peters, the Director of the Royal Museum of Berlin, with the loan of a little Flying Opossum, to which he has given the name of *notatus*, and which was procured by M. Gerard Krefft in the district of Victoria, generally known as Port Phillip.

Dr. Peters had doubtless duly compared this animal with the other members of the genus to which it is most nearly allied, and satisfied himself that it was distinct from either of them, otherwise I should have been inclined to regard it as identical with the *B. breviceps* of Waterhouse ; but in no specimen of *breviceps* that has come under my notice has the tail presented the peculiar marking which characterizes the present animal, the organ being rendered conspicuously different from that of every other member of the genus by the white stripe, bounded on each side by black, which passes down the centre, and by its snow-white tip ; and hence this remarkable deviation from all that has yet come to light certainly deserves to be figured in a work on the Mammals of Australia.

General colour of the upper surface grey, lightest on the head and back of the neck ; commencing on the forehead, and continuing down the centre of the head, neck, and back, is a narrow line of sooty black, which is deepest on the head, and gradually fades into the grey near the root of the tail ; a broad sooty-black mark also occupies the upper edge of the flying membrane ; the front part of the anterior limbs and the front and hinder part of the posterior limbs are also sooty black ; the ear, and the fur around its base, are black ; sides of the face and all the under surface greyish white ; tail grey, deepening into black towards the extremity, with a broad mark of light grey down the middle portion of the upper surface within the black ; the extreme tip snow-white.

The figures are of the natural size.

BELIDEUS ARIEL, Gould

T. Gould and H.C. Richter del. et lith.   Hullmandel & Walton Imp.

BELIDEUS ARIEL, *Gould.*

Ariel Belideus.

Belidea Ariel, Gould in Proc. of Zool. Soc., Part X. p. 11.
Petaurus (Belideus) Ariel, Waterh. Nat. Hist. of Mamm., p. 336.
Petaurus Ariel, Gray, List of Mamm. in Brit. Mus., p. 84.
Wor-gi, Aborigines of Port Essington.

THIS is the smallest and undoubtedly one of the most elegant species of the genus yet discovered : it is as much smaller in all its proportions than *B. breviceps* as that species is smaller than *B. sciureus* ; the tail is also much more cylindrical, less clothed with fur, and equally or more attenuated than that of *B. breviceps* ; the fur of the upper surface is also lighter and of a more delicate buffy grey, and the under surface either pale yellow or rich yellowish buff, as represented in the accompanying figure. The native country of this little animal is the northern portion of Australia ; all the specimens I have seen have been sent from Port Essington, where Mr. Gilbert states that, previous to the hurricane which visited that colony in 1839, it was exceedingly abundant, there being scarcely a hollow tree, shed, or hut, uninhabited by one or more pairs, but since that occurrence it had become exceedingly rare. Commander Ince, R.N., succeeded in bringing living examples from Port Essington to this country, and it is by his kindness in favouring me with the loan of his animals that I have been enabled to make the accompanying drawing from life.

In habits, economy and mode of life it assimilates so closely to the species inhabiting the south coast, that a separate description of them is unnecessary.

General colour of the upper surface pale ashy grey, faintly suffused with yellow ; a narrow black mark commences between the eyes, runs along the back, and extends nearly to the root of the tail ; eyes narrowly encircled with black, and a black ring surrounds the ear at the base, but is interrupted at the posterior angle where the hairs are pale yellow ; upper surface of the flank membrane blackish, especially near the margin, which is pale yellow ; anterior part of the arm, the wrist, and the posterior part of the hind-leg dusky ; under surface pale yellow.

The figures are of the natural size.

ACROBATES PYGMÆUS; *Desm.*

ACROBATES PYGMÆUS, *Desm.*

Pygmy Acrobates.

Didelphis pygmæa, Shaw, Zool. of New Holl., No. 1. pl. 2. p. 5.—Ib. Gen. Zool., vol. i. p. 501.
Petaurus pygmæus, Desm. Nouv. Dict. d'Hist. Nat., tom. xxv. p. 405.
Petaurista (Acrobata) pygmæa, Desm. Mamm., pt. 1. p. 270.
Petaurus (Acrobata) pygmæus, Waterh. Nat. Lib., vol. ix. (Marsupialia), p. 293. pl. 30.—Ib. Nat. Hist. of Mamm.,
    vol. i. p. 339.
Acrobates pygmæus, Gray, List of Mamm. in Brit. Mus., p. 83.

THE portion of Australia designated New South Wales is the only part of that great continent in which I have seen this elegant little Opossum; and it would appear that this is its sole habitat. In a letter recently received from my friend Ronald C. Gunn, Esq., he informs me that "The *Acrobates pygmæus* does not exist in Van Diemen's Land; nor in fact any of the Flying Opossums; but the *Belideus Sciureus*" (*B. breviceps?*) " is now not uncommon in the forests a few miles round Launceston: a number of individuals imported from Port Philip in the years 1835, 1836 and 1837, having escaped from confinement, are doubtless now breeding."

This pretty little animal, the " Opossum Mouse" of the colonists, is very common in every part of New South Wales; but from its nocturnal habits, its small size, and from the circumstance of its exclusively inhabiting the hollow limbs of the larger gum-trees, it rarely comes under the observation of ordinary travellers; it is in fact seen in considerable numbers only by those who really live in the bush, and to their notice it is seldom presented except under extraordinary circumstances, the most frequent of which are the blowing off of a large limb in which it is concealed: if this occurs in the daytime, the animal, being then in a torpid state, does not make its appearance; but if, as occurred several times during my explorations, the limb be thrown upon the traveller's fire, the little inhabitant is soon driven forth by the heat: occasionally as many as four or five are discovered by this means; it was thus that I obtained the specimens here figured, as well as numerous others which I kept alive for some time; and a more charming little pet cannot be imagined, an ordinary-sized pill-box forming a convenient domicile for the tiny creature, in which it lies coiled up during the day, becoming more and more active as night approaches. Its food consists of the saccharine matter which is so abundant in the flower-cups of the ever-blossoming *Eucalypti,* for which well-sweetened bread and milk forms an excellent substitute. The agility it displays among the branches in the night-time is very great; it not only passes over, around and beneath them, but, aided by the membrane attached to the sides and limbs, leaps from one bunch of flowers to another with the greatest facility.

The sexes are alike, but the female is somewhat smaller than her mate.

Fur short, dense, soft, glossy, and on the upper surface ashy greyish brown; under surface greyish white in some, yellowish white in others, this colour extending on to the lower part of the cheeks and the upper lip; circle surrounding and a space in front of each eye black; ears dusky towards the fore-part, and whitish behind; on the inner side of the ear near the apex and on the apical portion spring numerous long and extremely fine hairs; moustaches numerous, slender, and of a dusky brown; tail fringed on the sides with longer hairs than those clothing its upper and under surfaces.

The figures are of the natural size.

DROMICIA GLIRIFORMIS.

J. Gould and H.C. Richter del et lith.

Hullmandel & Walton Imp.

DROMICIA GLIRIFORMIS.

Thick-tailed Dromicia.

Phalangista gliriformis, Bell in Linn. Trans., vol. xvi. p. 121. pl. 13.
Dromicia gliriformis, List of Mamm. in Brit. Mus. Coll., p. 85.

THE *Dromicia gliriformis* is nowhere more abundant than in Van Diemen's Land, particularly the northern parts of the island ; and indeed it is very questionable if it is to be found in any other part of Australia ; but our present knowledge will not admit of the positive assertion.

I am sufficiently acquainted with the habits and economy of the *Dromicia gliriformis* to state that it is a strictly nocturnal animal, and that of all trees it prefers the Banksias, whose numerous blossoms supply it with a never-ceasing store of food, both of insects and sweets ; if I mistake not, it also feeds upon the tender buds and spikes of the flowers. During the day it generally slumbers coiled up in some hollow branch or fissure in the trees, whence if its retreat be discovered it is easily taken by the hand ; this state of inactivity is totally changed at night, when it runs over the smaller branches and leaps from flower to flower with the utmost ease and agility. This disposition is just as strongly displayed by it when kept in confinement ; being so drowsy during the daytime as to admit of its being handled without evincing the least anxiety to escape, while the contrary is the case as soon as night approaches. I have also observed that during the months of winter it is less active than in the summer ; undergoing in fact a kind of hyber-nation, somewhat similar, but not to the same extent, as the Dormouse.

That this interesting little animal bears confinement well and contentedly, is proved by the circumstance of the pair from which the accompanying drawing was taken being now alive in the possession of Her Most Gracious Majesty at Windsor Castle, where they are thriving as well as if they were in their native wilds. They were brought to this country by the Very Reverend the Archdeacon Marriott, who kindly permitted me to make drawings of them for the present work. If any difference is perceptible between examples in captivity and those in a state of nature, it is that the former are more sluggish in their actions and inclined to obesity.

Four individuals formed part of the collection in the Zoological Gardens, Regent's Park, and after living there for three years died, apparently without disease and probably from old age ; and my most estimable friend Thomas Bell, Esq., F.R.S., was in possession of living examples for four years, which furnished him with the materials for a paper on its habits and economy while in confinement, and I take the liberty of copying the following extract verbatim :—

"In their habits they are extremely like the Dormouse, feeding on nuts and other similar food, which they hold in their fore-paws, using them as hands. They are nocturnal, remaining asleep during the whole day, or, if disturbed, not easily roused to a state of activity, and coming forth late in the evening, and then assuming their natural rapid and vivacious habits ; they run about a small tree which is placed in their cage, using their paws to hold by the branches, and assisting themselves by their prehensile tail, which is always held in readiness to support them, especially when in a descending attitude. Sometimes the tail is thrown in a reverse direction, turned over the back, and at other times, when the weather is cold, it is rolled closely up towards the under part and coiled almost between the thighs. When eating they sit upon their hind-quarters, holding the food in their fore-paws, which, with the face, are the only parts apparently standing out from the ball of fur of which the body seems at that time to be composed. They are perfectly harmless and tame, permitting any one to hold and caress them without ever attempting to bite, but do not evince the least attachment either to persons about them or to each other."

Considerable diversity of colour exists in different individuals ; in some the upper surface is nearly uniform grey, while in others a fine tawny or rufous tint pervades the same parts ; and examples are constantly met with exhibiting every variety of intermediate shade.

The sexes are very nearly alike in size and colour.

The fur is very soft and thick ; all the upper surface either grey or yellowish grey, the yellow tint predominating on the sides, body, and the face ; under surface either greyish white or yellowish white ; base of the tail similar in colour to the upper surface of the body, but becoming purplish flesh-colour towards the tip.

The figures are of the natural size.

DROMICIA CONCINNA: Gould.

DROMICIA CONCINNA, *Gould.*

Beautiful Dromicia.

Dromicia concinna, Gould in Proc. of Zool. Soc., Jan. 14, 1845.
Man-duř-da, Aborigines of Western Australia.

THIS pretty little animal is abundantly and very generally distributed over the colony of Swan River. Its habits being strictly nocturnal, it secretes itself during the day in the hollows and chinks of trees, particularly those of the *Casuarinæ* and *Banksiæ* ; at night it leaves its retreat for the flowering branches of shrubby low trees in search of insects and sweets, of which, from the abundance of flowering plants, it easily obtains a never-ceasing supply.

It becomes very tame in confinement ; spending the daytime in sleep with its body rolled up in the form of a ball, and on the approach of evening throwing off its drowsiness and becoming animated in the extreme, leaping about from side to side of its cage in chase of insects, of which it is exceedingly fond.

The extent of its range over the continent of Australia, and all minute particulars respecting its habits and economy, have yet to be ascertained.

The sexes are so similar that they present but little difference either in size or colour ; in some specimens the under surface is slightly tinged with buff.

Before the eye a mark of black ; all the upper surface, the outer side of the limbs and the tail pale sandy brown ; all the under surface and the inner side of the limbs white ; the two colours distinctly separated, or not blending into each other.

The Plate represents a male and a female, on a branch of a very beautiful species of *Melaleuca*.

PHASCOGALE PENICILLATA.

PHASCOGALE PENICILLATA.

Brush-tailed Phascogale.

Tapoa-tafa, White's Journ., pl. in p. 281.
Didelphys penicillata, Shaw, Gen. Zool., vol. i. part 2. p. 502. pl. 113. fig. 1.
Dasyurus penicillatus, Geoff. Ann. du Mus., tom. iii. p. 361.
———— *Tafa,* Geoff., loc. cit.
Phascogale penicillata, Temm. Monogr. de Mamm., tom. i. p. 58.—Skull, pl. vii. figs. 9–12.—Waterh. Nat. Lib.
      Mamm., vol. ix. (Marsupialia), p. 136, pl. viii.—List of Mamm. in Brit. Mus. Coll., p. 98.
Tapoa-tafa, Aborigines of New South Wales.
Bul-loo-wa, Aborigines of the York district of Western Australia.
Bal-a-ga, Aborigines of Perth.
Bal-lă-wa-ra, Aborigines to the north of Perth.

As several specimens of this animal, contained in a collection lately received from Western Australia, offer on comparison no difference whatever from others procured in South Australia and New South Wales, it is evident that the Brush-tailed Phascogale has an unusually wide range of habitat. It probably does not extend so far south as the island of Van Diemen's Land, or northward of the twenty-fifth degree of south latitude.

The plain and the mountain districts appear to be equally inhabited by it, and from its destructive propensities is I fear likely to become a pest to the colonists. It has already been known to enter the stores of the settlers and commit severe depredations whenever they contained anything suited to its palate, and, whether justly or not I am unable to say, it has also been charged with killing the fowls and chickens of the hen-roost. In the stomachs of some that were dissected were found the remains of coleopterous insects, and what appeared to be a species of fungus. Nocturnal in its habits, it sleeps during the day in the hollows of decayed trees, from which retreat it emerges on the approach of evening, when it ascends the trees and displays the greatest activity among the branches. When captured it becomes quite ferocious and struggles hard to effect its escape, and so severe are the lacerations it inflicts, that even a native can rarely be induced to put his hand within reach of a living one. It breeds in the hollows of the gum-trees, but the precise number of its young has not yet been ascertained. The sexes differ but little in size and colour, but the male is somewhat the largest. The accompanying Plate represents the animal of the size it is when fully adult. It is necessary to mention this, because much difference exists in the relative size of specimens sent to this country, many individuals that I have seen not being more than half the size of those figured, and which is solely attributable to the youthful state of the animal, and not to a difference of species. It was first figured in White's "Journal of a Voyage to New South Wales," a work published in 1790, under the name *Tapoa-tafa*; the specimen there represented is still preserved in the Museum of the Royal College of Surgeons, so that we have the clearest evidence of its identity with the animal here figured.

The colouring of the Brush-tailed Phascogale may be thus described :—

Face, all the upper surface and the base of the tail grey; chin, throat, inside of the legs and feet greyish white; a darker grey mark commences at the tip of the nose and extends over the forehead to the nape; the fur is moderately long and loose, that which covers the back and upper surface being uniform blue-grey next the body, and grizzled grey and brown towards the surface; lengthened black hairs are also thinly scattered among the fur of the upper surface; the tail for about four-fifths of its length from the tip is clothed with long and stiff hairs of the finest black, giving that organ a brush-like appearance, whence its specific name; tip of the nose flesh-colour; ears purplish, very thinly covered with fine hairs.

The figures represent the two sexes of the size they attain when fully adult.

PHASCOGALE CALURUS: Gould.

PHASCOGALE CALURA, *Gould.*

Handsome-tailed Phascogale.

Phascogale calura, Gould in Proc. of Zool. Soc., Part XII. p. 104.

ALL mammalogists who are acquainted with the *Phascogale penicillata* will observe that a great similarity in form exists between that animal and the one here represented, of which a single individual has lately been forwarded to me from Western Australia, and which I believe to be the only specimen yet transmitted to Europe. I am led to consider it one of the most interesting of the Australian mammals lately discovered, not only from its forming the second species of the genus as now restricted, but from the extreme elegance of its form and the chaste but diversified character of its markings. The rich rust-red of the basal half of the upper surface of the tail is a very unusual mark in animals of this order. Mr. Gilbert procured the specimen above-mentioned while staying at the Military Station on William's River, and he merely says : " For this new animal I was indebted to a domestic cat who had captured it in the night. The soldiers informed me that they had often met with it in the store-room of the Station, but they could give me no other information respecting it, except that specimens with much larger or more brushy tails were some-times seen." The fact of its visiting the stores shows, that in habits and disposition it assimilates as closely to the *P. penicillata* as it does in form.

The fur is soft and moderately long ; its general colour is ashy grey externally and grey next the skin ; under surface of the body white, tinted with cream-colour, which becomes more distinct on the sides ; eyes surrounded by a narrow ring of black ; in front of the eye a blackish patch ; ears sparingly clothed for the most part with very minute dusky hairs, but at the base, both externally and internally, are some long yellowish hairs ; base of the tail for nearly half its length clothed with short hairs of a brilliant rusty red colour ; on the apical half of the tail the hairs are long and black, as is also the under surface of the base to near the root.

The Plate represents the animal, which is now in the British Museum, of the natural size.

PHASCOGALE LANIGERA, *Gould.*

PHASCOGALE LANIGERA, *Gould.*

Woolly Phascogale.

A SINGLE specimen of the little Phascogale figured on the accompanying Plate was discovered by Sir Thomas Mitchell, during one of his expeditions into the interior of Australia. This specimen, which is now in the British Museum, appears to be fully adult. In form it is precisely similar to the *Phascogale calura*, but in size and colouring it is very different, being a much smaller animal, and having no trace of the rufous colouring so conspicuous on the basal portion of the tail of that species. I regret to state that no account of the habits of this little animal accompanied the specimen.

I am indebted to Dr. Gray, of the British Museum, for permission to figure and describe this new and interesting addition to the genus *Phascogale*. The paucity of the information we possess respecting it affords further evidence of the little we know respecting the smaller animals of the interior of Australia, an acquaintance with which is rendered all the more difficult of acquisition from the circumstance of the whole, or nearly the whole of them being nocturnal in their habits.

The fur is soft, and of a character somewhat more woolly than that of *P. calura*; its general colour is brown externally and grey next the skin, becoming hoary on the hind-quarters; under surface of the body greyish-white; eyes surrounded by a narrow ring of black; ears sparingly clothed with minute brown hairs; fore- and hind-feet white, becoming brownish-grey on the toes; basal portion of the tail brown, like the body; hairs of the apical half of the tail long and black, as is the under surface of the base to near the root.

The figures are of the natural size.

ANTECHINUS SWAINSONI.

ANTECHINUS SWAINSONI.

Swainson's Antechinus.

Phascogale Swainsonii, Waterh. Mag. Nat. Hist., vol. iv. p. 300.
————— *(Antechinus) Swainsonii,* Waterh. Nat. Hist. of Mamm., vol. i. p. 411.—Gunn, Proc. of Roy. Soc. of
Van Diemen's Land, vol. ii. p. 82.—Zool. of Erebus and Terror, Beasts, pl. 25. fig. 1.

Of the animals comprising the restricted genus *Antechinus,* the present is the largest and the most darkly coloured species yet discovered. Van Diemen's Land, if not its sole habitat, is the country in which it is usually found, and I believe I am right in stating that up to the present time it has not been obtained elsewhere.

Mr. Waterhouse, after remarking that this species is of a much darker colour than any of the other *Antechini,* and is almost destitute of any grey hue, says, " The fur is long and moderately soft, and is of a deep grey colour next the skin; on the back the hairs are most of them annulated with rusty yellow or brownish rust-colour, the deeper tint being observable on the hinder parts. The hairs of the hinder parts of the body are grey, but tipped with yellowish. The tail is clothed throughout with small adpressed hairs of a dusky-brown colour, and a trifle paler on the under than on the upper surface. The feet are uniform dusky brown; the fleshy pads on their under surface are transversely striated, and the remaining naked portion of each foot is apparently smooth. The muzzle is narrower and more elongated than usual. The specimen from which the original description was taken measured from the tip of the nose to the root of the tail 5 inches and 2 lines in length, and its tail was 3 inches and 5 lines long." But that the animal attains a larger size is certain, there being an example in the British Museum which is 7 inches in length, and others of an equal size in the fine collection bequeathed to the town of Liverpool by the late munificent Earl of Derby.

The figures are of the size of life.

ANTECHINUS LEUCOPUS, *Gray*

ANTECHINUS LEUCOPUS, *Gray*.

White-footed Antechinus.

Phascogale leucopus, Gray in Ann. and Mag. of Nat. Hist., vol. x. p. 261.—Ib. List of Spec. of Mamm. in Coll. Brit. Mus., p. 100.
———— (*Antechinus*) *leucopus*, Waterh. Nat. Hist. of Mamm., vol. i. p. 423.

I HAVE figured this little Opossum as an inhabitant of Van Diemen's Land on the authority of Dr. Gray. The specimen from which he took his description is now in the British Museum, and appears to differ sufficiently from the other members of the group to warrant its being characterized as distinct; but on this point Mr. Waterhouse remarks,—

"The general tint of this animal is somewhat darker than that of *Phascogale albipes*; the upper surface of the tail is almost black, while in the species just mentioned it is greyish, and the ears are smaller. Beyond these, I can perceive no other points of distinction between the Van Diemen's Land animal and the Continental one (*P. albipes*). Of the former I have seen but one specimen, and I can scarcely satisfy myself, from such imperfect materials as are before me, that these White-footed *Phascogales* are specifically distinct.

"A small Phascogale is found at King George's Sound, which agrees very closely with the *P. leucopus*, being of the same dark colour, and having the tail black above, or nearly so. Two specimens in Mr. Gould's collection, thus resembling the Van Diemen's Land animal, differ, however, in having the chest of a dusky grey hue. A specimen from King George's Sound, and contained in the British Museum collection, differs in having the colouring less dark, and, indeed, very closely resembling that of *Phascogale albipes*.

"Fur very soft and rather long; general colour grey, much suffused with black on the back, and very finely pencilled with pale yellow, the yellow most distinct on the head and sides of the body; feet and under parts of the body white; ears tolerably large, and clothed with minute hairs, for the most part dusky, but pale at the basal portion of the ear externally; upper surface of the tail nearly black, under surface dirty white.

	Male.	
	inches.	lines.
"Length from tip of nose to root of tail	4	4
„ of tail	3	7
„ of ear		5
„ of hind-foot and nails		$8\frac{1}{2}$"

The figures are of the natural size.

ANTECHINUS FERRUGINIFRONS, Gould.

ANTECHINUS FERRUGINIFRONS, *Gould.*

Rusty-fronted Antechinus.

HAVING received two specimens of this animal direct from Sydney, I have little hesitation in stating New South Wales to be the true habitat of this new species; at the same time I am unable to say of what particular locality it is a native. Its yellowish rust-coloured face, more lengthened muzzle and larger size, at once distinguish it from *Antechinus flavipes* and *A. unicolor*, to both of which it is allied. It also differs from both in the more slender form and in the white colouring of its feet; points in which it offers some affinity to the smaller members of the genus, such as *Antechinus fuliginosus*, *A. albipes*, &c. In all probability this modification in the structure of the feet is accompanied by some diversity in the habits or economy of these slender-footed animals, but this is a point which can only be determined when we are able to obtain a more intimate knowledge of these singular little quadrupeds than we at present possess.

Fur moderately long and soft; face, head and occiput rusty fawn colour, interspersed with fine blackish hairs; general colour of the upper surface and sides of the body pale greyish brown, interspersed down the back with numerous fine blackish hairs; sides of the face washed with buff; throat and under surface of the body pale greyish white; all the four feet white; tail light brown.

Total length from the tip of the nose to the end of the tail 7¼ inches; of the tail 3½; of the nose to the ear 1¼ inch; of the ear ¼ an inch; of the tarsi and toes ¾ of an inch.

The figures represent the animal of the size of life.

ANTECHINUS UNICOLOR, *Gould.*

J.Gould. and H.C.Richter. del. et lith.   Hullmandel & Walton. Imp.

ANTECHINUS UNICOLOR, *Gould*.

Dusky Antechinus.

THIS animal is altogether larger and more robust than the *Antechinus ferruginifrons*, has a broader or more dilated hind foot, a shorter muzzle, and a more uniform style of colouring, the general tone being a rusty brown, with a somewhat heightened or brighter rusty hue on the lower part at the back and rump; both the fore and hind feet moreover are of a light brown.

Like the *A. ferruginifrons* this species is a native of New South Wales. The specimens I possess of both species were in fact received at one time by way of Sydney, without, unfortunately, any particulars as to the locality in which they had been obtained. Such then is all the information I am able to give respecting these rare species, of each of which two specimens were transmitted to me. Australia appears to abound in these small insectivorous animals, as evidenced by the numerous species described and figured in the present work, and when the forests of that great country are more closely searched, many others will doubtless be discovered. In its general structure this species must be associated with the broad-footed section, of which *A. flavipes* may be considered a typical example.

Fur moderately soft; general colour of the upper surface fulvous brown, interspersed with numerous black hairs; under surface paler fulvous brown; feet pale brown.

Length, from the tip of the nose to the end of the tail, 9¼ inches; of the tail, 3⅝; from the nose to the ear, 1¼; of the ear, ¼ an inch; of the tarsi, ¾ of an inch.

The Plate represents two animals of the size of life.

ANTECHINUS LEUCOGASTER. Gray.

Drawn and W? Baker del et lith

Hullmandel & Walton Imp.

ANTECHINUS LEUCOGASTER, *Gray.*

White-bellied Antechinus.

Phascogale leucogaster, Gray, App. to Grey's Journ., vol. ii. p. 407.
Antechinus leucogaster, Gray, List of Mamm. in Coll. Brit. Mus., p. 99.
Phascogale (Antechinus) leucogaster, Waterh. Nat. Hist of Mamm., p. 417.

"This animal so closely resembles the *Antechinus flavipes* in all its proportions, as well as in the structure of its skull and teeth, that it is with considerable hesitation I describe it as a distinct species. I have seen, however, several specimens from Western Australia which agree perfectly with the individual from which Mr. Gray drew up his original description, and which differ from the *A. flavipes* in having the under parts of the body white, and little or no rusty yellow on the sides of the body and on the feet. The general tint of the upper surface likewise differs somewhat, being less grey on the fore parts of the body, and on the hinder parts rich brown. The feet are brownish white, not unfrequently suffused with brown behind; the tail is brown above, pale brown beneath, and dusky towards the point.

"A skull removed from a specimen sent me from King George's Sound by Mr. Neill, differs from a skull of *A. flavipes* in the British Museum collection in having the muzzle (and consequently the nasal bones) a trifle shorter, but the difference is not more than is often found in individuals of the same species, and I think it by no means improbable that the *A. leucogaster* is a local variety of *A. flavipes*."

The above are Mr. Waterhouse's opinions respecting a Western Australian animal, of which my collection contains two or three examples obtained at King George's Sound. I figure it with the same degree of doubt as to its specific value that is entertained by Mr. Waterhouse, but I may state that I have seen hundreds of *A. flavipes* from Southern and Eastern Australia, all of which had the feet and under surface of a deep rusty colour, a hue I have never yet seen in any of the specimens of the Western Australian *Antechini*.

Fur rather soft, general colour dark brownish grey; the hind quarters tinted with rusty brown; all the upper surface beset with numerous fine black hairs; ears sparingly clothed with minute pale-coloured hairs; under surface greyish white; tail dusky, passing into blackish at the apex; feet light brown.

The figures are of the natural size.

ANTECHINUS APICALIS.

ANTECHINUS APICALIS.

Freckled Antechinus.

Phascogale apicalis, Gray, Ann. and Mag. Nat. Hist., vol. ix. p. 518.
Antechinus apicalis, List of Mamm. in Brit. Mus. Coll., p. 99.
Marh-dern, Aborigines in the neighbourhood of Moore's River.
Wÿ-a-lung, Aborigines of Perth.
Dib-bler, Aborigines of King George's Sound.

THIS animal is very generally distributed over every part of the colony of Western Australia, where it in-habits trees of various kinds, from the prostrate trunk of the once patriarchal gum of the dense forest to the living grass-trees of the more open districts. Mr. Gilbert's notes comprise all that is at present known of its habits, and these I give in his own words :—" The nest of this animal and the situation in which it is placed appear to vary in different parts of the country. The aborigines in the neighbourhood of Moore's River agree in stating that it is placed in a slight depression of the ground beneath the overhanging leaves of the *Xanthorrhœa*; on the other hand, the natives around Perth assured me that they always captured the animal either in a dead stump or among the grasses of the *Xanthorrhœa*; at King George's Sound it appears to differ from both the preceding, for there the natives always pointed out as the nest of this species, a raised structure of fine twigs and coarse grass, very closely resembling that of the common Perameles. The stomachs of those I dissected contained the remains of insects of various kinds. While at King George's Sound, I obtained a female with seven young attached; they were little more than half an inch in length, quite naked and blind. Above the mammæ of the mother is a very slight fold of skin, from which the long hairs of the under surface spread downwards and effectually cover and protect the young. The fold in the skin of the abdomen is the only approximation to a pouch that I have found in any member of this genus. The young are very tenacious of life; those above mentioned lived nearly two days, attached to the mammæ of the dead mother; and after being immersed in spirits of wine continued in motion for nearly two hours."

The sexes are precisely alike in colour; but the female is somewhat the smaller.

This little animal may be thus described :—All the upper surface reddish brown, interspersed with numerous longer hairs, which are black in the centre and white at the tip, giving the animal a peculiarly grizzled appearance; flanks and under surface buffy grey; outside of the fore and hind legs rufous; tail similar to the upper surface, passing into black at the tip which terminates in a fine point, whereas at the base it is thicker and the hairs more lengthened than in any other species of the genus; the hairs are also of a more stiff and wiry character.

The Plate represents both sexes of the natural size.

ANTECHINUS FLAVIPES.

ANTECHINUS FLAVIPES.

Rusty-footed Antechinus.

Phascogale flavipes, Waterh. in Proc. of Zool. Soc., part 5. p. 75.
————— *rufogaster*, Gray, App. to Grey's Journ., vol. ii. p. 407.
Antechinus flavipes, Gray, List of Mamm. in Coll. Brit. Mus., p. 99.
Phascogale (Antechinus) flavipes, Waterh. Nat. Hist. of Mamm., vol. i. p. 415.

———————————

THE specific term of *flavipes* is scarcely an appropriate appellation for this animal, for, as will be seen on reference to the accompanying drawing, which, if not taken from the original type, was made from precisely similarly coloured specimens, the feet are of a deep rust-red, the yellowish red hue which suggested the name being only found in some specimens. Of all the *Antechini* yet discovered, the present is the most common; I observed it to be very abundant both in New South Wales and in South Australia, and remarked that specimens from both these countries presented little or no difference either in size or colour. Like most of the other species of the genus, this little animal inhabits the hollow branches of the large *Eucalypti*. I frequently saw it running over the fallen logs by the creek sides of the plains of Adelaide, and remarked that in New South Wales it affected similar localities, and exhibited similar actions and habits. Its progression over the boles of trees is effected by a succession of very quick jumps like those of the Common Squirrel, and it passes round and beneath the branches with equal facility. Besides being conspicuous for its rusty coloured feet, this species is distinguished by the colouring of the face, back of the head, shoulders, and upper part of the back being dark grey with yellowish hairs interspersed, giving those parts a freckled appearance, while the lower part of the back and the thighs are more rufous. I could not observe any difference in the colouring of the sexes or of the young: in the relative size of the sexes, on the contrary, considerable difference exists, the adult female being always smaller than the male of the same age.

Fur moderately long and soft; general colour of the upper surface grey, tinted with fulvous on the lower part of the back; sides of the body washed with rusty yellow; under surface of the body and feet rusty yellow; chin, throat and chest whitish; tail black, freckled with yellow towards the base above, and rusty yellow beneath; tail clothed with short hairs, lengthened into a small tuft at the point.

The figures are of the natural size.

ANTECHINUS FULIGINOSUS, *Gould.*

Gould & Richter, del. et lith.
Hullmandel & Walton, Imp.

ANTECHINUS FULIGINOSUS, *Gould.*

Sooty Antechinus.

Antechinus fuliginosus, Gould in Proc. of Zool. Soc. 1852.
Twoor-dong, Aborigines of King George's Sound.

THIS species of *Antechinus*, which is a native of Western Australia, may be easily distinguished from the other members of the genus by its very dark colouring, a feature pervading both the upper and under surface of the animal. Up to this time, that is, after fourteen years' close attention to the natural productions of Australia, I have never seen an example of this species from any other than the western part of that country; there, however, it is very abundant, both at King George's Sound and in the vicinity of Perth. I am indebted to the researches of the late Mr. Gilbert for the following account, which, however, brief as it is, will I hope be read with interest by every true lover of zoology :—

" This is so much like the *Antechinus albipes,* that I considered it to be the same animal, until, by hunting for it myself, I found that it not only differs in habits, but is of a somewhat larger size and very much darker colour. Its favourite resorts are newly burnt spots, especially those adjacent to swamps and moist meadows. Among the clumps of the burnt stumps of coarse grass it burrows out the earth, and fills the cavity with short pieces of fine twigs and grass in the form of a round heap about two or three inches in height, the top being in most instances level with the surface of the surrounding earth; this structure is from six to twelve inches in diameter and from ten to fifteen in depth; in the top are several holes leading to galleries situated about half way down, which run horizontally among the roots of the surrounding scrub, and into one or other of which the animal escapes while the upper or loose portions of the sticks and grass are being removed. These structures are so precisely similar to the nests formed of pieces of grass and twigs of the same form and placed in similar situations by a small species of black ant, that I had passed hundreds without detecting them to be different, until the natives pointed them out to me as the nests of this animal, the only difference being the entrance-holes at the top and the absence of ants in the interior. I endeavoured to keep this species in captivity, but rarely succeeded in preserving it alive for more than a couple of days. It is exceedingly active in its habits, and when at rest the general contour of its body is short and ball-like; the eyes are black and prominent; the lower lip shows distinctly to the gape, and is of a pale lemon-yellow; it utters the singular hissing-like noise common to most of the Marsupials. It feeds at night, and appears to prey upon insects generally, as the stomachs of those I examined contained insects of various kinds."

The whole of the upper surface dark greyish brown, interspersed with numerous longer black hairs, giving it a fuliginous or sooty hue; face of a lighter tint; the whiskers and a narrow mark round the eyes black; sides of the chest sooty grey, separated down the centre by a narrow line of buffy grey extending from the chin to the insertion of the fore legs; under surface pale greyish white; fore feet and the hinder tarsi and feet white, slightly tinted with buff; tail dark reddish brown, becoming greyish beneath; ears inclined to silvery grey.

		inches.
Length from the nose to the root of the tail	. . . .	$3\frac{1}{4}$
,, of tail	. . . .	$3\frac{1}{4}$
,, ,, arm and hand	. . . .	$\frac{1}{2}$
,, ,, tarsi and toes	. . . .	$\frac{3}{4}$
,, ,, face from the tip of the nose to the base of the ear	. .	1
,, ,, ear	. . . .	$\frac{7}{16}$

The figures are of the natural size.

ANTECHINUS ALBIPES.

Gould & Richter, del et.
Hullmandel & Walton Imp.

ANTECHINUS ALBIPES.

White-footed Antechinus.

Phascogale albipes, Waterh. in Proc. of Zool. Soc., Part X. p. 48.
Phascogale (Antechinus) albipes, Waterh. Nat. Hist. of Mamm., vol. i. p. 421.
Otåm-in, Aborigines of Perth.

———————

THE accompanying Plate represents the *Phascogale (Antechinus) albipes* of Mr. Waterhouse, which appears to be almost universally distributed over the whole of the southern coast of Australia, from Swan River to New South Wales. I possess specimens collected by Mr. Gilbert in the vicinity of Perth, in the Swan River settlement, and others procured by him on the Darling Downs in New South Wales, while the specimen from which Mr. Waterhouse took his description had been sent from the intermediate district of Adelaide in South Australia. Some little difference is observable in specimens from the eastern and western coasts, particularly in the size of the ear, that organ being of a larger and rounder form in the individuals from New South Wales than in those from Western Australia; still this character is too slight to be regarded as indicative of anything but a mere local variety. I find the following notes respecting this animal among Mr. Gilbert's letters to me from Western Australia :—

"This species inhabits the dead stumps of the grass-trees (*Xanthorrhœa*). It makes no nest, but merely scrapes together a few of the dry fibrous parts: more than a single pair are rarely seen at one time. The stomachs of those I examined contained the remains of coleoptera. The length of the animal before skinning was seven inches from the tip of the nose to the extremity of the tail; the tail being three and three-eighths. This species is to be found among the grassy lands of the Toodyay district, as well as among the dense groves of *Xanthorrhœa* surrounding the swamps in the vicinity of Perth." When writing from Darling Downs in New South Wales he remarks : "This animal here inhabits clumps of grass in scrubby places : it may be readily distinguished from all the other members of the genus by its very large ears, the general lightness of its fur, and its long, slender tail."

Mr. Waterhouse's remarks on this species are as follows :—

"The White-footed Antechinus was discovered by the late J. B. Harvey, Esq., a very zealous corresponding member of the Zoological Society : in size and colouring it greatly resembles the Field Mouse of Europe ; its form is more robust than any of the other *Antechini*, its feet are more slender, and a greater portion of the palm of the hind foot is clothed with fur.

"The fur both on the upper and under parts of the body is of a deep slate-grey colour next the skin ; on the upper parts the hairs are of a very pale yellow colour near the point, and black at the point; those on the under surface are white at the point ; the eyes are encircled with black ; the large ears are clothed throughout with minute hairs, chiefly of a pale hue, but dusky on the outer surface near the anterior margin ; the tail is clothed with very small hairs of a dirty white colour on the under, and partly black and partly yellow-white on the upper surface."

The figures are of the natural size.

ANTECHINUS MURINUS, Waterh.

ANTECHINUS MURINUS.

Murine Antechinus.

Phascogale murina, Waterh. in Proc. of Zool. Soc., Part V. p. 75.—Ib. Nat. Lib., Marsupialia, p. 143. pl. 10.
Phascogale (Antechinus) murina, Waterh. Nat. Hist. of Mamm., p. 425.

———————

THE subject of the present Plate is another of the little Marsupials described by Mr. Waterhouse, of which my own collection contains two specimens from New South Wales, and which, on comparison with the original in the Museum of the Zoological Society of London, from which Mr. Waterhouse took his description, presents little or no difference. "The *A. murinus*," remarks Mr. Waterhouse, " is considerably smaller than *A. albipes*; its general colouring is paler, and its tail is uniformly white. The tarsi are almost entirely covered with hair on the under side, there being only a very narrow naked space running backwards from the digital pads, which are covered with small tubercles." With respect to the tail being entirely white, as remarked by Mr. Waterhouse, this does not appear to be a constant character, the tail of one of my specimens being wholly white, while in another it is washed with brown, particularly on the upper surface; and I am led to believe that the white tail is characteristic of immaturity, and not of the adult state. It must be admitted that it is a species, the distinctness of which from *A. albipes* is not very apparent. I have, however, no doubt of its being different, and in confirmation of this view I may state that it was sent to me as such by Mr. Gilbert, who, when writing from the Darling Downs in New South Wales, prior to joining Dr. Leichardt's expedition from Moreton Bay to Port Essington, in which his valuable life was unfortunately sacrificed, says, " I caught this species on the banks of the river Severn; the male is much larger in all its proportions than the female, and has a darker mark around and before the eye."

Fur very soft; upper parts of the body ashy grey; under parts and feet white; tail silvery white; ears clothed with minute pale hairs, becoming brownish in front on the outer surface; eyes encircled with black; fur of the under surface grey at the base.

The figures are of the natural size.

ANTECHINUS MACULATUS.

ANTECHINUS MACULATUS, *Gould.*

Spotted Antechinus.

Antechinus maculatus, Gould in Proc. of Zool. Soc., Dec. 9, 1851.

THE progress of civilization over the vast continent of Australia cannot fail to be the means of bringing to light many species of small quadrupeds at present entirely unknown to the zoologist; and the immense brushes which stretch along the southern and eastern coasts in particular, will, I feel confident, afford many treasures in this department of zoological science. During my short rambles in some of those extensive brushes, I frequently saw among the fallen leaves and thick herbage, many small, agile, mouse-like marsupials, which I found it impossible to procure, as they were too light to spring the ordinary traps, however lightly they might be set, and I was unprovided with any more suitable contrivances for capturing them; it must also be remembered that the difficulty of collecting them is much increased by their being all more or less nocturnal in their habits. Mr. Strange, however, from time to time sends me, among other objects, one of these little quadrupeds; and it is to him that we are indebted for our knowledge of the pretty species figured in the accompanying Plate, which was procured in the brushes near the river Clarence, a little to the southward of Moreton Bay. Contrarieties with regard to colouring and disposition of markings continually occur in the Australian Fauna, of which the present little animal offers another instance; since, contrary to the normal rule, we find it ornamented on the lower instead of the upper parts of the body.

The animal sent by Mr. Strange is a fully adult male and may be thus described :—

The fur is short, dense, and closely applied to the skin; the general tint of the upper surface is dark blackish brown, minutely grizzled with yellowish brown; lower part of the flanks and the under surface of the body dark brownish slate-grey, ornamented with oblong spots of white arranged in irregular rows in the direction of the body : there is also a streak of white down the centre of the throat.

The figures are of the natural size.

ANTECHINUS MINUTISSIMUS, *Gould.*

ANTECHINUS MINUTISSIMUS, *Gould.*

Minute Antechinus.

Antechinus minutissimus, Gould in Proc. of Zool. Soc., 1852.

This is by far the least of the Marsupials that have as yet been discovered in Australia. The accompanying figures which were taken from fully adult specimens represent the animal of the natural size. The little *Antechini* of Australia constitute two very distinct groups or subsections; the form of one of which is characterized by a more elegant and lengthened contour, a sharper and more attenuated muzzle, larger ears and longer, more slender and mouse-like formed tarsi, such as is seen in *Antechinus albipes, A. fuliginosus* and *A. murinus*; and the other by a shorter and bluffer head, smaller rounded ears, and extremely short and broad tarsi, as is perceptible in *A. flavipes, A. maculatus,* and the present species, *A. minutissmus.* I am quite sure that this difference in structure is accompanied by an equally marked difference in the habits and actions of the animals constituting these two groups: I had many opportunities of observing the *A. flavipes* in a state of nature, and noticed that it exhibited some very curious actions while traversing the large boles and limbs of the trees, both those that were prostrate as well as those still erect: its mode of progression was more like that of the squirrel than of any other animal with which I can compare it; as it traversed the limbs in every direction by leaps, with widely spread legs, sometimes sideways, at others with the head downwards; indeed in any position in which it wished to move.

The native habitat of the *A. minutissimus* is the districts on the eastern coast of Australia, in the neighbourhood of Moreton Bay. I have specimens collected by Mr. Strange labelled with the native name of *Pimburam.*

Fur short and closely applied to the skin; the whole of the upper surface, including the tail, greyish brown, the latter being paler beneath; chin and throat pale buff; feet buffy brown; under surface of the body and legs greyish buff gradually blending with the brown of the upper surface, but the buffy tint predominating on the centre of the abdomen and vent.

	inches.
Length from the nose to the root of the tail	$2\frac{1}{4}$
„ of tail	$2\frac{3}{8}$
„ „ arm and hand	$\frac{3}{8}$
„ „ tarsi and toes	$\frac{7}{16}$
„ „ face from the tip of the nose to the base of the ear	$\frac{9}{16}$
„ „ ear	$\frac{1}{4}$

The figures are of the natural size.

PODABRUS MACROURUS. Gould.

J. Gould and H.C.Richter del. et lith.

Hullmandel & Walton Imp.

PODABRUS MACROURUS, *Gould.*

Large-tailed Podabrus.

Podabrus macrourus, Gould in Proc. of Zool. Soc., Part XIII. p. 79.
Phascogale (Antechinus) macroura, Waterh. Nat. Hist. of Mamm., vol. i. p. 426.

ALL that I have to record respecting this new and interesting little animal, is that several specimens were procured and sent to me, accompanied with the following remarks, by Mr. Gilbert, just prior to starting on his ill-fated expedition from Moreton Bay to Port Essington.

" This is an interesting species, inasmuch as it assimilates in the large size of its tail to the little thick-tailed species (*P. crassicaudatus*) of the western coast. I found it inhabiting clumps of grass on the open plains in pairs, but I am told by an intelligent native of the Namoi that he has frequently found as many as four or five in a nest beneath a large stone, or in a dead hollow log lying on the ground. It is the *Toon-moŏ-ra-lă-ga* of the natives of the Namoi. All my specimens were obtained in the Darling Downs district. Like many others of the genus, the sexes differ very much in size ; the size of the tail also varies much in different individuals, and was always very much smaller in the females."

The fur in this animal is very soft, and both on the upper and under parts of the body of a slate-grey colour next the skin ; the general hue of the upper parts of the body is ashy grey, but the fur is much pencilled with black ; on the sides of the body there is but little of the black pencilling, and hence the general hue is paler ; and on these parts, as well as on the sides of the head, is a faint yellow tint ; the under parts of the body are white, very indistinctly suffused with yellow on the mesial portion of the abdomen ; between the white of the under parts and the greyish hue of the sides of the body is a narrowish space which is of an almost uniform pale yellow hue, and the same tint is observed on the outer side of the legs ; the feet are white, obscurely tinted with pale yellow ; on the upper surface of the head is a mark, narrow on the muzzle, but becoming expanded behind, which is almost entirely black, and immediately around the eyes the hairs are also black. The ears are of moderate size, have the posterior margin nearly straight, and are clothed internally with small pale yellowish hairs, and externally with black hairs, excepting on the hinder part, where they are pale. The tail is very thick at the base, being about 3¼ lines in diameter at this part, and becomes gradually slender to the apex ; it is clothed throughout with very minute hairs, between which the scaly skin is visible ; on the upper parts and sides of the tail the hairs are partly black and partly yellow, and on the under surface they are dirty white.

The figures represent the two sexes of the natural size.

PODABRUS CRASSICAUDATUS, Gould.

PODABRUS CRASSICAUDATUS, *Gould.*

Thick-tailed Podabrus.

Phascogale crassicaudata, Gould in Proc. of Zool. Soc., Part XII. p. 105.

A SINGLE specimen only of this curious little Marsupial has as yet come under my notice. This was sent me from Western Australia by Mr. Gilbert, who procured it at the Military Station on William's River. The following notes which accompanied the specimen comprise all that is at present known respecting the animal :—

"I regret to say that I have been unable to procure any information whatever respecting the habits and economy of this very curious species. It was brought into the station by a domestic cat, which is constantly in the habit of going into the bush and returning several times during the night with a small mammal or bird in her mouth; and by this means I obtained it fortunately uninjured. The most striking and singular character of this pretty little animal is the form of the tail, which it was quite impossible to skin without making an incision throughout its entire length; when the skin was removed the fat presented precisely the same appearance as that of the tail of the Beaver. From the circumstance of none of the natives recognising it, I am induced to believe it to be a very rare species. Before it was skinned its entire length was 5⅘ inches; tail, 1¹⁸⁄₁₆; from the nose to the ear, ⅞; from the nose to the eye, ½; ear, ¾. The eyes were black, full and prominent."

Upper surface grey with a wash of yellow, and on the sides of the body distinctly tinted with reddish buff; under parts and feet pure white; tail much swollen, especially in the middle, and clothed throughout with very minute pale hairs; ears externally dark brown, with a patch of buff at the tip; internally flesh-colour clothed with minute pale hairs.

The figures represent the animal in two positions of the natural size; the specimen from which they were drawn now forms part of the collection at the British Museum.

The beautiful little flowering plant (*Calectaria cyanea*) represented in the drawing is very common in many parts of Western Australia.

SARCOPHILUS URSINUS.

SARCOPHILUS URSINUS.

Ursine Sarcophilus.

Didelphis ursina, Harris in Linn. Trans., vol. ix. p. 176. pl. 19. fig. 2.
Dasyurus ursinus, Geoff. Ann. du Mus., tom. xv. p. 305.—Temm. Mon. de Mamm., tom. i. p. 69.—Waterh. Nat. Lib.
  Marsupialia, p. 128.
Diabolus ursinus, Gray in App. to Grey's Journ. of Two Exp. to N.W. and W. Australia, p. 400. No. 12.—List of
  Mamm. in Brit. Mus., p. 97.
Dasyurus (Sarcophilus) ursinus, Waterh. Nat. Hist. Mamm., vol. i. p. 448.
Devil and *Native Devil* of the Colonists of Van Diemen's Land.

THE Ursine Sarcophilus was one of the first of the native quadrupeds encountered by the early settlers in Van Diemen's Land, from whom its black colouring and unsightly appearance obtained for it the trivial names of Devil and Native Devil. It has now become so scarce in all the cultivated districts, that it is rarely, if ever, seen there in a state of nature; there are yet, however, large districts in Van Diemen's Land untrodden by man; and such localities, particularly the rocky gullies and vast forests on the western side of the island, afford it a secure retreat. During my visit to the continent of Australia I met with no evidence that the animal is to be found in any of its colonies, consequently Tasmania alone must be regarded as its native habitat.

In its disposition it is untameable and savage in the extreme, and is not only destructive to the smaller kangaroos and other native quadrupeds, but assails the sheep-folds and hen-roosts whenever an opportunity occurs for its entering upon its destructive errand.

Although the animal has been well known for so many years, little or nothing more has been recorded respecting it than that which appeared in the ninth volume of the Linnean Society's Transactions from the pen of Mr. Harris, who states :—

"These animals were very common on our first settling at Hobart Town, and were particularly destructive to poultry, &c. They, however, furnished the convicts with a fresh meal, and the flesh was said to be not unlike veal. As the settlement increased, and the ground became cleared, they were driven from their haunts near the town to the deeper recesses of the forests yet unexplored. They are, however, easily procured by setting a trap in the most unfrequented parts of the woods, baited with raw flesh, all kinds of which they eat indiscriminately and voraciously; they also, it is probable, prey on dead fish, blubber, &c., as their tracks are frequently found on the sands of the sea-shore.

"In a state of confinement they appear to be untameably savage, biting severely, and uttering at the same time a low yelling growl. A male and a female which I kept for a couple of months chained together in an empty cask, were continually fighting; their quarrels began as soon as it was dark (as they slept all day), and continued throughout the night almost without intermission, accompanied by a kind of hollow barking not unlike that of a dog, and sometimes a sudden kind of snorting, as if the breath was retained a considerable time and then suddenly expelled. They frequently sat on their hind parts, and used their fore paws to convey food to their mouths. The muscles of the jaws were very strong, as they cracked the largest bones asunder with ease."

Mr. Gunn remarks, that notwithstanding their comparatively small size, they are so fierce and bite so severely, that they are a match for any ordinary-sized dog.

The fur is coarse, moderate in length, and black, excepting on the head, tail and under parts of the body; a broad white band usually crosses the chest, and extends backwards on either side, more or less, over the base of the fore leg; and a second crosses the back near the root of the tail; the nose, the ears and the soles of the feet are of a fleshy pink.

Much diversity exists in the colouring of different individuals; in fact, scarcely two are found precisely alike; some being uniformly black, while others are crossed with bands of pure white; some having a white patch on the chest only, while others have a band of white stretching round the neck; and others again patches of the same hue across the fore or hind legs, or both.

Mr. Waterhouse states that a very fine specimen, which died in the menagerie of the Zoological Society, measured from the tip of the nose to the root of the tail 23 inches 9 lines; from the root to the tip of the tail 11 inches, and round the body at the chest 20 inches.

The drawing on the accompanying Plate was made by Mr. H. C. Richter, from a fine living specimen in the menagerie of the Zoological Society, and represents the animal about two-thirds of the natural size.

DASYURUS MACULATUS.

DASYURUS MACULATUS.

Spotted-tailed Dasyurus.

The Spotted Martin, Phillip's Voy. to Bot. Bay, p. 276.—Martin, Cat., pl. 46.

Viverra maculata, Shaw, Gen. Zool., vol. i. pt. ii. p. 433.

Mustela Novæ-Hollandiæ, Meyen.

Dasyurus macrourus, Geoff. Ann. du Mus., tom. iii. p. 358.—Peron et Lesueur, Voy. aux Terr. Australes, pl. 33.—
    Temm. Mon. de Mamm., tom. i. p. 69.—Waterh. Nat. Lib. *Marsupialia*, vol. xi. p. 139. pl. 6.

——— *maculatus*, Gray, List of Mamm. in Brit. Mus., p. 98.—Waterh. Nat. Hist. Mamm., vol. i. p. 439.

THE Spotted-tailed Dasyurus is universally dispersed over every portion of Van Diemen's Land suitable to its habits and mode of life; I have also received specimens from the Liverpool Range and similar districts of New South Wales; but from no other portion of Australia have I seen examples. Rocky gullies trending from the mountain ranges through primitive forests are the favourite abode of this animal, and here, like the Pole and Martin Cats of Europe, it skulks beneath large stones and in holes of the ground; it also ascends trees with the greatest facility in pursuit of birds, which, with bandicoots and other small quadrupeds, afford it an abundant supply of food. It is a strictly nocturnal animal, and, as might be supposed, a most dreaded enemy to poultry: it is consequently regarded by the settler as one of his greatest pests.

The sexes are not distinguishable in colour, neither do the young, which are from four to six in number, materially differ in this respect; the female, however, never attains the large size of the male. It is the largest species of the genus yet discovered, and differs from all its known congeners in the spotted markings of its tail.

Mr. Waterhouse having most carefully described the colour and markings of all the members of this genus, and in many instances from specimens in my own collection, I take the liberty of transcribing the following description from his valuable work :—

"The fur is harsh to the touch, and rather short; its colour varies from a very deep brown to a rich red brown; the head is always paler than the back, and sometimes assumes a yellowish hue, being much pencilled with this pale tint; other parts of the body are more or less pencilled with yellowish, and the whole under parts of the body, as well as the fore-legs and feet, are of a dirty yellow; the upper lip, chin and throat are of a more pure yellow tint; the toes of the fore feet are yellowish; the hind legs externally, and the hind feet, scarcely differ in tint from the upper parts of the body; the tail is nearly equal in length to the head and body, cylindrical, and clothed with tolerably long and harsh hairs; its general colour is the same as that of the body, or nearly so; the ears are short, clothed internally for the most part with small yellowish hairs, but at the margin the hairs are longer, and near the anterior angle they are tolerably long; on the outer side the ears are of the same colour as the crown of the head. With regard to the white spots with which this animal is adorned, they vary considerably in different individuals, and are very irregular in size and form; they are observed on the whole of the upper parts and sides of the body; some few are also visible on the under parts and on the legs; the head is usually immaculate, or presents but two or three very small spots; the spots on the tail are often large, but never numerous."

The Plate represents a male of the natural size.

DASYURUS VIVERRINUS.

DASYURUS VIVERRINUS.

Variable Dasyurus.

The Spotted Opossum, Phillip's Voy. to Bot. Bay, p. 147. pl. 15.
Tapoa Tafa, White, Journ. of a Voy. to New South Wales, p. 285 and pl.
Didelphis Viverrina, Shaw, Gen. Zool., vol. i. pt. ii. p. 491. pl. 111.
———— *guttatus*, Desm. Nouv. Dict. d'Hist. Nat.
Dasyurus Viverrinus et *D. Maugei*, Geoff. Ann. du Mus., tom. iii. pp. 359, 360.—Temm. Mon. de Mamm., pp. 71,
72. pl. 7. figs. 1–8, skull and lower jaw.—Waterh. Nat. Hist. Mamm., vol. i. pp. 442, 444.—Ib. Nat.
Lib. *Marsupialia*, pp. 133, 135, pl. 7.—Cat. of Mamm. in Brit. Mus. p. 97.

THAT the specific terms *Viverrinus* and *Maugei* have reference to one and the same animal, I had abundant opportunities of ascertaining during my residence in Van Diemen's Land; where not unfrequently litters came under my notice in which the prevailing colour of some of the young was black, and in others grey: to the former the old specific term of *Viverrinus* was applied, and to the latter the more modern one of *Maugei*.

The habits and economy of the present species are very similar to those of the Spotted-tailed Dasyurus; it also inhabits almost exclusively the same countries—Van Diemen's Land and New South Wales. During the daytime, hollow trees, holes in the rocks, and stony places form the retreats of this pretty animal; as night approaches, it becomes alert and active, and seeks for its living prey, which consists of small quadrupeds and birds without restriction to any particular species.

I believe that six will be found to be the normal number of the young, since that was the number contained in a litter I obtained in Van Diemen's Land, of which three were black, and three grey-coloured animals: the former, I am assured, are not, however, so regularly met with, and must be considered the variety. Mr. Waterhouse remarks, in his "Natural History of the Mammalia"—

" As no individuals presenting an intermediate condition of colouring are found, I at first suspected that the difference might be sexual; but such is not the case, since I have seen male specimens, both of the black and grey varieties. The former vary only from brown black to black; the under parts of the body and the feet are generally brownish. The fur on the back is grey next the skin, and that on the abdomen is also grey, but of a paler hue; the white spots on the body vary in size, some being very small, and others more than half an inch in diameter; on the head there are a few small white spots; the tail is bushy, being provided with long hairs, averaging on the basal portion about an inch in length, but of double that length at the point; on the under surface they are, however, comparatively short; in length the tail is about equal to the body; the ears are tolerably large and somewhat attenuated at the apex; they are clothed with short black hairs, which are most abundant on the outer surface, but are also plentiful on the inner surface at the point and near the anterior angle, in which latter part the hairs are considerably longer than elsewhere; the back of the ear is of a pale pink colour in the living animal, as is also the naked tip of the nose and the soles of the feet, the latter being also destitute of hair, but covered with small fleshy tubercles."

In the light-coloured animals :—" The general colour of the fur is greyish, but much suffused with yellow; each hair of the ordinary fur on the upper parts of the body is of a pale grey colour at the root, pale yellow near the point, and black at the point, and the coarser interspersed hairs have their visible portions almost entirely black; on the feet and under parts of the body the hairs are of an uniform yellowish white tint; the sides of the face are almost of an uniform pale yellow; the ears are for the most part rather sparingly clothed with pale hairs; at their base externally is a white spot; the tail is bushy, of the same general hue as the body at the base, but becomes gradually paler towards the opposite extremity, and is terminated with white or dirty yellow-white hairs."

The figures represent both states of the animal of the natural size.

DASYURUS GEOFFROYI.

J. Gould and H. C. Richter del. et lith.

Hullmandel & Walton, Imp.

DASYURUS GEOFFROYI, *Gould.*

Geoffroy's Dasyurus.

Dasyurus Geoffroyi, Gould in Proc. of Zool. Soc., part viii. p. 151.—Waterh. Nat. Lib. *Marsupialia*, p. 132.—Ib., Nat. Hist. Mamm., vol. i. p. 437.—Cat. of Mamm. in Brit. Mus., p. 98.

No other species of the present genus is so widely distributed over the continent of Australia as the *Dasyurus Geoffroyi*, which inhabits the whole of the southern portion of the country from Moreton Bay on the east to Swan River on the west. Unlike the *D. Viverrinus* and *D. maculatus* which frequent the country lying between the mountain ranges and the sea, the present animal appears to be exclusively confined to the regions on the interior side of the hills, the specimens I have seen having been procured on the Liverpool Plains in New South Wales, the Murray Scrub in South Australia, and beyond the ranges of Swan River on the western coast. I have stated of the other members of this genus that they are nocturnal in their habits, but that the present is not strictly so is shown by my having encountered one at midday while silently wandering in the Murray Scrub in South Australia, which, squirrel-like, ran up to the topmost branches of a neighbouring gum-tree, whence I immediately dislodged it with my gun in order to obtain a knowledge of the species.

I believe that the *Dasyurus Geoffroyi* is never subject to those variations of colour so conspicuous in *D. Viverrinus.*

Its brown tail clothed with much shorter hairs than in any of its congeners is a character by which it may at all times be distinguished from either of them.

I have named this species in honour of M. Geoffroy de St. Hilaire, the eminent French naturalist, in token of respect for his valuable labours in the field of science.

"The fur is moderately long, rather soft, and on the upper part of the body of greyish hue, but much suffused with yellow, and pencilled with black; and these parts moreover, as well as the sides of the body, are adorned with numerous irregular white spots; the head has a few small white spots only, and is often of a greyer hue than other parts, but the muzzle is somewhat tinted with brownish, and in front of the eye is a dusky patch; the ears are dusky brown and clothed externally with minute blackish brown hairs; internally with longish pencilled black and grey hairs, at or near the anterior margin; but towards the apex, and on the hinder parts, the hairs are minute and brownish; the under parts of the body are white, or very nearly so; the fore feet are brownish, sometimes brownish white; the hind feet are nearly white, or greyish suffused with yellow; the tail is yellowish at the base, but much pencilled with black, the ends of the hairs being of that colour; the black gradually increases towards the tip of the tail, and usually about one-third is entirely black."

The Plate represents the two sexes of the natural size.

DASYURUS HALLUCATUS.

DASYURUS HALLUCATUS, *Gould.*

North Australian Dasyurus.

Dasyurus hallucatus, Gould in Proc. of Zool. Soc., part x. p. 41.—Waterh. Nat. Hist. Mamm., vol. i. p. 434.
Cat. of Mamm. in Brit. Mus., p. 98.

The small number of specimens of the *Dasyurus hallucatus* that have come under my notice renders it unsafe for me to affirm that it is or is not subject to the variations in colour which are observable in *D. Viverrinus* ; but I have reason to believe that such is the case. All the examples that have yet been sent to Europe have been procured in the extreme northern portion of the Australian continent, and the greater number of them from the Port Essington Settlement on the Cobourg Peninsula.

Mr. Waterhouse having instituted a very careful examination and comparison of this animal with the other members of the genus, I cannot do better than give his remarks verbatim from his "Natural History of the Mammalia," above referred to.

"This is the smallest species of the true Dasyures, being a trifle less than *D. Viverrinus* or *D. Geoffroyi* ; with the latter animal it might be confounded, having like it a thumb to the hind feet ; upon a close examination, however, I discovered several characters by which it might be easily distinguished. It is of smaller size than *D. Geoffroyi*, of a darker colour ; with the ears of a paler colour and clothed with pale hairs ; the longer hairs which cover the root of the ear externally are whitish ; the toes of the hind foot are longer, since I found them to be seven lines in length in *hallucatus*, and only six and a quarter in a specimen of *Geoffroyi*, which was of the same sex and of considerably larger size ; and, lastly, I find the whole sole, both of the fore and hind feet in *D. Geoffroyi*, covered with minute but distinct fleshy tubercles, as is also the case in *D. Viverrinus* ; while in *D. hallucatus* I could scarcely perceive a trace of tubercles ; and the fleshy pads at the base of the toes and elsewhere, on which the tubercles were most distinct in *Geoffroyi*, are covered with numerous oblique or transverse grooves ; the pads, moreover, at the base of the toes, were much narrower and proportionately longer."

"The fur is less dense and harsher than in *D. Geoffroyi* ; the upper parts of the body dusky brown, inclining to black, but pretty freely pencilled with yellowish, and having numerous, irregular and moderate-sized white spots, which extend likewise on the sides of the body ; on the crown of the head are a few very small white spots ; the under parts of the body are white, but suffused with yellowish ; most distinctly so about the throat ; the cheeks, a large patch above the eye, and the sides of the body are greyish ; ears pinkish flesh colour, thinly clothed with small pale-coloured hairs ; immediately at the base externally the hairs are longer and dense, and of a yellowish white colour, and the part of the head immediately adjoining the root of the ear has similar pale hairs ; the tail is immaculate, cylindrical, clothed throughout with longish harsh hairs, but by no means bushy ; the basal third is brownish, but considerably pencilled with black, and the remaining two-thirds almost entirely black ; the feet are brownish, and the region of the pouch is clothed with very dark red hairs appearing as if stained with blood."

The figures are of the natural size.

THYLACINUS CYNOCEPHALUS.

H C Richter, del et lith.   Hullmandel & Walton, Imp.

THYLACINUS CYNOCEPHALUS.

Thylacinus.

HEAD, OF THE SIZE OF LIFE.

WHEN the comparatively small island of Tasmania becomes more densely populated, and its primitive forests are intersected with roads from the eastern to the western coast, the numbers of this singular animal will speedily diminish, extermination will have its full sway, and it will then, like the Wolf in England and Scotland, be recorded as an animal of the past: although this will be a source of much regret, neither the shepherd nor the farmer can be blamed for wishing to rid the island of so troublesome a creature. A price is already put upon the head of the native Tiger, as it is called; but the fastnesses of the Tasmanian rocky gullies, clothed with impenetrable forests, will, for the present, preserve it from destruction.

I trust my readers will duly estimate the life-sized head, taken from the living animal.

For all that is known respecting the *Thylacinus cynocephalus*, the reader is referred to the account given with the reduced figures.

THYLACINUS CYNOCEPHALUS.

THYLACINUS CYNOCEPHALUS.

Thylacinus.

Didelphis cynocephala, Harris, Trans. Linn. Soc., vol. ix. p. 174. pl. 19.
Dasyurus cynocephalus, Geoff. Ann. du Mus., tom. xv. p. 304.
Thylacinus cynocephalus, Fisch. Syn. Mamm., p. 270.—Wagn. in Schreb. Saug. Supp., 109-110 Heft, p. 19.—
    Waterh. Nat. Lib. (Marsupialia), vol. xi. p. 123. pl. 5.—Ib. Nat. Hist. Mamm., vol. i. p. 456. pl.
    16. fig. 2.
Thylacinus Harrisii, Temm. Mon. de Mamm., vol. i. p. 63. pl. 7. figs. 1-4.
Peracyon cynocephalus, Gray, List of Mamm. in Brit. Mus., p. 97.
Tiger, Hyæna, Zebra-Opossum, Zebra-Wolf, and *Dog-headed Opossum* of the Colonists.

THE circumstance of a fine pair, male and female, of the *Thylacinus cynocephalus* being now living in the Gardens of the Zoological Society in the Regent's Park, enables me to give the best figure of the animal that has yet appeared; and so great is the interest which attaches to this singular species, that I have been induced to give a representation of its head of the natural size, in addition to that of the entire animal on a reduced scale. Tasmania, better known as Van Diemen's Land, is the country it inhabits, and so strictly is it confined to that island, that I believe no instance is on record of its having been found on the neighbouring continent of Australia. It must be regarded as the most formidable, both of the Marsupialia and of the indigenous mammals of Australia; for although too feeble to make a successful attack on man, it commits sad havoc among the smaller quadrupeds of the country, and among the poultry, and other domestic animals of the settler; even sheep are not secure from its attacks, which are the more difficult to be guarded against, as the habits of the animal being nocturnal, they are always made at night. The destruction it deals around has, as a matter of course, called forth the enmity of the settler, and hence in all cultivated districts the animal is nearly extirpated; on the other hand, so much of Tasmania still remains in a state of nature, and so much of its forest land yet uncleared, that an abundance of covert still remains in which the animal is secure from the attacks of man; many years must therefore elapse before it can become entirely extinct; in these remote districts it preys upon *Halmaturus Billardieri* and *H. Bennetti, Bandicoots, Echidnæ,* and all the smaller animals.

In confinement it is excessively shy, and on being alarmed dashes and leaps about its cage in the most violent manner, uttering at the same time a short guttural cry resembling a bark; but whether this sound is also emitted in a state of nature, has not been observed. Mr. Ronald C. Gunn, who has had better opportunities than any other scientific man of observing the animal in its native wilds, states that it is common in the more remote parts of the colony, and that it is often caught at Woolnooth and the Hampshire Hills. He has seen some so large and powerful, that a number of dogs would not face one of them. Its attacks on sheep are usually made by night, but it also prowls about in the day-time, when, perhaps from its imperfect vision by day, its pace is very slow.

Mr. Harris, to whom we are indebted for our first knowledge of the animal, states that it dwells among caverns and rocks in the deep and almost impenetrable glens in the neighbourhood of the highest mountains of Van Diemen's Land. The specimen from which his description was taken, was caught in a trap baited with kangaroo's flesh; it remained alive but a few hours, having received some internal hurt while being secured; it appeared exceedingly inactive and stupid, and, like the Owl, kept almost continually moving the nictitating membrane with which the eye is furnished: the remains of an Echidna were found in its stomach.

In a letter lately received from Mr. Gunn by D. W. Mitchell, Esq., Sec. Zool. Soc., dated Launceston, Nov. 12, 1850, the following note occurs respecting the specimens in the Society's menagerie :—

"I feel little doubt but that the Thylacines will do well and very probably breed; the number of young is four at a litter—at least I have seen four in the female's pouch, but there may often be fewer. They inhabit the summits of the western mountains (alt. 3500 feet), where, occasionally, snow falls for many months of the year, where the ground is sometimes covered with snow for weeks, and where frosts are severe; therefore I can imagine nothing in the climate of London likely to injure them very materially."

The fur is short, and closely applied to the skin, though of a somewhat woolly texture, owing to each of the hairs of which it is composed being waved; the general tint is greyish brown, but faintly suffused with yellowish; on the under parts of the body of a paler hue than the upper; the fur of the back is of a deep brown colour next the skin, and each hair, excepting those which form the transverse black bands, is yellowish brown towards and dusky at the point; on the abdomen the hairs are of a paler brown at the root, and brown white externally; the black bands are usually about fourteen in number; they commence immediately behind the shoulders and are at first narrow and confined to the back, but, proceeding towards the tail, they become gradually broader and are more extended on the sides; those on the haunches are longest and often forked at their extremities; the general tint of the head is rather paler than that of the body, and the region of the eye is of a whitish hue, but a dark spot is observable at the anterior angle of the eye, and a narrow dark line runs over the eye; the muzzle is dusky, the edge of the upper lip white; the eye is large, full, and of a blackish brown; long black bristles spring from the upper lip, a few also occur on the cheeks, and above the eyes; the limbs externally and the feet scarcely differ in colour from the body; the tail is clothed at the base with a somewhat woolly fur like that of the body, crossed by three or four black bands, but about the commencement of the second fourth of the tail the hairs become short and harsh, closely applied to the skin, brown on the upper surface and pale brown beneath; on the under surface of the apical portion of the tail the hairs are comparatively long, as well as at the point where they are blackish.

The animals are figured in life-like positions, but necessarily much reduced; the figure of the head represents that of the male of the natural size.

PHASCOLOMYS WOMBAT, *Per. et Les.*

J.Gould and H.C.Richter, del. et lith.

Hullmandel & Walton, Imp.

PHASCOLOMYS WOMBAT, *Pér. et Les.*

Wombat.

HEAD, OF THE SIZE OF LIFE.

I WISH it to be understood that, an interval of eighteen years having passed away between the commencement and termination of the present work, there may be some instances in which opinions expressed in years gone by now require modification. When I published the reduced figures of this animal, I remarked that it was uncertain whether there was more than one species of the genus *Phascolomys*. I now, in 1863, feel confident that there are three, if not four, quite distinct Wombats—one inhabiting Tasmania, or Van Diemen's Land, and certainly two, if not three, the opposite portion of the Australian continent.

The life-sized portrait given on the opposite Plate was taken from a Van Diemen's Land animal. It will be seen that it is very dark in colour—a feature common, I believe, to most of the specimens in that island; I have, however, heard of lighter-coloured examples being occasionally seen, but none have come under my own observation. It will be observed that, independently of the difference of colour, it is a small animal when compared with *P. latifrons*. I would call the attention of Professor M'Coy and others who have opportunities of studying the Wombats in their native country to the importance of investigating their history, since it is to them that the mammalogists of Europe must look for accurate information on the subject: and this should be done speedily; for, like the Badger in England, these large and singular Marsupials will soon become scarce.

My figure was taken from an individual which lived for many years in the menagerie of the Zoological Society of London.

PHASCOLOMYS WOMBAT, *Per et Lee*

PHASCOLOMYS WOMBAT, *Per.* et *Les.*

Wombat.

Phascolomys Wombat, Peron et Lesueur, Voy. aux Terres Australes, Atlas, tab. 28.—Desm. Mamm., part i. p. 276.—
    Waterh. in Jard. Nat. Lib. Mamm., vol. x. p. 300.—Ib. Nat. Hist. of Mamm., vol. i. p. 246.—Gunn
    in Proc. of Roy. Soc. of Van Diem. Land, vol. ii. p. 85.
——————— *fossor*, Sevastianoff in Mém. de l'Acad. Imp. de St. Pétersb., tom. i. p. 444.
——————— *wombatus*, Leach, Zool. Misc., vol. ii. p. 101. pl. 96.
——————— *fusca*, Desm. Dict. des Sci. Nat., tom. xxv. p. 500. tab. G 44. fig. 1.
——————— *Bassii*, Less. Man. du Mamm., p. 229.
——————— *ursinus*, Gray, List of Mamm. in Coll. Brit. Mus., p. 95.
Didelphis ursina, Shaw, Gen. Zool., vol. i. part ii. p. 504.
Wombatus fossor, Geoff.
Opossum hirsutum, Perry, Arcana.
Amblotis fossor, Ill. Prod., p. 77.
Perameles fossor, Peron.
Womback, Bewick's Quadr., 6th Edit. p. 522.
Wombat, Collins's Account of New South Wales, vol. ii. p. 153, and plate at p. 157.
Badger, of the Colonists.

THE Wombat may be regarded as one of the most curious of the Australian Mammals, ranking as it does, in respect to its anomalous structure and appearance, with the Koala and Ornithorhynchus. In no other part of the world is the form to be found, and it is difficult to say of which of the great groups of placental animals it is the representative in its own class—the *Marsupiatæ*. I obtained several examples in Van Diemen's Land, but failed in procuring continental specimens, which I regret, because it leaves the question as to there being more than one species of this form still undecided; nor can this point be determined until specimens from South Australia have been sent to Europe, or until comparisons have been made in that country by a naturalist competent to set the question at rest. Professor Owen informs us that a skull in the Collection of the Royal College of Surgeons, sent from South Australia, offers sufficient differences from skulls from Van Diemen's Land to convince him that there are at least two species; and when such a statement is made by so high an authority, the doubt that exists on the subject is much diminished. Mr. Waterhouse also states, that in his opinion the continental species will prove to be distinct from the animal found in Van Diemen's Land and the islands in Bass's Straits. I may mention also that His Excellency Sir George Grey has placed in my hands a pencil drawing of the head of a specimen killed in South Australia, to which, from the great breadth of the head, the name of *latifrons*, proposed by Professor Owen for the continental animal, might apply. There appears, therefore, good reasons for concluding that the continental animal is really distinct; but the question still remains an open one, and it is much to be regretted that both skins and skeletons have not been sent home, for its proper elucidation. After what has been stated, it is almost superfluous to say, that my figure was taken from a specimen procured in Van Diemen's Land, where the animal, particularly in certain districts, is extremely common. I met with it myself in the neighbourhood of Port Arthur, in the sterile districts behind Mount Wellington, and in many other situations where a similar character of country prevails. It is also found in the islands in Bass's Straits, where the specimen first described, in "Collins's Voyage," vol. ii. p. 153, was procured. In its habits it is nocturnal, living in the deep stony burrows, excavated by itself, during the day, and emerging on the approach of evening, but seldom trusting itself far from its stronghold, to which it immediately runs for safety on the appearance of an intruder. The natives state, however, that it sometimes indulges in a long ramble, and, if a river should cross its course, quietly walks into the water and traverses the bottom of the stream until it reaches the other side; but I am unable to confirm this statement from personal observation. In its disposition it is quiet and docile in the extreme, soon becoming familiar with and apparently attached to those who feed it; as an evidence of which, I may mention that the two specimens which are now (1855), and have been for a long period, living in the Gardens of the Zoological Society in the Regent's Park, not only admit the closest inspection, but may be handled and scratched by all who choose to make so intimate an acquaintance with them. The following notes are from the pens of various authors who have written on the Wombat; the earliest of whom was Mr. Bass, in "Collins's Voyage" above referred to.

"The *Wombat*," says Mr. Bass, "is a squat, thick, short-legged, and rather inactive quadruped. Its figure and movements strongly remind one of those of a Bear; its pace, too, is hobbling or shuffling, and not unlike the awkward gait of that animal. In disposition it is mild and gentle, but it bites hard and becomes furious when provoked, and then utters a low cry between a hissing and a whizzing sound, which cannot be heard at a greater distance than thirty or forty yards." Mr. Bass chased one of these animals, lifted it off the ground and laid it along his arm, as if carrying a child. It made no noise, nor any effort to escape, not even a struggle. Its countenance was placid and undisturbed, and it exhibited no discomposure, although in the course of a mile walk it was frequently shifted from arm to arm, and sometimes laid over the shoulder; when, however, he proceeded to secure it by tying its legs, while he left it to cut a specimen of a new wood, it became irritated, whizzed, kicked and scratched most furiously, and snapped off a piece from the elbow of Mr. Bass's jacket with its powerful incisors. Its temper being now ruffled, it remained implacable all the way to the boat, ceasing to kick and struggle only when quite exhausted.

Mr. G. Bennett in his "Wanderings," speaking of one of these animals, kept in a state of domestication at Been in the Tumat country, states that "it would remain in its habitation till dark; it would then come out and seek for the milk-vessels, and should none be uncovered, it would contrive to get off the covers and bathe itself in the milk, drinking at the same time. It would also enter the little vegetable garden attached to the station in search of lettuces, for which it evinced much partiality. If none could be found, it would gnaw the cabbage stalks, without touching the leaves. Although this animal is very numerous in the more distant parts of the colony, it is difficult to procure from the great depth to which it burrows."

"The specimen dissected by Sir Everard Home," says Mr. Waterhouse, "and which was brought from one of the islands in Bass's Straits by Mr. Brown, the eminent botanist, lived as a domestic pet in the house of Mr. Clift for two years. This animal was a male, measured two feet and two inches in length, and weighed about twenty pounds. The observations made by Sir Everard Home on the habits of this animal whilst in confinement, correspond pretty closely with those already given. 'It burrowed in the ground whenever it had an opportunity, and covered itself in the earth with surprising quickness; it was very quiet during the day, but constantly in motion in the night; was very sensible to cold; ate all kinds of vegetables, but was particularly fond of new hay, which it ate stalk by stalk, taking it into its mouth, like a Beaver, by small bits at a time. It was not wanting in intelligence, and appeared attached to those to whom it was accustomed, and who were kind to it. When it saw them it would put up its fore paws on their knees, and when taken up would sleep in the lap. It allowed children to pull and carry it about, and when it bit them, it did not appear to do it in anger or with violence.'"

This animal, like almost every other of the Australian quadrupeds, is eaten by the natives, but as an article of food it must give place to the Kangaroo and its affines. I partook of it myself, but always found its flesh tough, with a musky flavour, and not altogether agreeable.

Mr. Bass remarks that the size of the two sexes is nearly the same, but that the female is somewhat the heavier, and such appears to be the case; the weight, whenever ascertained, being always in favour of the female.

In Mr. Gunn's paper on the Mammals indigenous to Tasmania, published in the "Proceedings of the Royal Society of Van Diemen's Land," above referred to, that gentleman states that—"The Wombats of Tasmania differ much in colour in different localities, some being dirty black, and others of a silvery grey. They are found on the tops of the mountains, and thence to the sea-coasts; and are very numerous in some localities, 234 having been killed in less than a year upon a farm, at present occupied by me, on the St. Patrick's River."

For the details of the internal structure of this curious animal, I must refer my readers to the "Leçons d'Anatomie Comparée" of the celebrated Cuvier, and to the writings of our equally well-known countrymen, Sir Everard Home, in the "Philosophical Transactions" for 1808, and Professor Owen, in the "Proceedings of the Zoological Society" for 1836. The original memoir of the latter author, on *Phascolomys latifrons*, will be found in the "Proceedings" of the same Society for 1845.

The general hue of the tolerably long and very coarse fur of this animal is grey-brown; next the skin, the hairs of the ordinary fur of the upper surface are dusky brown, with the exposed portion of a dirty white, but the longer and coarser hairs are black at the point; on the under surface the hairs are dusky at the root, and dirty white for the remainder of their length, the general hue being paler than that of the upper surface; the muffle is naked and black; the small pointed ears are well clothed with hairs; the legs are short and strong, and the feet broad, naked beneath, and covered with minute, round, fleshy tubercles; the claws are large; those of the fore feet solid, or not concave beneath, slightly curved and depressed; those of the hind feet are curved, slightly compressed, and concave beneath; the hairs of the moustaches are numerous, strong and black, as are also some long bristly hairs which spring from the cheeks; the tail is a mere tubercle, and is hidden by the fur.

"The skeleton," says Mr. Waterhouse, "presents certain peculiarities well worthy of attention: the number of its ribs, and consequently of its dorsal vertebræ, is unusually large, being fifteen, whilst twelve or thirteen are usually found in the Marsupialia; the body of the atlas vertebra remains permanently cartilaginous; the humerus, besides having the inner condyle perforated, has an opening between the condyles; and the patella, or knee-bone, is wanting."

The Plates represent the head of the natural size, and the entire animal considerably reduced.

PHASCOLOMYS LATIFRONS, *Owen*.

J.Gould and H.C.Richter del. et lith.

Hullmandel & Walton, Imp.

PHASCOLOMYS LATIFRONS, *Owen.*

Broad-fronted Wombat.

HEAD AND FORE FEET, OF THE SIZE OF LIFE.

FOR many years the skull of a distinct species of Wombat, from the southern portion of Australia, has formed part of the collection at the Royal College of Surgeons of England, in Lincoln's Inn Fields. To the animal to which this skull belonged Professor Owen applied the name of *Phascolomys latifrons*. On the receipt by the British Museum of a skin of an apparently second species of Wombat from Victoria or Adelaide, I came to the conclusion that it was the skin of the animal characterized by Professor Owen,—a point which could have been easily determined had the skull been sent with the skin, but unfortunately it was destitute of this very important appendage: in other words, the skull at the College of Surgeons arrived without the skin, and the skin at the British Museum without the skull. I have little doubt, however, of my having applied the term *latifrons* to the right animal.

The accompanying head was taken from the skin in the British Museum, above alluded to, and is of the size of life; it will be seen that it differs very considerably in colour from the *P. Wombat* so common in Tasmania.

PHASCOLOMYS LATIFRONS, *Owen*

J.Gould and H.C Richter, del. et lith.

Hullmandel & Walton, Imp.

PHASCOLOMYS LATIFRONS, *Owen.*

Broad-fronted Wombat.

Phascolomys latifrons, Owen in Proc. of Zool. Soc., part xiii. p. 82.—Waterh. Nat. Hist. of Mamm., vol. i. p. 252.

WHEN I attempted to write the history of the *Phascolomys Wombat* in the preceding page of this work, no authenticated specimen from the continent of Australia existed in our museums, and I then stated that the question of there being more than one species of this form must remain in doubt for the present, this remark having special reference to the species characterized from a skull by Professor Owen, under the name of *latifrons*; but during the present year, the skin of a large Wombat from the southern parts of the continent of Australia has arrived in this country, unfortunately "sans teeth, sans skull, sans everything" that could have settled the question as to there being one or two species; if, however, we may judge from the skin, much difference exists between the island and continental animals. In size, this skin, which is now in the British Museum, far exceeds all others I have seen; in colour too it is equally distinct; for while most of the specimens from Tasmania are either black, brown, or greyish brown, that from Victoria is of a light sandy buff or isabelline colour. I cannot then do otherwise than give a figure of this skin, which I presume to be an example of Professor Owen's *Phascolomys latifrons*. Surely some of the collectors in South Australia or Victoria will forward specimens to Europe, and not leave zoologists still in doubt respecting the existence or non-existence of a second species.

"Of the Broad-fronted Wombat," says Mr. Waterhouse, "all that is known is a skull sent from South Australia to Professor Owen. This skull presents so many marked differences when compared with that of the *Phasc. Wombat*, that no doubt can be entertained of the existence of two distinct species of Wombats. I have sought in vain, however, amongst the specimens of Wombats contained in our museums, for an animal which might be identified with Professor Owen's new species. In none have I found the incisor teeth presenting the broadest surface in front, a peculiarity in which the *P. latifrons* differs from *P. Wombat*, where the broadest part of the incisor is at the side. The new species differs moreover in having the upper incisors distinctly broader than the lower, whilst in the Common Wombat the upper and lower incisors are very nearly equal in width, when viewed in front. The following points of distinction presented by the skull of *P. latifrons*, when compared with that of *P. Wombat*, are for the most part pointed out in some notes from the pen of Professor Owen, who has kindly placed them at my disposal :—

"'The skull of *Phasc. latifrons* is rather smaller and broader in proportion to its length; the upper incisors have a semi-oval, transverse section; the convex enamelled surface directed more forward, and longitudinally substriated. The lower incisors narrower, trihedral, with the enamelled outer surface flat. The first lower molar tooth relatively larger, the last relatively smaller. The lower jaw is shorter, more suddenly curved behind, and has the symphysis deeper; the intermaxillary part of the skull is higher in proportion to the width, and less convex externally; the palate is less contracted between the foremost molars, and the palatine portion of the intermaxillaries is wider and very concave. The nasal bones are broader, forming the whole upper surface of the anterior third of the skull. The interorbital part of the cranium is much broader, and presents a well-marked supraorbital ridge and postorbital processes, both of which are almost obsolete in *Phasc. Wombat*. The temporal fossæ are not bounded, as in the last-mentioned animal, by two nearly parallel ridges, but are continued by a convex tract to the upper surface of the cranium; and the supratympanic depression is much larger.'"

This, like most other Australian quadrupeds, forms an article of food, its flesh being brought to market for the supply at least of the Celestial part of the mixed population of that country, as will be seen by the following extract from the "Ovens and Murray Advertiser":—

"A NOVELTY.—We happened a few days ago to walk through one of the Chinese camps, and were attracted by a crowd of mixed people standing outside the shop of a Celestial butcher who lives in Joss-house-street, main encampment. Taking a place among the assemblage, we beheld, tethered to the door-frame, a full-grown Wombat, which was ever and anon turned and hauled about by some one of the bystanders. One Chinaman, more curious than the rest of his fellows, put the animal through its 'facings;' and after spending some time in stroking down its back, examining its hair and pinching its sides, he lifted the round plump body of the Wombat on its fore legs, and viewed it all over. The act was received by his countrymen as a capital joke, setting them all laughing, accompanied by a sing-song sonorous 'yabber' that we did not understand. When it had subsided, we moved out of the heterogeneous group, and observed on an adjoining table pieces of strange-looking meat. We made inquiry, and learned that the flesh was pieces of Wombat offered for sale by the Chinese victualler."

The accompanying Plates represent the head of the natural size, and the entire animal much reduced.

PHASCOLOMYS LASIORHINUS, Gould

PHASCOLOMYS LASIORHINUS, *Gould.*

Hairy-nosed Wombat.

HEAD, OF THE SIZE OF LIFE.

THIS full-sized representation of the head of this new and highly interesting species of Wombat has been executed partly from a fine drawing sent to me by Professor M'Coy, of Melbourne, and partly from an example now in the Gardens of the Zoological Society. This living animal exhibits a fleshy muzzle, partially covered with fine white hairs, through which the pink flesh-tint shows very conspicuously. Mr. Bartlett, the superintendent, informs me that, after carefully comparing Professor M'Coy's drawing with the Society's animal, he has no doubt of its having been made from an example identical in species.

For further particulars respecting the *Phascolomys lasiorhinus,* the reader is referred to the interesting notes by Mr. Angas and Professor M'Coy, which will be found in the next page.

PHASCOLOMYS LASIORHINUS, Gould.

PHASCOLOMYS LASIORHINUS, *Gould.*

Hairy-nosed Wombat.

THE arrival in this country of two or three living examples of a species of Wombat with the very remarkable character of a hairy muzzle has naturally excited as much interest among our own naturalists as its recent discovery had done among those of Australia. Both Mr. Angas and Professor M'Coy have forwarded lengthened papers respecting it for publication in the 'Proceedings of the Zoological Society of London,' and both those gentlemen were of opinion that the new Wombat was identical with the *P. latifrons* of Owen, who many years since had applied that specific term to a skull in the Museum of the Royal College of Surgeons in Lincoln's Inn Fields, but of which no skin had ever been received; and not-withstanding what I have said in my account of the preceding species, I should have considered that Mr. Angas and Professor M'Coy were correct in their conclusion, had not one of the animals sent to this country died, and thus afforded an opportunity of comparing its skull with that in the College Museum above mentioned. On this being done, it was found that the two skulls did not agree; and I believe I am at liberty to say that Mr. Flower, who has charge of the collection, is of opinion that they could never be considered as belonging to the same species. Under these circumstances, I had no alternative but to give the Hairy-nosed Wombat a distinctive specific appellation; and, at the suggestion of Dr. Sclater, I have assigned to it that of *lasiorhinus.* This course, however, will not prevent me from giving the remarks of my friends M'Coy and Angas, which indeed will have the more interest as descriptive of this new and extraordinary animal.

" I have lately," says Mr. Angas, " had an opportunity of examining a living full-grown male example of a Wombat, in the Botanical Gardens in Adelaide, which, on comparing it with adult specimens of the Tas-manian Wombat, I find to be quite distinct from that species."

" The fur of the latter is very rough and coarse, of a dark grizzly grey; ears quite small, blackish brown outside, whitish internally; nose nearly black, and more pointed than that of the former, giving to the face an expression slightly resembling the 'Koala,' whereas the other presents a bold, bull-dog-like aspect from the greater expansion of his face and width of nostrils. The general aspect of *P. Wombat* is more bear-like: in standing, it arches its back considerably, and does not hold its head so erect; the expression of the eye, too, is decidedly fierce, and lacks the good-natured twinkle of the South Australian species.

" The specimen in the Adelaide Botanical Gardens was caught some twelve months since near the Gawler River, about thirty miles north of Adelaide. It is kept in an enclosure, where it is secured with a strong chain and collar to prevent its escape by burrowing; it is perfectly docile, and never attempts to bite like the Common Wombat. It is fed on bran and weeds, and drinks freely of water. The only sound it emits is a short, quick grunt when annoyed; it sleeps a good deal during the day, rolled up almost into a ball, with its flesh-coloured nose buried between its fore paws; and appears impatient of heat and rain, as in its wild state it is entirely a burrowing animal, living in large holes in the limestone districts, and only leaving its habitation towards dusk for the purpose of obtaining food. It is fond of lying on its back like a bear, will burrow three or four feet into the soft ground of its enclosure, and scratches alternately with its fore paws. When worried, it presents its hind quarters to the enemy, and, suddenly turning round, makes a charge at his legs, evidently for the purpose of throwing him down; otherwise it is perfectly harmless. He runs fast for a short distance in a sort of gallop, but soon tires, and is easily caught. Although in some parts of the colony, especially on Yorke's Peninsula and about Port Lincoln, the holes of these Wombats are very numerous, yet the animals are but rarely seen. Many of the oldest colonists have informed me that they never saw a Wombat alive. The blacks on the Murray describe two kinds of Wombats: one (evidently *P. latifrons*) they speak of as ' big yellow fellow,' the other as being smaller and dark; they also say that the impressions of their feet in the sand-tracks leading to their burrows bear a striking resemblance to those of the footprints of a young child. The flesh they describe as being like pork, and excellent eating. They are extremely difficult to obtain, on account of their great timidity. The usual plan is to make a screen of boughs in the vicinity of their haunts, behind which the natives conceal themselves. If not killed on the spot, they will scramble to their holes, whence it is utterly impossible to dislodge them."—Proc. of Zool. Soc. 1861, pp. 268–271.

The following is Professor M'Coy's account of the animal examined by him in Melbourne:—

" As the description given by Mr. Angas in the 'Proceedings of the Zoological Society' for June 1861, of what he supposed to be the living *P. latifrons,* and as the first skin of this Wombat could only be identified with that species by an examination of the bones of the skull, which Mr. Angas had not seen, I felt much doubt as to the identity of his species and that of Professor Owen; and when, a few weeks ago, the Acclimatization Society of Melbourne received two specimens of a Wombat from South Australia of an entirely different species from that of Tasmania or Victoria, I gladly embraced the opportunity of determining the species in the only way in which it could for the first time be done from the skin, namely, from an examination of the skull; and in this way I am able to furnish a description of the external characters of what I believe to be the *Phascolomys latifrons*. Mr. Angas not noticing the extraordinary character of the hairy muffle, I am quite uncertain as to what his species may really be. His differs from mine also in having the feet of the same colour as the body, instead of a rich brown.

"The two specimens examined were quite alike; both were males. Their general size and shape is somewhat like that of *P. Wombat*, but the body is considerably longer and narrower in proportion; the neck is also so much more slender that the animal may easily be confined by a dog-collar round the neck, which cannot be done with *P. Wombat*, from the neck being nearly as thick as the head. The head, instead of being broader, is narrower and deeper in proportion to the length of the body; the forehead is much more elevated across the eyes, forming a very obtuse but distinct angle with the line of the nose; but the most extraordinary difference in the character of the head is produced by the great width and flatness of the nose, which is covered with dense coarse white hair, and is much wider than high, instead of being narrow, black, naked, and longitudinally ovate. Another marked peculiarity is the great comparative length of the ears, which are long and elliptically pointed, differing remarkably from the very short rounded ears of *P. Wombat*. The general outline, too, of *P. latifrons* is rendered remarkably different by the broad, flat, oblique form of the body behind the crest of the hips, and the hair forming two circular rosettes on that part, uniting to form a short transverse crest across the back by meeting the ordinary hair of the back coming down in the opposite direction. The claws are shorter and broader than those of *P. Wombat*. The tail, also, instead of being a mere tubercle, is slender and cylindrical.

"The fur is totally unlike the coarse, harsh, densely adpressed fur of *P. Wombat*, being soft, and in length, texture, and feel resembling more nearly that of an English wild rabbit. It is closer on the feet and toes than on the legs, differing thus from the long bristly covering of the toes of *P. Wombat*.

"Upper part of head, back, sides, and legs brownish grey; a semicircular spot under the nostrils, one in front of the eye, a broad spot on the chin, the back of the ears, and the feet dark brown; the crest of the lower part of the back is dusky brown; the tail is blackish and naked; the under part of the body is whitish or dull grey, and the sides under the head ochraceous or yellowish; whiskers black, with a few white hairs; hairs on muffle silvery-whitish grey.

"From Mr. Angas omitting to notice the broad, white *hairy muffle*, and from the narrowness of the head and great width of the nostrils, I supposed the species I have above described would prove to be the *Phascolomys platyrhinus* of Owen, which seems to have been overlooked by most subsequent writers, but an examination of the skull proved the above-given description to belong truly to the *P. latifrons*; possibly Mr. Gould and Mr. Angas may have had *P. platyrhinus* under their eyes."

It will be remarked that both Professor M'Coy and Mr. Angas consider the animal they respectively describe to be the true *P. latifrons* of Owen; and from an examination of examples received from South Australia, I have no doubt that the remarks of both gentlemen refer to one and the same animal: the omission of the hairy muzzle by Mr. Angas may have arisen from the hairs having been eroded in consequence of confinement, as is the case to a certain extent with the animal in our Gardens. With regard to the *P. vlatyrhinus* I may mention, that on a further examination of the skull in the Museum of the College of Surgeons to which that name has been applied, it so much resembles several skulls of *P. Wombat*, that it is questionable if it be really distinct.

While writing these remarks, another *Phascolomys* has just arrived at the Zoological Society's Gardens in the Regent's Park, which certainly differs from all the rest, its colour being uniform jet-black, even to the plated bare shield on the nose. It is allied to the *P. lasiorhinus* in its long pointed ears, which at once separates it from *P. Wombat* and the animal I have figured as *P. latifrons*. For this new species I propose the name of *P. niger*. Its native locality is unknown: can it be the dark animal spoken of by the blacks to Mr. Angas as inhabiting the Murray scrub?

In concluding these remarks, I must express a hope that mammalogists will adopt the names I have applied to the four species of Wombat. I admit that there is still some little difficulty as to the identity of the *P. latifrons* of Owen, whether it be or be not a species still unknown to us, or whether it be the animal I have figured under that name. I must also in fairness state that the skull of *P. lasiorhinus* sent to the Museum of the Royal College of Surgeons for comparison appeared not to be fully adult; at the same time it exhibited so many striking differences from the skull to which the name of *P. latifrons* was assigned, that no anatomist would for a moment consider them to be identical; and we can scarcely suppose that the progress of age would produce so great a change in the character of the skull that ultimately they would be alike.

My figures were taken from a drawing made by Mr. Wolf, from the animal in the Zoological Society's Gardens, which was received from South Australia.

THE

MAMMALS OF AUSTRALIA.

BY

JOHN GOULD, F.R.S.,

F.L.S., F.Z.S., M.E.S., F.R.GEOG.S., M.RAY S.: HON. MEMB. OF THE ROYAL ACADEMY OF SCIENCES OF TURIN; OF THE ROYAL
ZOOL. SOC. OF IRELAND; OF THE PENZANCE NAT. HIST. SOC.; OF THE WORCESTER NAT. HIST. SOC.; OF THE
NORTHUMBERLAND, DURHAM, AND NEWCASTLE NAT. HIST. SOC.; OF THE NAT. HIST. SOC.
OF DARMSTADT; OF THE TASMANIAN SOC. OF VAN DIEMEN'S LAND; OF THE NAT.
HIST. SOC. OF STRASBOURG; OF THE NAT. HIST. SOC. OF IPSWICH; AND
CORR. MEMB. SOC. OF NAT. HIST. OF WÜRTEMBERG.

IN THREE VOLUMES.

VOL. II.

LONDON:

PRINTED BY TAYLOR AND FRANCIS, RED LION COURT, FLEET STREET.

PUBLISHED BY THE AUTHOR, 26 CHARLOTTE STREET, BEDFORD SQUARE.

1863.

LIST OF PLATES.

VOL. II.

Macropus major	Great Grey Kangaroo		1, 2
———— ocydromus	West-Australian Great Kangaroo		3, 4
———— fuliginosus	Sooty Kangaroo		5
Osphranter rufus	Great Red Kangaroo		6, 7
———— Antilopinus	Red Wallaroo		8, 9
———— robustus	Black Wallaroo		10, 11
———— ? Parryi	Parry's Wallaroo		12, 13
Halmaturus ruficollis	Rufous-necked Wallaby		14, 15
———— Bennettii	Bennett's Wallaby		16, 17
———— Greyi	Grey's Wallaby		18, 19
———— manicatus	Black-gloved Wallaby		20, 21
———— Ualabatus	Black Wallaby		22, 23
———— agilis	Agile Wallaby		24, 25
———— dorsalis	Black-striped Wallaby		26, 27
———— Parma	Parma Wallaby		28
———— Derbianus	Derby's Wallaby		29, 30
———— Thetidis	Pademelon Wallaby		31, 32
———— stigmaticus	Branded Wallaby		33, 34
———— Billardieri	Tasmanian Wallaby		35, 36
———— brachyurus	Short-tailed Wallaby		37, 38
Petrogale penicillata	Brush-tailed Rock-Wallaby		39, 40
———— lateralis	Stripe-sided Rock-Wallaby		41, 42
———— xanthopus	Yellow-footed Rock-Wallaby		43, 44
———— inornata	Unadorned Rock-Wallaby		45, 46
———— brachyotis	Short-eared Rock-Wallaby		47
———— concinna	Little Rock-Wallaby		48
Dendrolagus ursinus	Black Tree-Kangaroo		49
———— inustus	Brown Tree-Kangaroo		50
Dorcopsis Bruni	Filander		51
Onychogalea unguifer	Nail-tailed Kangaroo		52, 53
———— frænata	Bridled Nail-tailed Kangaroo		54
———— lunata	Lunated Nail-tailed Kangaroo		55
Lagorchestes fasciatus	Banded Hare-Kangaroo		56
———— Leporoïdes	Hare-Kangaroo		57
———— hirsutus	Rufous Hare-Kangaroo		58
———— conspicillatus	Spectacled Hare-Kangaroo		59
———— Leichardti	Leichardt's Hare-Kangaroo		60
Bettongia penicillata	Jerboa-Kangaroo		61
———— Ogilbyi	Ogilby's Jerboa-Kangaroo		62
———— Cuniculus	Tasmanian Jerboa-Kangaroo		63
———— Grayii	Gray's Jerboa-Kangaroo		64
———— rufescens	Rufous Jerboa-Kangaroo		65
———— campestris	Plain-loving Jerboa-Kangaroo		66
Hypsiprymnus murinus	New South Wales Rat-Kangaroo		67
———— apicalis	Tasmanian Rat-Kangaroo		68
———— Gilberti	Gilbert's Rat-Kangaroo		69
———— platyops	Broad-faced Rat-Kangaroo		70

MACROPUS MAJOR, *Shaw*

Gould and H.C. Richter, del. et lith.   Hullmandel & Walton Imp.

MACROPUS MAJOR, *Shaw.*

Great Grey Kangaroo.

HEAD OF A MALE, LIFE-SIZE.

As reduced figures furnish but an inadequate idea of the size and facial expression of the larger species of Kangaroos, I have been constrained in many instances to publish double plates of these important Marsupials. The accompanying illustration will convey a just conception of the appearance of the animal at the moment of surprise, when it stands bolt upright on its hind legs, its huge tail completing the tripod which sustains its body. Twitching of the nose and the upcurling of the lips shown in the drawing are indications of anger often exhibited by these animals when their haunts are intruded upon. The hairy muffle, common to all the members of the restricted genus *Macropus*, is far better shown in the opposite Plate than it possibly could be in the reduced figures; and this forms an additional reason for giving the life-sized illustration.

A detailed account of the species will be found on the succeeding page.

MACROPUS MAJOR. _Shaw._

MACROPUS MAJOR, Shaw.

Great Grey Kangaroo.

Yerbua gigantea, Zimm. Zool. Geog. Quadr., p. 526.
Didelphis gigantea, Schreb. Saug., tom. iii. p. 552. tab. 154.—Gmel. Edit. Linn. Syst. Nat., tom. i. p. 109.
Macropus giganteus, Shaw, Nat. Misc., pl. 33.—Waterh. Nat. Hist. of Mamm., vol. i. p. 62.
Kanguroo, Cook's Voy., vol. iii. p. 577. pl. 20.—Phill. Voy., pl. in p. 106.—White's Voy., pl. in p. 272.
Macropus major, Shaw, Gen. Zool., vol. i. p. 305. pl. 115.—Cook's First Voy., vol. iv. p. 45. pl. 2.—Desm. Nouv.
      Dict. d'Hist. Nat., tom. xvii. p. 33.—Gould, Mon. of Macropodidæ, pl.
Kangurus labiatus, Geoff. Encycl., pl. 21. fig. 4.—Desm. Ency. Méth. Mamm., p. 273.
Halmaturus griseo-fuscus, Goldf. in Oken's Isis, 1819, p. 266 (Waterhouse).
Macropus ocydromus, Gould in Ann. and Mag. Nat. Hist., vol. x. p. 1? (Waterhouse).
—————— *melanops*, Gould in Proc. of Zool. Soc., part x. p. 10?
—————— *fuliginosus* and its synonyms?
Boomer, *Forester*, *Old Man Kangaroo* of the Colonists, *Bundaary* of the Aborigines of the Liverpool range.

THERE can be little doubt of the present species being that noticed by our celebrated navigator Cook, in his voyage round the world in 1770; and as I conceive all information connected with this early-known species will be interesting, I shall commence my account of its history with a quotation from the above-mentioned work.

"On Friday, June the twenty-second, while stationed for a short time on the south-east coast of Australia," says Captain Cook, "a party, who were engaged in shooting pigeons for the use of the sick of the ship, saw an animal, which they described to be 'as large as a greyhound, of a slender make, of a mouse-colour, and extremely swift.' The following day the same kind of animal was again seen by a great many other people. On the twenty-fourth it was seen by Captain Cook himself, who, walking at a little distance from the shore, observed a quadruped, which he thought bore some resemblance to a greyhound, and was of a light mouse-colour, with a long tail, and which he should have taken for a kind of wild dog, had not its extraordinary manner of leaping, instead of running, convinced him of the contrary. Mr. Banks also obtained a transient view of it, and immediately concluded it to be an animal perfectly new and undescribed.

"The sight of a creature so extraordinary could not fail to excite, in the mind of a philosophic observer, the most ardent wishes for a complete examination. These were at length gratified; Mr. Gore, one of the associates in the expedition of Captain Cook, having been so fortunate as to shoot one in the course of a few days."

Such is the earliest notice to be found relative to this fine species, of which living examples were a few years afterwards brought to Europe, and have from time to time formed an interesting addition to our menageries. It is however remarkable, that though it has now been introduced for so long a period, all attempts at naturalizing it have hitherto proved futile; still, from my own observations of the animal in a state of nature, I am led to believe that a small degree of perseverance is alone requisite to effect so desirable an object. Should I be so fortunate as to interest any who have the means, as well as the inclination, in the furtherance of this object, we may yet hope to see our large parks and forests graced with the presence of this highly ornamental and singular animal. That it would bear the severities of our winters is almost beyond a doubt, since in Van Diemen's Land, among other places, it resorts to the bleak, wet, and frequently snow-capped summit of Mount Wellington. The kind of country which appears most suitable to its nature, consists of low grassy hills and plains, skirted by thin open forests of brushwood, to the latter of which, especially on the continent of Australia, it resorts for shelter from the oppressive heat of the mid-day sun. Although the numbers of this large species are becoming greatly reduced in consequence of the intrusion of civilized man, and though it has disappeared from those localities where he has taken up his abode, accompanied by his vast flocks and herds, still the immense tracts of sterile unwatered country which characterize Australia, and present physical obstacles to cultivation, will, in my opinion, for a long period afford a sufficient asylum for the preservation of the race.

The Great Grey Kangaroo enjoys a wide range of habitat, being spread over the colony of New South Wales, South Australia, and the intervening countries; and if the animal from Western Australia, to which I have given the name of *Macropus ocydromus*, should ultimately prove to be merely a local variety, then its range may be said to extend throughout the whole of the southern portion of the continent from east to west.

I should not consider it, strictly speaking, a gregarious animal, as I have never seen more than six or eight together, and have more frequently met with it singly or in pairs; and this view of its habits is confirmed by R. C. Gunn, Esq., who states, that "although, from the circumstance of its food being abundant on certain spots, as on recently burnt land, it may be seen in flocks, it is not gregarious; their food brings them to one spot; but on no occasion have I ever known them in flocks, owning a leader and proceeding *en masse* as all other wild animals do." Yet Mr. Gilbert, speaking of the animal observed by him in Western Australia (*M. ocydromus*, if distinct), says—"Mr. Gunn's remarks will not at all apply to the Kangaroos of this part of the country, for I have seen hundreds of instances in which the whole herd has followed the leading one *en masse*, unless divided by the dogs: it is true there is no regular leader, but when one is disturbed, the whole herd immediately take alarm, and one bounding off is the signal for the whole to follow: when running in this way, the does soon take the lead, while the males from their greater weight are unable to keep up with them, and often bring up the rear a long distance behind, but they all follow in the same track as the leading does, and when the latter stop the entire herd stop also: this habit I have noticed so frequently, that I have always considered the Kangaroo as a gregarious animal. Occasionally an old and very large male will take possession of a valley, and there remain for years without moving a mile from the spot, leading in fact a perfectly misanthropic life; such instances, however, are not very common; still, two or three spots are known to me which have been thus tenanted for years, many of the settlers and aborigines, now young men, remembering these particular animals from their childhood. Some of the most experienced Kangaroo-hunters have endeavoured to capture them, but have invariably failed, at the cost of much injury to their dogs: with the exception of cases like these, it is rare to meet with a single Kangaroo."

Mr. Gunn states that in Van Diemen's Land the Kangaroos "lodge during the heat of the day amongst high ferns, such as *Pteris esculenta*, high-grass, and in underwood, commonly called *scrub*, that is, dense patches of *Melaleuca*, *Leptospermum*, &c. on the margins of streams; and although almost all the forest trees (*Eucalypti*) are hollow at the butt, and innumerable dead and hollow trees cover the ground, I have never known them used as sleeping-places: the space under a dead tree is much more likely to be resorted to for this purpose than the hollow of a living one."

The senses of smelling and hearing are so exquisite in this animal that it is extremely difficult of approach without detection, and to effect this it is always necessary to advance against the wind. It browses upon various kinds of grasses, herbs and low shrubs, a kind of food which renders its flesh well-tasted and nutritive. The early dawn and evening are the periods at which it feeds, and at which it is most certain to be met with.

Although hunted and frequently killed by the Dingo, or native dog, its most formidable antagonist has hitherto been the Aborigine, who employs several modes of obtaining it; sometimes stealing upon it with the utmost caution under covert of the trees and bushes, until it is within the range of his spear, which is generally thrown with unerring aim; at other times, having discovered their retreat, the natives unite in a party, and, forming a large circle, gradually close in upon them with shouts and yells, by which the animals are so terrified and confused, that they easily become victims to the bommerengs, clubs and spears which are directed against them from all sides.

Still, however formidable an enemy the Aborigine may have been, the Great Grey Kangaroo finds, at the present time, a far greater one in the white man, whose superior knowledge enables him to employ, for its destruction, much more efficient weapons and assailants than those of the more simple son of nature. Independently of the gun, he brings to his aid dogs of superior breed, and of so savage a nature, that the timid Kangaroo has but little chance when opposed to them. These dogs, which run entirely by sight, partake of the nature of the greyhound and deerhound, and from their great strength and fleetness are so well adapted for the duties to which they are trained, that its escape, when this occurs, is owing to peculiar and favourable circumstances, as, for example, the oppressive heat of the day, or the nature of the ground; the former incapacitating the dogs for a severe chase, and the hard ridges which the Kangaroo invariably endeavours to gain giving him a great advantage over his pursuers. On such grounds the females in particular will frequently outstrip the fleetest greyhound, while, on the contrary, heavy old males, on soft ground, are easily overtaken. Many of these fine Kangaroo-dogs are kept at the stock-stations of the interior for the sole purpose of running the Kangaroo and the Emu, the latter being killed solely for the supply of oil which it yields, and the former for mere sport, or for food for the dogs. Although I have killed the largest males with a single dog, it is not generally advisable to attempt this, as they possess great power, and frequently rip up the dogs, and sometimes even cut them to the heart with a single stroke of the hind leg. Three or four dogs are more generally laid on, one of superior fleetness to "pull" the Kangaroo, while the others rush in upon and kill it. It sometimes adopts a singular mode of defending itself by clasping its short powerful fore-limbs around its antagonist, leaping away with it to the nearest water-hole, and there keeping it beneath the surface until drowned: with dogs the old males will do this whenever they have an opportunity, and it is said that they will also attempt to do the same with man. In Van Diemen's Land the *Macropus major* forms an object of chase, and like the Deer and Fox in England, is hunted with hounds; and twice a week, during the season, the Nimrods of this distant land may be seen, mounted on their fleet steeds, crossing the ferry of the Derwent, at Hobart Town, on their way to the hunting-ground, where they seldom meet without "finding." The following particulars of the "hunt" have been obligingly forwarded to me by the Honourable Henry Elliot, late aide-de-camp to His Excellency Sir John Franklin, and one of its chief patrons.

"I have much pleasure in telling you all I know of the Kangaroo-hunting in Van Diemen's Land. The hounds are kept by Mr. Gregson, and have been bred by him from foxhounds imported from England; and though not so fast as most hounds here now are, they are quite as fast as it is possible to ride to in that country.

"The 'Boomer' is the only Kangaroo which shows good sport, for the strongest 'Brush Kangaroo' cannot live above twenty minutes before the hounds; but as the two kinds are always found in perfectly different situations, we never were at a loss to find a Boomer, and I must say that they seldom failed to show us good sport. We generally 'found' in a high cover of young wattles; but sometimes we 'found' in the open forest, and then it was really pretty to see the style in which a good Kangaroo would go away. I recollect one day in particular, when a very fine Boomer jumped up in the very middle of the hounds, in the 'open'; he at first took a few high jumps with his head up, looking about him to see on which side the coast was clearest, and then, without a moment's hesitation, he stooped forward and shot away from the hounds, apparently without an effort, and gave us the longest run I ever saw after a Kangaroo. He ran fourteen miles by the map from point to point, and if he had had fair play, I have very little doubt but that he would then have beat us; but he had taken along a tongue of land which ran into the sea, so that, on being pressed, he was forced to try to swim across the arm of the sea, which, at the place where he took the water, cannot have been less than two miles broad; in spite of a fresh breeze and a head sea against him, he got fully half-way over, but he could not make head against the waves any further, and was obliged to turn back, when, being quite exhausted, he was soon killed.

"The distance he ran, taking in the different bends in the line, cannot have been less than eighteen miles, and he certainly swam more than two. I can give no idea of the length of time it took him to run this distance, but it took us something more than two hours; and it was evident, from the way in which the hounds were running, that he was a long way before us; and it was also plain that he was still fresh, as, quite at the end of the run, he went over the top of a very high hill, which a tired Kangaroo never will attempt to do, as dogs gain so much on them in going up-hill. His hind quarters weighed within a pound or two of seventy pounds, which is large for the Van Diemen's Land Kangaroo, though I have seen larger.

"We did not measure the length of the hop of this Kangaroo; but on another occasion, when the Boomer had taken along the beach, and left his prints in the sand, the length of each jump was found to be just fifteen feet, and as regular as if they had been stepped by a serjeant. When a Boomer is pressed, he is very apt to take to the water, and then it requires several good dogs to kill him; for he stands waiting for them, and as soon as they swim up to the attack, he takes hold of them with his fore-feet, and holds them under water. The buck is altogether very bold, and will generally make a stout resistance; for if he cannot get to the water, he will place his back against a tree, so that he cannot be attacked from behind, and then the best dog will find in him a formidable antagonist.

"The doe, on the contrary, is a very timid creature; and I have even seen one die of fear. It was in a place where we wished to preserve them, and as soon as we found that we were running a doe, we stopped the hounds just at the moment they were running into her. She had not received the slightest injury, but she lay down and died in about ten minutes. When a doe is beat she generally makes several sharp doubles, and then gets among the branches, or close to the trunk of a fallen tree, and remains so perfectly still, that she will allow you almost to ride over her without moving, and in this way she often escapes. A tolerably good Kangaroo will generally give a run of from six to ten miles; but in general they do not run that distance in a straight line, but make one large ring back to the place where they were found, though the larger ones often go straight away."

An extraordinary difference is observable in the size of the sexes of this species, the female being not more than half the size of the male: she brings forth one young at a time, which, as soon as it is clothed with hair, assumes the colouring of the adult.

A slight variation is found to exist in specimens from different localities, some being much darker than those represented in the Plate, and others of a foxy-red. Albinoes are occasionally, but very rarely, to be met with. As might reasonably be expected also, the fur is much thicker and more woolly in winter than in summer.

All the fur on the upper surface uniform greyish-brown above, passing into grisly-grey on the arm and under surface; a faint line of greyish-white above the upper lip and along the sides of the face; hands, feet, and tip of the tail black.

	Male.		Female.	
	feet.	inches.	feet.	inches.
Length from the nose to the extremity of the tail	7	10	5	11½
„ of tail	3	2	2	4½
„ „ tarsus and toes, including the nail	1	3	1	½
„ „ arm and hand, including the nails	1	6		10½
„ „ face from the tip of the nose to the base of the ear		9		8
„ „ ear		5½		5

The accompanying Plates represent a head of the male animal of the natural size, and reduced figures of adult examples of both sexes.

Since the publication of my Monograph of the Kangaroos, in which my account of this species first appeared, my friend Mr. G. R. Waterhouse has paid particular attention to the Marsupialia, and has recorded his opinion in the work above referred to, that the Kangaroos described by me as distinct, under the names of *Macropus ocydromus* and *M. melanops*, are merely local varieties of the present animal; whether my own or Mr. Waterhouse's view of the subject be the correct one, time and future research can alone determine; in the meanwhile it will be as well to append my descriptions, and the information I have received in reference to the animals to which I have assigned the two names above mentioned. Mr. Waterhouse is also of opinion that the animals in the Paris Museum, described under the name of *Kangurus* and *Macropus fuliginosus*, will probably prove to be merely a variety of *M. major*, in which I believe he is correct, for I have never seen any other examples than those mentioned, and consequently have no direct evidence of their being distinct.

The animal from Western Australia, which I have called *Macropus ocydromus,* is a fine large species closely resembling the *M. major,* but differs in being of a more slender form, and in having the fur of a more woolly texture and of a darker colour on the upper surface, particularly at the base of the ears and back of the neck; the cheeks are destitute also of the usual white stripe.

"This animal," says Mr. Gilbert, "the male of which is called *Yoon-gur,* and the female *Work,* by the aborigines of Western Australia, is tolerably abundant over the whole colony from King George's Sound south, to forty miles north of Moore's River, the farthest point in that direction I have yet explored. It does not appear to be confined to any particular description of country, being as often seen in gum-forests among the mountains as on the open plains and clear grassy hills; but it is certainly most numerous in the more open parts of the country, where it is not liable to surprise. In travelling along the road from Guilford to York, from two to four or five may occasionally be met with; but farther in the interior, particularly on the Guvangun Plains, herds of thirty or forty in number may often be seen; and still farther north, beyond Kojenup, it is still more numerous; indeed, I have never seen, in any part of Australia, so large a herd as I met with on the Gordon Plains in 1840; it could not, at the most moderate computation, have comprised less than five hundred individuals; and several of my party in their astonishment considered there were even a much greater number than I have stated. The full-grown male, termed a buck or Boomer, attains a large size, and is a most formidable opponent to many of the best dogs in the country; indeed, there are few dogs that will even attempt to run him; this may in some measure account for the few recorded instances of very large ones being killed by the hunters; it is not that their speed enables them to escape, for, on the contrary, their great weight incapacitates them for running fast or to any distance, and almost any dog may overtake them; instead, therefore, of running away, the Boomer invariably turns round and faces his pursuers, erects himself to his full height, and, if possible, supports his back against a tree, and thus awaits the approach and rush of the dogs, endeavouring to strike them with his powerful hind-toe or to catch them in his fore-arms, and, while holding them, to inflict dreadful and often fatal wounds with the same weapon of offence. Old dogs, well broken in, and accustomed to keep a Boomer at bay, never attempt to run in, but by barking keep the Kangaroo at bay until the hunter comes up, when a blow or two on the head with a short heavy stick soon brings him down. The hunter himself, however, often runs great hazard, for the Boomer will frequently on his approach leave the dogs and attack him most fiercely; and it is no easy matter for him to avoid being severely cut while attempting to kill the animal. When closely pressed, it takes to the water, and as the dogs approach, catches them in its arms and holds them beneath the surface till drowned; but if the water be too shallow for drowning them, it has been known to catch one dog and place him beneath its foot while courageously awaiting the approach of a second.

"The female of the first year, before having young, and during the second year, with her first young, is termed the 'flying doe,' her speed being so great, if she obtains anything like a fair start, that she gives the fleetest dogs a very long and severe run, and frequently succeeds in outstripping them; upon finding herself too closely pursued, she usually attempts to evade the dogs by making a sudden leap almost at right angles, when the dogs, being at full speed, bound past her to such a distance, that before they can recover the track, the Kangaroo has gained so much ground that it is enabled to escape; this stratagem, however, often accelerates its death, for, in turning off so suddenly, its whole weight is thrown upon one limb, which being broken by the pressure, the animal falls to the ground and becomes an easy prey. Even the large bucks are often taken in this way, and, in their fright and anxiety to escape from the dogs, they not unfrequently run against a tree or stump with such violence as to be killed on the spot. It would scarcely be supposed by any one who has only seen this animal in confinement, where it appears so quiet and harmless, that it can be excited to rage and ferocity, yet such is the case in a state of nature; for upon finding itself without a chance of escape, it summons up all its dormant energies for a last struggle, and would doubtless often come off victor if it had dogs alone to contend with; but the moment it observes the approach of man, it seems intuitively to know that its most formidable opponent is before it; its lips are then twisted and contracted, its eyes become brilliant, and almost start from their sockets with rage, its ears are in constant motion, and it emits a peculiar, low, smothered grunt, half hiss or hard breathing-like sound; in fact, when man approaches, it seems altogether to forget the dogs, and regardless of the consequences of withdrawing its attention from them to him, soon loses its former advantage, and the dogs being enabled to obtain a secure hold, soon bring it down.

"The individuals inhabiting the forests are invariably much darker, and have a somewhat thicker coat, than those frequenting the plains. The young at first are of a very light fawn colour, which deepens in tint until they are two years old; after that age it gradually fades until, in the old males, it becomes of a very light grey. In summer their coat assumes a light and hairy character, while in winter it approaches more nearly to the texture of wool. It is very common to find them with white marks, or spots of white about the head, more particularly a white spot between the eyes or on the forehead. On one occasion I met with a very curiously marked individual, having the whole of the throat, cheeks, and the upper part of the head spotted with yellowish-white; and perfect albinoes have been observed by the hunters. The largest and heaviest Kangaroo I have been able to obtain any authentic account of, was killed at the Murray; it weighed 160 lbs."

MACROPUS OCYDROMUS, Gould.

Gould and H.C. Richter del. et lith.   Hullmandel & Walton Imp.

MACROPUS OCYDROMUS, *Gould.*

West-Australian Great Kangaroo.

HEADS OF A MALE AND A FEMALE, LIFE-SIZE.

If the letterpress annexed to the succeeding Plate, containing reduced figures of this species, be referred to, sufficient reasons will be found for figuring life-sized heads of the two sexes of the West-Australian Kangaroo. On comparing *Macropus major* and *M. Ocydromus*, it will be seen that a very considerable difference exists between the two animals—the deep vinous colouring of the entire body, deep brown hue of the nose, and the black mark at the base of the ears, which are peculiar to the latter, being very striking, and rendering it conspicuously distinct from its near ally: the opinion that they are really different species is moreover strengthened by the circumstance of the one being an inhabitant of the western, and the other of the eastern parts of the great continent of Australia; and from what we have seen in so many other instances of representative species, we might naturally expect this would be the case.

For a more detailed account of the *Macropus Ocydromus*, the reader is referred to the pages given with the entire figures of the animal.

MACROPUS OFYDROMUS, Gould

MACROPUS OCYDROMUS, *Gould.*

West-Australian Great Kangaroo.

Macropus ocydromus, Gould in Ann. and Mag. of Nat. Hist., vol. x. p. 1.—Gray, List of Spec. of Mamm. in Coll.
  Brit. Mus., p. 86.
Yoŏn-gur, the male : *Work,* the female : of the Aborigines of Western Australia.

SINCE my account of the Great Grey Kangaroo (*Macropus major*) was printed, some additional examples of the West-Australian animal, to which I had assigned the specific term *ocydromus,* have reached this country; a careful and accurate comparison of which induces the belief that my original opinion of its being a distinct species is really correct, and that I was quite right in conferring upon it a distinctive appellation.

A very young individual which I now possess has not only a thicker and more woolly fur than the young of *Macropus major,* but has a much more vinaceous colour pervading the whole of the body; and I find that this peculiar woolly texture as well as the colour are retained to the extreme adult age, and that the cheeks are of a nearly uniform vinous brown, while the cheeks of *M. major* are brown with a stripe of white. The bases of the ears in *ocydromus* are of a rich hair-brown for the extent of about an inch, when that colour abruptly terminates, and the remainder of the outside of the ear is white; the interior is also white, and has the basal portion thickly clothed with long white hairs,—a style of colouring, so far as regards the outer part of the ears, which is never found in specimens of the true *Macropus major*; there is also a lesser amount of white about the under surface of the body of *ocydromus* than in that of *major,* the vinous tint pervading the lower part of the chest and a great part of the abdomen.

That a great similarity exists in the anatomy of the two animals there can be no doubt; but the same may be said with regard to many other quadrupeds and birds which are considered distinct species. It must be recollected that the Western and Eastern Australian species, both of quadrupeds and birds, differ in almost every instance, and that but few cases occur of a species ranging across the entire continent—an extent of three thousand miles, more or less. Time, and a greater acquaintance with the mammals of Australia, will be necessary before we can say with certainty over what portion of the country this species may range.

Upon looking over my MSS. of West-Australian animals, I find a note from the pen of the late Mr. Gilbert, from which I learn that not only had he observed the difference in the colouring of the animals there found, but he had noticed that a still darker one inhabits the brushes, and that this darker-coloured animal has a more woolly coat. This may be the same as the animal I have called *Macropus melanops,* and is probably distinct from both the others, although I have placed that name as a synonym of *M. major.*

The following is Mr. Gilbert's note above referred to :—

" *Macropus ocydromus.*—You will receive herewith a very large male and two mature females, from different localities; the two latter showing the extremes of the dark and light variations of colouring. The largest Kangaroo yet killed in the colony (the weight of which is well authenticated) was shot on the Canning by Mr. Phillips, the Resident at King George's Sound; it was ascertained to weigh 180 lbs., its unusual size having induced that gentleman to weigh it before any part was removed."

The male has the face and forehead dull cinnamon-brown, becoming darker over the nose and forehead; cheeks without a white stripe; upper lip and chin beset with a number of long and short fine black hairs, those on the edge of the upper lip being rigid; base of the ears and occiput dark vinous brown, the remainder of the ears clothed externally with short grizzled hairs, the tips of which are white, and the base brown, offering a strong contrast to the dark colouring of the lower part of the ear; internally the ears are clothed with long white hairs; the vinous brown colouring of the occiput is continued down the back of the neck and over the middle of the back, becoming lighter towards the tail; throat, fore part of the neck, and chest brownish white; sides of the body, flanks, and under surface dull cinnamon-brown; arms and hands grizzled brown, externally becoming lighter on the inner surface and darker towards the extremities; thighs, legs, and feet similar; a deep vinous brown mark extends along the ridge of the tail, gradually passing into black at the tip, the remainder of the tail cinnamon-brown.

The female is similar in colour, but lighter in every part.

The accompanying Plates represent the heads of an adult male and female of the size of life, and reduced figures of the entire animals.

MACROPUS FULIGINOSUS.

MACROPUS FULIGINOSUS.

Sooty Kangaroo.

Kangurus fuliginosus, Desm. Nouv. Dict. d'Hist. Nat., tom. xvii. p. 35. pl. E. 22 (*K. géant*).—Ib. Ency. Méth. Mamm., part i. p. 263.

Kangurou géant, F. Cuv. et Geoff. Hist. Nat. des Mamm., fasc. 2.—F. Cuv. Dict. Sci. Nat., tom. xxiv. p. 347.

Macropus fuliginosus, Less. Man. de Mamm., p. 225.—Gould, Mon. of Macrop., pl.   .—Waterh. in Jard. Nat. Lib. *Marsupialia*, p. 200.—Gray, List of Mamm. in Coll. Brit. Mus., p. 88.—Waterh. Nat. Hist. of Mamm.,vol. i. p. 73.

ALTHOUGH I have mentioned, in my account of *Macropus major*, the probability that the animal to which the name of *M. fuliginosus* has been assigned is merely a variety of that species, I have thought it advisable to give reduced figures from the original specimens in the Paris and Leyden Museums, because these specimens differ considerably in the colouring of their fur from the ordinary examples of *M. major*, and because I believe the animal is no longer to be found on Kangaroo Island, where, according to Desmarest, the specimens above referred to were procured. The peculiar dull red colouring they exhibit may or may not be due to some unusual mode of preparing the skin before mounting; but one thing is certain: whatever may have been the original colouring of their fur, the term *fuliginosus* is now by no means descriptive of it; in all probability the change is due to the long exposure to light and dust to which they have been subjected,—an agency which has not only had a deleterious effect upon the specimens in question, but upon all those I brought from Australia. To become acquainted with the natural colouring of the various species of Kangaroo, it is positively necessary to observe them in their native country, where the newly-killed animals present colours which no art on the part of the Taxidermist or care on the part of Museum curators has the power of preserving, and to give a faithful portraiture of which, coloured drawings should then and there be made. I see the necessity of this more and more, whenever I look at specimens in our museums, from all of which the colours have more or less faded, until a general sameness of tint pervades the whole.

Fur of the body rather long and inclining to a woolly texture; general colour rusty yellowish brown, darker and inclining to sooty on the shoulders and centre of the back; hairs of the throat, back and abdomen grisly; sides of the face and muzzle uniform, and of the same colour as the body; inner surface of the ear furnished with long white hairs; external surface blackish brown; toes and apical half of the tail blackish brown.

	Male.		Female.	
	feet.	inches.	feet.	inches.
Length from the nose to the extremity of the tail	7	3	4	9¼
„ of the tail	2	6	1	9
„ „ tarsus and toes, without the nail	1	0		9¼
„ „ ear		4⅛		3¼
„ „ head		9¾		7

The Plate represents both sexes, necessarily greatly reduced.

OSPHRANTER RUFUS.

J. Gould and H.C. Richter, del. et lith.   Hullmandel & Walton, Imp.

OSPHRANTER RUFUS.

Great Red Kangaroo.

HEAD OF A MALE, LIFE-SIZE.

I REGRET very much to say that the time may not be far distant when an opportunity of giving a full-sized drawing of the head of this noble animal, taken from life, will not be possible.   The larger and more conspicuous productions of an island are often, as a natural consequence, the first that become extirpated; and this result takes place more speedily where no protection is afforded to them.   Short-sighted indeed are the Anglo-Australians, or they would long ere this have made laws for the preservation of their highly singular, and in many instances noble, indigenous animals; and doubly short-sighted are they for wishing to introduce into Australia the productions of other climes, whose forms and nature are not adapted to that country.   Let me then urge them to bestir themselves, ere it be too late, to establish laws for the preservation of the large Kangaroos, the Emeu, and other conspicuous indigenous animals: without some such protection, the remnant that is left will soon disappear, to be followed by unavailing regret for the apathy with which they had been previously regarded.   I make no apology, therefore, for publishing a life-sized head of the Great Red Kangaroo of the plains, a detailed history of which will be found accompanying the reduced figures.

OSPHRANTER RUFUS.

OSPHRANTER RUFUS.

Great Red Kangaroo.

Kangurus rufus, Desm. Mamm. Supp., p. 541.—Gray, in Griff. Anim. Kingd., vol. v. p. 202.
———— *laniger,* Gaim. Bull. des Sci. par la Soc. Philom., année 1822, p. 138.—Quoy et Gaim. Voy. de l'Uranie, p. 65. pl. 9.
Macropus lanigerus, Gray, in Griff. Anim. Kingd., vol. iii. p. 49. pl. opposite p. 48.
———— *laniger,* G. Benn. Cat. of Australian Museum, Sydney, p. 6. no. 28.—Gould, Mon. of Macropodidæ.
—Gray, List of Mamm. in Coll. Brit. Mus., p. 88.
———— *(Halmaturus) rufus,* Waterh. Nat. Hist. Mamm., vol. i. p. 104.

Not only is this species the most beautiful member of the family to which it belongs, but it may also be regarded as the finest of the indigenous Mammals of Australia yet discovered ; its large size, great elegance of form, and rich and conspicuous colouring all tending to warrant such an opinion. A splendid male, which in health and colour fully equals any examples I have personally observed in their native wilds, is now (1853) living in the Gardens of the Zoological Society in the Regent's Park, and, although it has not yet attained the stature of a fully adult animal, forms an object of great attraction to the visitors, and particularly to those naturalists who take an interest in the singular Mammals of Australia. This fine example is the first that I have seen alive in Europe ; and it will be much to be regretted if a female cannot be procured, for in all probability the success which has attended the introduction of the Common Kangaroo and other members of the family would also wait upon the domestication of this noble animal, and ultimately lead to the perpetuation of its race in Europe ; an object of the highest importance, since from the limited extent of its native habitat, daily encroached upon by civilized man, and the wanton manner in which it is unrelentingly killed, it is constantly becoming more and more scarce in the open plains and low grassy hills of its native land. The kind of country it frequents being of the utmost value to the pastoral portion of the Australian community, it is diligently sought for and occupied as soon as found, for depasturing their immense flocks and herds, in the stockmen and keepers of which, aided by their fleet, powerful, and well-trained dogs, the Red Kangaroo finds an enemy which at once drives it from all newly occupied districts, and which will ultimately lead to its entire extirpation, unless some law be enacted for its preservation; and to this point I would direct the attention of the present enlightened Governor and Assembly of New South Wales, who surely will not hesitate to make some provision for the protection of this noble animal, as well as for some other fine species of the family still inhabiting that Colony ; in fact, if this be not done, a few years will see them expunged from the Fauna of Australia.

The range of the Great Red Kangaroo, so far as it is yet known, extends over the plains of the interior of the Colonies of New South Wales, Port Philip, and South Australia ; I have never seen a specimen from the country to the westward of the latter colony, or from the northward of the latitude of Moreton Bay ; the plains bordering the rivers Gwydir, Namoi, Morumbidgee, Darling and Murray, and the grassy hills of South Australia, particularly those to the northward of Adelaide, are the districts over which it formerly ranged in abundance, and in which, notwithstanding the persecution to which it has been subjected, it may still be found, though in much smaller numbers. It does not so strictly affect the rich grassy plains as the Common Kangaroo (*Macropus major*), but evinces a greater partiality for the sides of the low stony hills and patches of hard ground clothed with box, intersecting those alluvial flats. In this part of its economy, as well as in the structure of its hinder feet, the greater length of its arms, the comparative nakedness of its muzzle, and in the much smaller size of the female compared with the females of the true *Macropi*, and in the difference in the colouring of the sexes, it is most intimately allied to the Great Rock Walleroos, to which I have given the generic name of *Osphranter*, and hence I have been induced to associate it with the members of that genus, and to call it *Osphranter rufus*, which latter or specific name has the priority over that of *laniger* assigned to it by M. Gaimard, and under which it appears in my "Monograph of the Macropodidæ." It is to be regretted that the colouring of the fur of this fine animal cannot by any means be preserved after death if exposed to light ; nothing can be more different than its colour on the living animal and that of the mounted specimens in the National Museum, which were procured by myself while in Australia ; so great in fact is the difference, that they might readily be mistaken for two different animals. The beautiful pink hue of the throat and chest appears to be due to some peculiar exudation from the skin rather than to the colouring of the hair itself; for if those parts be rubbed with a white handkerchief, a pinky pollen-like substance will be found adhering to it : this tint is deeper at some seasons than at others, and is probably developed under some particular condition of the animal.

The female is still more gracefully and elegantly formed than the male, and has a very different style of colouring, delicate blue being the prevailing tint in those parts which in the male are red, whence the colonial names for the two sexes of Red Buck and Blue Doe; the female has also been called the Flying Doe, from her extraordinary fleetness, which is in fact so great, that I have no hesitation in saying that on hard ground and under favourable circumstances she would outstrip the fastest dogs. Occasionally both sexes are run successfully, either from the chase being over soft muddy soil, or from the female being encumbered with a large and heavy young one, which she has not been able to eject from the pouch, as she always will do if possible when hardly pressed; the female specimen in the British Museum above alluded to was procured under these circumstances. Observing a pair sheltering from the heat of the sun under a small group of Myalls (*Acacia pendula*) on the plains near the Namoi, I succeeded in leading a fine dog to within seventy yards without being perceived; the dog was so quickly at the heels of the female, which was carrying a large young one, that her escape was impossible: the male in the British Museum was also secured by a single dog, which, after a short chase, "pulled" and kept him at bay until I came up and despatched him, after a fearful resistance. It weighed above two hundred pounds, and was killed while I was making a forced march between the River Murray and the City of Adelaide, at a time when our provisions were exhausted, and I can therefore speak with a lasting recollection of its flesh, which supported me and my party for four days.

The male has the head, all the upper surface and flanks rich orange-red; a wash of grey on the outer side of the thigh; sides of the muzzle as far as the angle of the mouth and the chin white; intermingled with the white of the muzzle some interrupted rows of black hairs; ears white at the base, the remainder greyish brown, fringed with white; throat and chest delicate pink, deeper at some seasons than at others; arms and legs tawny white; hands and toes blackish brown; under surface of the body and tail white, tinged with tawny.

The female is blue-grey where the male is red, but has a wash of red on the sides of the body and the haunches tinged with vinous; has a broad white mark extending from the angle of the mouth under the eye, and the under surface of the body and the limbs pure white.

In the young animal the upper surface is nearly of a uniform blue-grey.

As a reduced figure can give but a faint idea of the size of this fine animal, I give the measurements of the male killed near the Namoi.

Total length from the nose to the end of the tail eight feet two inches; of the tail three feet; of the arm, hand and nails eighteen inches and a half; of the tarsi, toes and nail fourteen inches; of the face from the tip of the nose to the base of the ear eight inches; of the ear five inches and a half.

The drawing of the head, taken by Mr. Richter from the animal in the Gardens of the Zoological Society, is about two-thirds of the fully adult size.

OSPHRANTER ANTILOPINUS, Gould.

J. Gould and H. C. Richter, del. et lith.   Hullmandel & Walton, Imp.

OSPHRANTER ANTILOPINUS, *Gould.*

Red Wallaroo.

HEAD OF A MALE, LIFE-SIZE, AND OF A FEMALE, REDUCED.

THE Red Wallaroo, of the Cobourg Peninsula, a noble species, second only in colour and structure to the *Osphranter rufus*, must for ever form a conspicuous object among the indigenous quadrupeds of Australia. Its bare mufflle at once indicates it to be a less browzing animal than the *Macropus major*; while the structure of its feet and toes equally indicate that stony and rocky districts are the situations in which it is destined to dwell. Much disparity occurs in the size of the sexes, the female being very much smaller than the male; the accompanying illustration, however, does not portray the head of the female so large as it really is; on the other hand, the head of the adult male is the size of life. Fierce, bold, and even dangerous is this powerful animal. Its native rocks afford it partial protection; but it is one of the species which will soon be extirpated when Northern Australia becomes peopled by miners or stockholders.

The following Plate gives reduced figures of this fine species, and the accompanying letter-press a detailed account of its history and economy.

OSPHRANTER ANTILOPINUS: Gould.

OSPHRANTER ANTILOPINUS, *Gould.*

Red Wallaroo.

Osphranter Antilopinus, Gould in Proc. of Zool. Soc., Part ix. p. 80.—Ib. Mon. of Macrop., pl.   .—Gray, List
of Mamm. in Coll. Brit. Mus., p. 91.
Macropus (Halmaturus) Antilopinus, Waterh. Nat. Hist. of Mamm., vol. i. p. 95.
Mar-ra-a-woke of the Aborigines of Port Essington.

FROM the period at which Australia was first visited by our enterprising navigators to the present time, our knowledge of its natural productions has been almost entirely confined to those of the narrow and limited tract of land bordering its eastern and southern shores; and it may fairly be said, that the whole of the zoology of the vast range of country washed by the seas of Torres' Straits is as much or more unknown than that of any similar extent of country in the world. In exemplification of what I have here asserted of our ignorance of the productions of that region, I may mention that the noble Kangaroo here figured is only one of many new and interesting animals I have lately received from these parts. It is very abundant on the Cobourg Peninsula, and I have no doubt that, when the country towards the interior is explored, it will there be found in great numbers.

Two very fine specimens, from which my figures and dimensions are taken, were collected by Mr. Gilbert while at Port Essington, and these in all probability are the only perfect specimens in Europe: the weight of the male was about one hundred and twenty pounds. Captain Chambers, however, late of H.M.S. Pelorus, has placed at my disposal, for the purpose of comparison, &c., several imperfect skins of this species, which clearly indicate that the animal frequently attains a much larger size; and that gentleman also assured me that he has himself seen examples weighing one hundred and seventy pounds; few species therefore exceed it in size, and certainly, with the exception of *Macropus laniger*, none in the richness of its colour and markings. Captain Chambers further informed me, that when hard pressed in the chase it becomes exceedingly fierce and bold, and while among the rocks a most dangerous animal to encounter, one of his finest dogs being tumbled over a precipice and killed by an old male: in this fierceness of disposition it exhibits a striking resemblance to the Black Wallaroo; they also closely assimilate in the diminutive size of their females.

Although fifteen years have elapsed since the above remarks were published in my monograph of the Kangaroos, no additional information or examples have been transmitted to this country. As I have given life-sized drawings of the heads of the other large Kangaroos, I have thought it necessary, for the sake of uniformity, to give a similar illustration of this noble species, of which specimens are to be seen in the National Collection.

The male has the fur of the body rigid and adpressed; general colour rusty red, becoming paler on the face and shoulders, and white or yellowish white on the throat, chest, abdomen and inside of the limbs; hands and feet dark reddish brown, passing into black on the toes; tip of the tail reddish brown.

The female has the fur less rigid and more loose than the male; general colour reddish sandy brown, passing into vinous grey on the shoulders, back of the neck and face; base of the ear externally dark brownish grey, passing into yellowish white towards the tips; immediately in front of the ear a conspicuous patch of yellowish buff; a light buff mark also extends from beneath the eye along the upper lip; throat, chest, abdomen and inside of the limbs pale yellowish white; hands and feet dark brown, becoming black towards the nails.

	Male.		Female.	
	feet.	inches.	feet.	inches.
Length from the nose to the extremity of the tail	7	3	5	6
„ of tail	2	9	2	3
„ „ tarsus and toes, including the nail	1	1		11
„ „ arm and hand, including the nails	1	2		11
„ „ face from the tip of the nose to the base of the ear		7¼		6
„ „ ear		4¼		3¼

The first Plate represents the head of the male of the natural size; the second contains reduced figures of both sexes.

OSPHRANTER ROBUSTUS, Gould.

J. Gould and H. C. Richter, del. et lith.   Hullmandel & Walton, Imp.

OSPHRANTER ROBUSTUS, *Gould.*

Black Wallaroo.

HEAD OF A MALE AND OF A FEMALE, LIFE-SIZE.

IF there be any one of the Great Kangaroos the discovery of which afforded me more pleasure than another during my sojourn in Australia, it is the Great Black Wallaroo of the mountain-districts of New South Wales. Surprising, indeed, it was that so large and conspicuous an animal had not been previously made known; and still more surprising is the fact that, from the period of my visit in 1838–39 to the present time, 1863, few if any skins of the animal have been sent to Europe. Still I can assure my readers that the existence of the Black Wallaroo is not a myth; for specimens of both sexes grace the collections at the British Museum and at Leyden. Like the *O. antilopinus*, the *O. robustus* becomes dangerous both to man and dogs when the rocky and sterile mountain elevations it frequents are traversed; for, like the *Ibex* of the mountain-ranges of the northern hemisphere, the old males will make a determined stand when assaulted and escape is impossible.

As is the case with the sexes of all the other members of this section of the *Macropodidæ*, the male and female of *O. robustus* differ considerably in size, the latter being much smaller and weaker than the former.

As the districts inhabited by this fine species are fully described in the succeeding pages, it is unnecessary to mention them here.

A glance at the accompanying illustration, which represents a head of each sex of the size of life, will furnish a just conception of the features of these animals.

OSPHRANTER ROBUSTUS, Gould

OSPHRANTER ROBUSTUS, *Gould.*

Black Wallaroo.

Macropus (Petrogale) robustus, Gould in Proc. of Zool. Soc., part viii. p. 92.
Osphranter robustus, Gould, Mon. of Macrop., pl.   .—Gray, List of Mamm. in Coll. Brit. Mus., p. 91.
Macropus robustus, Waterh. in Jard. Nat. Lib. *Marsupialia*, p. 241.
Macropus (Halmaturus) robustus, Waterh. Nat. Hist. of Mamm., vol. i. p. 100.
Black Wallaroo of the Colonists.

SINCE my return from Australia in 1840, I have in vain requested my numerous friends and correspondents to procure and transmit examples of this large and truly fine animal. I believe I was the first scientific man who visited the locality in which it dwells, as well as the first who made it known to science; and I may ask, is it not surprising that during the interval of fifteen years which has elapsed since the account of this species was published in my " Monograph of the Kangaroos," no examples besides those I myself brought home should have been procured, and that no attempts to secure living examples of so conspicuous an animal should have been attempted? Indeed, were it not for my visit to its native haunts, it might have remained unknown to us even to the present time. This is the more to be wondered at, since the animal is found within the colony of New South Wales. Surely the exterminating hand of civilized man, so fatal to the animal productions of a new country, cannot have dealt out destruction so unsparingly as to have destroyed the entire race.

The following account of this species appeared in my Monograph of the group, and I regret that I have nothing to add to it.

The Black Wallaroo inhabits the summits of sterile and rocky mountains, seldom descending to the coverts of their sides, and never to their base; few, therefore, have had an opportunity of observing it in a state of nature; indeed there are thousands of persons in Australia who are not even aware of its existence. Although the south-eastern portion of the continent is, I believe, the only part of the country in which it has yet been observed, in all probability it has an extensive range northwards. It is tolerably abundant on the Liverpool Range, and I ascertained that it inhabited many of those hills that branch off on either side of this great mountain-chain, towards the interior as well as towards the coast. Its retreats are so well chosen among the crags and overhanging ledges of rocks, that it is nearly useless to attempt its capture with dogs. It is a formidable and even dangerous animal to approach, for if so closely pressed that it has no other chance of escape, it will rush at and force the invader over the edge of the rocks, as the Ibex is said to do under similar circumstances. Independently of its great muscular power, this animal is rendered still more formidable by the manner in which it makes use of its teeth, biting its antagonist with great severity.

The Black Wallaroo may be regarded as a gregarious animal, four, six, and even more being frequently seen in company. On one of the mountains near Turi, to the eastward of the Liverpool Plains, it was very numerous; and from the nature of this and the other localities in which I observed it, it must possess the power of existing for long periods without water, that element being rarely to be met with in such situations.

The summits of the hills to which this species resorts soon become intersected by numerous roads and well-trodden tracks, caused by its repeatedly traversing from one part to the other; its food consists of grasses and the shoots and leaves of the low scrubby trees which clothe the hills it frequents.

Although much shorter in stature, and consequently less elegant in form, the fully adult male of this species equals in weight the largest specimens of *Macropus major*; and so remarkable is the difference in the colour and size of the sexes, that, had I not seen them together in a state of nature, I should have considered them to be different species, the black and powerful male offering so great a contrast to the small and delicate female.

The male has the fur harsh and somewhat shaggy; general colour slate-grey, obscurely washed with brownish, and tinted with vinous on the outer sides of the thighs; feet dark brown, gradually passing into black on the fore part; upper part of the arm brownish; hands and wrists black; inner surface of the ear white, the exterior brown; muzzle and a patch on the chin blackish; a line round the angle of the mouth and the lower lip white; throat and fore part of the neck white, the hairs being grey at the base; under surface like the upper, but paler; tail blackish brown above, paler beneath.

The female has the general colour silvery grey, obscurely tinted with purplish or vinous on the back; under surface nearly white; cheeks hoary, with a blackish patch on the chin; tail dirty white, slightly tinged with brown on the upper side; legs paler than the body; hands brown, becoming nearly black on the fingers; toes brownish black above.

	Male.		Female.	
	feet.	inches.	feet.	inches.
Length from the nose to the extremity of the tail	7	0 . . .	5	10
„ of the tail	2	10¼ . . .	2	6
„ „ tarsus and toes, including the nail.	1	0 . . .	1	10¼
„ „ arm and hand, including the nails		13¼ . . .		9¼
„ „ face from the tip of the nose to the base of the ear .		8 . . .		7
„ „ ear		3¼ . . .		3

The accompanying Plates represent the head of the male of the natural size, and whole but greatly reduced figures of both sexes.

OSPHRANTER? PARRYI.

J. Gould and H.C. Richter del. et lith.

Hullmandel & Walton, Imp.

OSPHRANTER? PARRYI.

Parry's Wallaroo.

LIFE-SIZED HEAD AND FORE-ARMS OF BOTH SEXES.

If this be the least of the Wallaroos, it is one of the most elegant and chastely coloured species yet discovered. The rocky districts of the eastern portion of New South Wales are its true, and probably its restricted habitat; there it dwells among the precipitous rocks and sterile crowns of the mountains, feeding upon the vegetation peculiar to such situations. In most instances the Osphranters differ considerably in the colouring of the male and female, but in the *O. Parryi* this feature is less conspicuous than in its congeners. The more diminutive size of this species enables me to figure not only the head, but also a considerable portion of the fore quarters, of the size of life. On reference to the Plate, it will be seen that, as with the other members of the genus, there is much difference in the size of the sexes. Its muzzle, like that of *O. rufus*, is more hairy than that of *O. robustus* or *O. antilopinus*.

Like some of the larger Kangaroos, this fine animal requires protection, otherwise it will be speedily extirpated; its extreme agility among the rocks, and the sterile nature of the districts it frequents, will, however, tend somewhat to its preservation.

For a history of the species the reader is referred to the succeeding page.

OSPHRANTER ? PARRYI.

J. Gould and H.C. Richter, del. et lith.

Hullmandel & Walton, Imp.

OSPHRANTER? PARRYI.

Parry's Wallaroo.

Macropus Parryi, Bennett in Proc. of Zool. Soc., Part II. p. 151.—Ib. Trans. Zool. Soc., vol. i. p. 295. pl. 37.

Macropus (Halmaturus) Parryi, Waterh. Nat. Lib. Marsupialia, p. 206. pl. 18.—Ib. Nat. Hist. of Mamm , vol. i. p. 113.

Halmaturus Parryii, Gould, Mon. of Macropodidæ, Part II.—Gray, List of Mamm. in Brit. Mus., p. 89.

Macropus elegans, Lambert, Trans. of Linn. Soc., vol. viii. p. 318. pl. 16 ?

THE known range of this fine species extends along the east coast from Port Stephens to Wide Bay, a newly opened district to the northward of Moreton Bay. Mr. Strange informs me that it inhabits the rocky ranges of the Clarence, occasionally descending into the more open broken country, where it frequents the ledges of rocks at an elevation of 2000 feet; it is also met with between the open grassy hills trending upward to the main range. So fleet is this animal, that it is only with the assistance of the finest dogs that there is any chance of procuring examples; it surpasses in fact every other animal in speed, and when fairly on the swing no dog can catch it. Their general contour, short and stout hind limbs and short blunt nails are all in accordance with their habit of frequenting rocks. Like most other members of its race, it is easily tamed, readily becoming familiar and docile.

A living specimen, presented to the Zoological Society of London by Captain Sir Edward W. Parry, R.N., after whom the animal has been named, " was obtained at Stroud, near Port Stephens, in the latitude of about 30° south. It was caught by the natives, having been thrown out of its mother's pouch when the latter was hunted. At that time it was somewhat less than a rabbit, but was full-grown on its arrival in England. It was never kept in confinement until it was embarked for England, but lived in the kitchen, and ran about the house and grounds like a dog, going out every night after dark in the bush or forest to feed, and usually returning to its friend the man-cook, in whose bed it slept, about two o'clock in the morning. Besides what it might obtain in these excursions, it ate meat, bread, vegetables, in short everything given to it by the cook, with whom it was extremely tame, but would allow nobody else to take liberties with it. It expressed its anger when very closely approached by others, by a sort of half-grunting, half-hissing, very discordant sound, which appeared to come from the throat, without altering the expression of the coun. tenance. In the daytime it would occasionally, but not often, venture out to a considerable distance from home, in which it would sometimes be chased back by strange dogs, especially those belonging to the natives. From these, however, it had no difficulty in escaping, through its extreme swiftness; and it was curious to see it bounding up a hill and over the garden fence, until it had placed itself under the protection of the dogs belonging to the house, especially two of the Newfoundland breed to which it was attached, and which never failed to afford it their assistance, by sallying forth in pursuit of its adversaries."

But little doubt exists in my mind that Lambert's characters of his *Macropus elegans* were taken from an animal of this species, although neither his figure nor his description is sufficiently correct to determine this point with certainty.

Fur moderately long and soft; general colour silvery grey, the lower part of the back tinged with purplish brown; muzzle deep brown inclining to black, gradually becoming paler on the forehead until it passes into the grey of the upper surface; a broad pure white mark extends from near the tip of the muzzle along the cheeks, and terminates a little beyond the posterior angle of the eye; below this a faint grey line; ears nearly naked within, but having a few small white hairs on the apical portion; externally they are clothed with blackish brown fur at the base, with adpressed white hairs in the middle, and with black hairs at the tip; chin, throat, inner side of the limbs, under surface of the body and under side of the basal half of the tail white; the tips of the hairs on the chest faintly tinged with grey; arms hoary grey; hands black; tarsi and two inner toes white; the other toes black at the extremity, and with a mixture of black and white hairs at the base; tail nearly white, with the exception of the tip, which, with a fringe of long hairs on the under surface of the extremity, are black.

	Male.	
	feet.	inches.
Length from the nose to the extremity of the tail	5	5
„ of tail	2	7
„ „ tarsus and toes, including the nail		10
„ „ arm and hand, including the nails		8
„ „ face from the tip of the nose to the base of the ear		5¼
„ „ ear		3¼

The head and fore-arm represented in the accompanying Plate are of the natural size, while the entire figures in the other are much reduced.

HALMATURUS RUFICOLLIS.

J.Gould and H.C.Richter, del et lith.

Hullmandel & Walton, Imp.

HALMATURUS RUFICOLLIS.

Rufous-necked Wallaby.

HEAD AND ARMS OF BOTH SEXES, OF THE SIZE OF LIFE.

AN opinion exists among zoologists that the *Halmaturus Bennetti*, which is exclusively confined to Tasmania, and the *H. ruficollis* of the opposite portion of the Australian continent, are one and the same species, and that the difference in the colouring of the two animals is merely the effect of climate; the full-sized heads, then, which are given as accessories to the reduced figures on the succeeding Plate will be of value as illustrative of the difference in question. Certain it is, that all the specimens from the continent (the *H. ruficollis* of this work and of previous authors) are much redder in colour, and have the white of the cheeks extending further on the breast, than the *H. Bennetti* procured in Tasmania. As these differences are carefully detailed on the succeeding page, it will not be necessary to enter into them here.

The accompanying Illustration represents the head and forearms of both sexes, as near the size of life as possible.

HALMATURUS RUFICOLLIS.

HALMATURUS RUFICOLLIS.

Rufous-necked Wallaby.

Kangurus ruficollis, Desm. Dict. d'Hist. Nat., tom. xvii. p. 37. —Ib. Ency. Méth. Mammalogie, p. 274.
Kangaroo à cou roux, F. Cuv. Dict. des Sci. Nat., tom. xxiv. p. 348.
Macropus ruficollis, Less. Man. de Mamm. p. 228.—Gould, Mon. of Macropodidæ, pl.   .—Waterh. in Jard. Nat.
        Lib., Marsupialia, p. 216.
Kangurus ruficollis, Peron.
Macropus rufo-griseus, Waterh. in Jard. Nat. Lib., Marsupialia, p. 217.
Kangurus rufo-griseus, Desm. Nouv. Dict. d'Hist. Nat., tom. xvii. p. 36.
Halmaturus griseo-rufus, Goldf. in Oken's Isis, 1819, p. 267
Macropus (Halmaturus) ruficollis, Waterh. Nat. Hist. of Mamm., vol. i. p. 125.
Warroon, of the Aborigines of the Illawarra district.

THE low table-lands of New South Wales, particularly those on which the *Daveysia* scrub abounds, are the favourite localities of this species of *Halmaturus*. I found it especially abundant on the fine estate of Charles Throsby, Esq., at Bongbong, immediately behind Illawarra, and ascertained that it ranges westward from thence nearly to Port Philip and eastward to Moreton Bay ; it is also said to inhabit the larger islands in Bass's Straits. Since writing the account of this species given in my Monograph of the Kangaroos, referred to above, numerous New South Wales specimens have been sent to me by my collectors, and many living examples have been forwarded to this country, one of which is now (1854) living in the Menagerie of the Zoological Society in the Regent's Park. A careful examination of all these examples tends to strengthen the supposition that the present animal and the *Halmaturus Bennetti* of Van Diemen's Land are quite distinct ; at the same time I cannot but admit that I am still in doubt as to whether this is the case, or if the differences they exhibit are due to local causes ; under these circumstances, I have thought it best to figure both animals under the names by which they are respectively known, and leave the determination of these points to future research.

The specimens contained in the great collections of the Continent, particularly those of Paris and Leyden, are from the mainland, and not from Van Diemen's Land, and have the names of *ruficollis* and *rufo-griseus* attached to them,—appellations which are not applicable to the Tasmanian specimens. I have observed that the mainland animals not only differ in colour, but are larger than those from the islands.

As is the case with most of the other species of the family, the male of the present animal much exceeds the female both in the size of the body and in the strength of the fore-arm.

Fur moderate as to length and softness of texture ; general colour rusty brown pencilled with white, brownish grey at the base succeeded by rusty, broadly annulated with white near the extremity and black at the point ; neck, shoulders and arms bright rust-red, pencilled with white ; muzzle brownish black ; on the upper lip a tolerably distinct white mark, running backward and terminating beneath the eye ; apical half of the external surface of the ear blackish, internal surface of the ear white, narrowly margined at the tip with black ; on the chin a patch of brownish black ; throat whitish ; under surface greyish white, the hairs being grey at the base and white at the extremity ; hand black ; tarsi clothed with hairs, which are brownish black at the base and white at the tip ; toes covered with black hairs ; tail hoary grey, with a pencil of black hairs at the tip.

The accompanying Plates represent the head of each sex of the natural size, and reduced figures of the entire animals.

HALMATURUS BENNETTI.

J. Gould and H.C. Richter, del. et lith.   Hullmandel & Walton, Imp.

HALMATURUS BENNETTI.

Bennett's Wallaby.

HEAD AND FOREARM, OF THE SIZE OF LIFE.

THREE species of the larger Kangaroos are indigenous to the Island of Tasmania, or Van Diemen's Land, viz. *Macropus major, Halmaturus Bennetti*, and *H. Billardieri*. Of these the *H. Bennetti* is intermediate in size, and is the most important, since its flesh is a staple commodity as an article of food, and its skin affords no inconsiderable profit to the settlers, vast numbers of skins being annually sold.

The life-sized portrait of the upper portion of this animal will convey a just conception of its physiognomy, while the reduced figures, which are drawn to scale, will show how disproportionate in size are its hind quarters: it is these latter parts of the animal which are eaten by all classes, from the Governor of the colony to the stockmen. The relative weights and admeasurements of the two sexes are given on the succeeding page, to which I must therefore refer my readers for further particulars respecting the species.

HALMATURUS BENNETTII.

HALMATURUS BENNETTI.

Bennett's Wallaby.

Macropus Bennettii, Waterh. in Proc. of Zool. Soc., part v. p. 103.—Ib. in Jard. Nat. Lib. Marsupialia, p. 211.
  pl. 19.
———— (*Halmaturus*) *fruticus*, Ogilby, Ann. of Nat. Hist. 1838, vol. i. p. 219.
Halmaturus Ualabatus, Gray, Mag. of Nat. Hist. new ser. 1837, vol. i. p. 583.
———— *Bennettii*, Gould, Mon. of Macropodidæ, pl.   .—Wagn. in Schreb. Saug. Suppl. 111, 112. Heft,
  p. 115.—Gray, List of Mamm. in Coll. Brit. Mus., p. 89.
Macropus ruficollis, var. *Bennettii*, Waterh. Nat. Hist. of Mamm., vol. i. p. 130.—Gunn in Proc. Roy. Soc. of Van
  Diem. Land, vol. ii. p. 88.
Halmaturus leptonyx, Wagn.
Brush Kangaroo, Colonists of Van Diemen's Land.

THE native habitat of the animal here represented is Van Diemen's Land and the larger islands in
Bass's Straits. It would be useless to attempt an enumeration of the localities in which it may be found,
since its dispersion may be said to be general over the islands named, from the snowy summits of Mount
Wellington and the hills of lesser elevation to the forests in the lowest valleys; it evinces, however,
a decided preference to situations of a humid character, being seldom, if ever, seen on the hot and sandy
plains : the localities it affects afford it a retreat, so secure as to preclude all chance of its extermination,
although many thousands are killed annually for the sake of its flesh, which is very generally eaten and
highly esteemed, being delicate, juicy and well-flavoured; its skin also forms a considerable article of
commerce, being largely exported from Van Diemen's Land into England for the manufacture of the upper
parts of boots and shoes, for which it is admirably adapted, besides being extensively used for the same
purpose in the colony. I have read advertisements in the Hobart Town newspapers, stating that three
thousand skins were immediately wanted, and which were quickly supplied by the settlers, servants and
shepherds at the out-stations : the skins are generally taken off on the spot where the animal is killed,
and afterwards stretched on the ground to dry; they are then sold for about fourpence or sixpence each
to persons who visit the stock stations of the interior for the purpose of collecting them, and who retail
them again in Hobart Town or Launceston to the advertiser or others for colonial consumption or for
exportation.

  The Bennett's Wallaby is gregarious in its habits, and although truly a brush animal, does not confine
itself so strictly to localities of that description as the smaller members of the genus, but frequently resorts
to the thinly-timbered forests and the crowns of the low grassy hills, always, however, seeking security in
the thick brush when pursued, or such steep rocky acclivities as present almost insurmountable obstacles
to the pursuit with dogs. It is one of the most hardy members of its family, and would doubtless readily
become acclimatized in this country, since the temperature of Van Diemen's Land more nearly resembles
that of the British Islands than does any other part of Australia, in proof of which I may mention that
numbers have been bred in the Menagerie of the Zoological Society, in that of the late Earl of Derby, and
others.

  "In a large piece of enclosed ground in his Lordship's park," says Mr. Waterhouse, "I had the pleasure
of seeing many individuals of the Brush Kangaroo in a state of comparative freedom, and where they
appeared to thrive well. When I entered the paddock in which they were kept, being all concealed beneath
some heath, I was not aware of their presence until, on approaching their place of shelter, they suddenly
elevated the fore part of their bodies and then darted off to a distant spot with great swiftness. When at
rest they frequently assume a singular position ; the fore feet are applied to the ground, and they at the
same time sit upon their haunches, having the hind legs stretched forwards, and perfectly straight, as well
as the tail, which lies between them. The young animal does not finally quit the pouch of the mother until
some time after it has attained the size of a full-grown rabbit; at which time it does not differ in colouring
from the parent."

  The full-grown male varies in weight from forty to sixty pounds : the haunch and loins are the only parts that are eaten, and these are constantly exposed for sale in Hobart Town, Launceston, and other parts of the country. The female closely resembles the male in colour, but is about one-third less in size.

Mr. Waterhouse, who gave the specific appellation of *Bennettii* to this animal, in honour of a late talented Secretary of the Zoological Society, is now inclined to consider it to be merely a local variety of the *Halmaturus ruficollis* of New South Wales, an animal which does not accord with it in colour, and which is of a somewhat larger size; it will be seen that I have treated them as distinct: in either case it becomes necessary, in order duly to illustrate the subject, to figure both.

Fur rather long and moderately soft; general tint a very deep grey, inclining to black on the back; somewhat paler on the sides of the body, with a rust-like tint on the back of the neck, base of ears, the haunches, shoulders, and in the region of the eye; under surface of the body and the inner side and fore part of the hinder legs greyish-white; muzzle black; crown of the head brownish-black; an obscure whitish line extends backwards from the corners of the mouth, and becomes obliterated on the cheeks; lips dirty-white; chin blackish; ears white internally, black externally; hands, toes, and outer side of the heel black; hairs of the tail (excepting at the base, where they are of the same colours and character as those of the body) black, broadly annulated with white near the apex; tip of the tail black, under side of the tail white; the hairs on the upper part of the body are of a deep slate-colour at the base, the remaining portion of each hair is black, annulated with white, or more generally with pale rust-colour; on the under parts of the body the hairs are of a deep slate-colour, with the apical portion white.

The figure of the head is of the natural size; that of the entire animal is much reduced.

J.Gould and H.C Richter, del. et lith.

HALMATURUS GREYI, *Gray.*

Hullmandel & Walton, Imp.

HALMATURUS GREYI, *Gray.*

Grey's Wallaby.

UPPER HALF OF A MALE AND HEAD OF A FEMALE, OF LIFE-SIZE.

THE name of Sir George Grey must always be conspicuous in the annals of Australian history, whether we regard this enlightened and valuable public servant as an explorer or a ruler; and, for my own part, I am much gratified that so fine a species as the present should have been named in honour of the present Governor of New Zealand.

The *Halmaturus Greyi*, if not so beautiful as the *H. manicatus*, is very little inferior in this respect to that species. It will be seen that, while the forearm is as short as in that animal, the black colouring of the fore feet is not so sharply defined, and that on the tips of the ears this colour is wholly or nearly absent. South Australia is the native habitat of this fine animal; it therefore inhabits that part of the country lying between the eastern and western parts of the continent. Those who are not well versed in the Mammals of Australia may perhaps consider the variation in the shades of colour above mentioned insufficient to constitute a species; but I can assure them that such is not the case. The *H. Greyi* is further distinguished from both the *H. ruficollis* and *H. manicatus* by having a more-lengthened-hairy or shaggy coat, by its nearly white tail, and generally lighter colouring.

HALMATURUS GREYI, Gray.

HALMATURUS GREYI, *Gray.*

Grey's Wallaby.

Halmaturus Greyii, Gray, List of Mamm. in Brit. Mus., p. 90.
Macropus (Halmaturus) Greyi, Waterh. Nat. Hist. of Mamm., vol. i. p. 122.

THIS fine Wallaby was first sent to this country from South Australia by His Excellency Sir George Grey, after whom it was named. It is a species quite distinct from every other, but is perhaps most nearly allied to the *Halmaturus manicatus*, an animal inhabiting the country farther to the westward. Its powerful and finely proportioned hinder extremities, contrasted as they are with its slender and diminutive fore-arms, are indicative of a structure adapted for rapid movements, and, in strict accordance with this view, we find that it is one of the most fleet and agile members of its race. Its favourite places of resort are flats near the sea-shore, particularly low sand-hills and open grounds, where the surface is bare and unbroken, to which is doubtless to be attributed the circumstance of its claws being more attenuated and spine-like than those of any other species. In size the *H. Greyi* rather exceeds the *H. manicatus*, but it is less than *H. ruficollis* and *H. Bennetti*.

Mr. Strange informs me that he met with this animal " between Lake Albert and the Glenelg. The kind of country in which it is found consists of large open plains intersected by extensive salt lagoons and bordered by pine ridges. On fine sunny days it is to be found in the salt-water scrub around the lagoons and amid the long grass of the plains. I never saw anything so swift of foot as is this species : it does not appear to hurry itself until the dogs have got pretty close, when it bounds away like an antelope, with first a short jump and then a long one, leaving the dogs far behind it. In wet weather it confines itself to the sand-hills. I have had twenty runs in a day with four swift dogs and not succeeded in getting one."

The description of this animal by Mr. Waterhouse from Sir George Grey's specimens so closely accords with my own, that I cannot do better than give it in his own words :—

" General colour pale ashy brown, slightly tinted with yellowish; the pale tint of the upper parts of the body is produced by the mixture of white with pale rust-colour and black, the visible portion of each hair exhibiting these colours; on the under parts of the body the hairs are of a pale buff-yellow colour externally, and pale grey at the root; the head is grey above, obscurely tinted with rufous, and this latter tint is also observable on the back of the ears, as well as on the neck; immediately behind the naked tip, the muzzle is dusky black above, but the black hue is almost immediately blended into the general grey tint; on the sides of the muzzle are three longitudinal bands, of which the middle one, representing the ordinary pale cheek-mark, is pale yellow; the upper one almost black, but slightly pencilled with whitish, and the lower one is somewhat suffused with brownish; ears well clothed internally with rich yellow hairs, but they are rather narrowly margined with black at the apex; externally, the black extends downwards from the point for about half an inch; behind the eye is a yellowish spot; the chin and throat are tinted with fulvous, and there is a greyish spot on the former; the chest is greyish; below the chest the fur has a pale rusty grey hue; the arms are grey-white at the base, and of a very pale fulvous colour, or fulvous white beyond, and the hands are of the same colour, but the fingers are black, and the black extends slightly beyond the base of the fingers; the hind legs and feet are coloured in the same manner; the thighs are somewhat greyish externally at the base, and the toes are black, with the exception of the long hairs which cover the nails, which are brownish; tail well clothed with hairs of a very pale grey colour, washed as it were with yellow on the upper parts and brown-white beneath; a considerable space at the apex covered with long dirty yellowish hairs."

	feet.	inches.
Length from the nose to the root of the tail	2	8
„ of tail	2	5
„ „ tarsus and toes, including the nail		9¼
„ „ arm and hand, including the nails		6¼
„ „ face from the tip of the nose to the base of the ear . .		5¼
„ „ ear		3

The head and fore-arm represented in the accompanying Plate are of the natural size, while the entire figures in the other are much reduced.

HALMATURUS MANICATUS, *Gould.*

J. Gould and H.C.Richter, del. et lith.   Hullmandel & Walton, Imp.

HALMATURUS MANICATUS, *Gould*.

Black-gloved Wallaby.

HEAD, NECK, AND FOREARM OF BOTH SEXES, OF THE SIZE OF LIFE.

EVERY naturalist who has diligently worked out a monograph of any group of animals must have observed that while some conspicuous feature, either of colouring or marking, pervades all the species, it is much more strongly developed in some of them than in others; in one, perhaps, it is only faintly indicated, while in another it is bold and decided. Now, there is a tendency in all the Wallabys to a blackish brown or black colouring on the hands or the tips of the ears: in some this colouring occurs on both; in others it is confined to the hands alone. The present animal, which is a native of Western Australia, may be cited as the species in which this character is carried to its maximum; for if its fore feet and the tips of its ears had been carefully dipped in ink, they could not be of a blacker hue, nor could this colouring terminate more abruptly. That there is no special end or purpose for the fantastic markings of the Kangaroos and many other animals, beyond mere ornament, I think there cannot be a doubt. Nature revels in variety, as may be seen in the stripings of the various species of Zebra, the fantastic markings of the Antelopes, the banding of the Perameles, and a thousand other creatures. I make no apology for giving full-sized heads of this very pretty species, the peculiarity of whose markings is not so apparent in the reduced figures.

HALMATURUS MANICATUS, *Gould*.

HALMATURUS MANICATUS, *Gould.*

Black-gloved Wallaby.

Macropus (Halmaturus) manicatus, Gould in Proc. of Zool. Soc., Part VIII. p. 127.
Halmaturus manicatus, Ib. Mon. of Macropodidæ.
Macropus (Halmaturus) Irma, Waterh. Nat. Hist. of Mamm., vol. i. p. 117 *.
Goŏrh-a, Aborigines of Perth, and
Quăr-ra, Aborigines of the interior of Western Australia.
Brush and Blue Kangaroo of the Colonists of Western Australia.

IT must, I think, be admitted, that generally a degree of elegance and beauty reigns among the indigenous animals of Australia, and the present species may be cited as an instance in point; the size, form and colouring of this Kangaroo presenting a combination of elegance and beauty; while its jet-black hands and feet render it so conspicuous, that there is no other species with which it can be confounded, except its near ally the *Halmaturus Greyi,* from which, however, it may at a glance be distinguished by its darker-coloured face and nape. To what extent this pretty animal ranges over Western Australia has not been ascertained, but we know that it is very generally diffused over every part of the colony of Swan River, wherever sterile and scrubby districts interspersed with belts of dwarf *Eucalypti* exist; from these retreats it occasionally advances to more open grounds, to feed upon the grasses which there occur in greater abundance than in the glades of the forest.

Mr. Gilbert informs us that it may be ranked among the fleetest of its race; that it requires dogs of the highest breed to capture it, and that a full-grown male weighs nearly twenty pounds. The flesh forms an excellent viand for the table, and the skins manufactured into rugs are extensively used by those whose avocations and mode of life lead them to spend much of their time in the bush.

The sexes are alike in colour and similarly marked about the hands and feet, but the female is always much smaller than the male.

General colour of the upper surface of the body deep grey, produced by the admixture of black and white, the hairs being black at the tip, and annulated with white near the tip; sides and under surface of the body paler grey, tinted with buff-yellow; this yellow tint is almost pure on the abdomen between the hind legs, on the feet and inner side of the ears: the upper surface of the head and muzzle are of a soot-like colour, and the occiput and back of the ears, as well as the apical portion in front, are pure black; a yellowish white line is observable on each side of the muzzle, commencing at the tip, and running backwards beneath the eye; the fore half of the hands and feet are pure black, appearing as if they had been dipped in ink or some other black liquid, the black not blending, as usual, with the pale colour of the hind part of the feet, but terminating in an abrupt line; the greater portion of the tail (which is well clothed with harsh hairs) is of the same black colour; at the base, however, it is coloured as the body; and on the upper surface, for a considerable distance from the base, the black hairs are more or less annulated with whitish, producing a grizzled appearance: on the chin is a small black patch.

	Female.	
	feet.	inches.
Length from the nose to the extremity of the tail	5	0
„ of tail	2	3
„ „ tarsus and toes, including the nail		8¾
„ „ arm and hand, including the nails		5
„ „ face from the tip of the nose to the base of the ear . . .		5¼
„ „ ear		3¼

In one of the accompanying Plates the head and fore-arm are represented of the natural size; while the reduced figures represent the entire animal.

* Mr. Waterhouse considers that the *Macropus Irma* of M. Jourdan may be synonymous with this species, but this is by no means certain; for upon purposely visiting Lyons to clear up this point, I did not find the animal in the collection of that city, and M. Jourdan informed me that his description was taken from a specimen in the Museum at Paris, where also I could not find it.

HALMATURUS UALABATUS.

J. Gould, and H.C. Richter, del. et lith.   Hullmandel & Walton, Imp.

HALMATURUS UALABATUS.

Black Wallaby.

HEAD AND FORE QUARTERS, OF THE NATURAL SIZE.

THIS huge Wallaby is an inhabitant of morasses, mangrove-swamps, and humid woods; and, so far as I am aware, New South Wales is the only part of Australia it frequents. Its characteristics are its black and rich rusty-red colouring, its shaggy thick coat, short ears, and long swinging tail. It stands quite alone among the great family of Kangaroos, there being no other species with which it can be confounded. The weight of an old male is about sixty pounds, while that of the female is considerably less. Its flesh is eaten both by the natives and settlers, but, so far as I recollect, is not so palatable as that of *H. ruficollis* or *H. Bennetti*.

Very correct reduced figures will be found on the next Plate, and a full description on the opposite page.

HALMATURUS UALABATUS.

HALMATURUS UALABATUS.

Black Wallaby.

Kangurus Ualabatus, Less. et Garn. Zool. de la Coquille, tom. i. p. 161. pl. 7.
Macropus Ualabatus, Less. Man. de Mamm., p. 227.—Waterh. in Jard. Nat. Lib. Marsupialia, p. 219.
Kangurus Brunii, Desm. Ency. Méth. Mamm., p. 275.
Halmaturus Lessonii, Gray, Mag. Nat. Hist., vol. i. new ser. p. 583.
———— *nemoralis*, Wagn. in Schreb. Saug. Suppl., part 111–112. p. 114 (Waterhouse).
Macropus (Halmaturus) Ualabatus, Waterh. Nat. Hist. of Mamm., vol. i. p. 136.

———————

This well-marked species inhabits, with but few exceptions, all the thick brushes of New South Wales, especially such as are wet or humid. I hunted it successfully at Illawarra, on the small islands at the mouth of the Hunter, and on the Liverpool ranges. In the former localities it was frequently found in the wettest places, either among the high grass and other dense vegetation, or among the thick mangroves, whose roots are washed by each succeeding tide. The islands at the mouth of the Hunter, particularly Mosquito and Ash Islands, are not unfrequently flooded to a great extent, yet it leaps through the shallow parts with apparent enjoyment, and even crosses the river from one island to the other. On the Liverpool range it as strictly keeps to such parts as are most humid—often near the crowns of mountains, which are frequently enveloped in fogs and dews. Over what extent of country this species will be found to range, it is impossible to say; as yet, I have only observed it in the localities above mentioned; the dense brushes of the Clarence, Manning, and, in fact, all the brushes from Western Port to Moreton Bay, are probably inhabited by it.

Independently of its dark colouring, lengthened tail, and stiff wiry hair, it may be readily distinguished from every other species by the jet-black spot immediately beneath the insertion of the arm. When full-grown, this animal is about the size of *H. Bennettii* and *H. ruficollis*.

Fur long, harsh to the touch; general colour blackish-brown, pencilled with a lighter hue; under surface yellowish in some specimens, in others deep sandy- or rusty-red; ears clothed with dirty-white hairs internally; a rusty patch surrounds their base, and is extended on the neck; cheeks pale brown, mingled with dirty-white; upper part of the muzzle and round the eye blackish; lips and chin whitish; wrists and hand black; immediately beneath the insertion of the fore-arm a jet-black patch; tarsi black; basal third of the tail like the body, the remainder black.

	Male.	
	feet.	inches.
Length from the tip of the nose to the extremity of the tail . . .	4	4
,, of tail	1	4
,, ,, tarsus and toes, including the nail		8¼
,, ,, arm and hand, including the nails		6¼
,, ,, face from the tip of the nose to the base of the ear . . .		4¼
,, ,, ear		2¼

The first Plate represents the head of the animal the size of life; the second, entire figures necessarily much reduced.

HALMATURUS AGILIS, *Gould.*

J. Gould and H.C. Richter del. et lith. Hullmandel & Walton, Imp.

HALMATURUS AGILIS, *Gould.*

Agile Wallaby.

HEAD AND FORE PART OF THE BODY, OF THE NATURAL SIZE.

IT will be seen that Mr. Richter has indulged in a little variation as regards the opposite illustration; the animal, however, is faithfully portrayed, both as regards its form and colouring and the texture of the short adpressed hair which covers its body. All these points are seen to much greater advantage than in the Jerboa-like reduced figures on the next Plate: it is, indeed, impossible to do justice to the appearance of these animals in such small representations of them.

It will be seen that the full-sized head and fore quarters are more darkly coloured than the reduced figures; but as such differences really exist in the various specimens, no apology is necessary for the seeming discrepancy. The time, we may suppose, is not far distant when the northern part of the great southern continent will be peopled by our enterprising settlers. That country contains the bones of my worthy assistant Gilbert, who fell a sacrifice to the treachery of the natives, while arduously prosecuting his researches for the advancement of science and the furtherance of the present work. It is well known that he was in company with the celebrated explorer Dr. Leichardt, who, in like manner, found a resting-place in that *terra incognita*; but it is still unknown in what precise locality his fate was sealed. This country of the Kangaroos is second to none in the sacrifice of valuable lives in the various attempts which have been made to unfold the hidden recesses of its treasures.

HALMATURUS AGILIS; Gould.

HALMATURUS AGILIS, *Gould.*

Agile Wallaby.

Halmaturus agilis, Gould in Proc. of Zool. Soc., part ix. p. 81.—Ib. Monograph of the Macropodidæ, pl.
 Hamb. et Jacq. Voy. au Pole Sud, pl. 19.—Gould, Mon. of Macropodidæ, pl.
——————— *Binoë*, Gould in Proc. of Zool. Soc., part x. p. 58.
Macropus (Halmaturus) agilis, Waterh. Nat. Hist. of Mamm., vol. i. p. 108.

This species of Wallaby may be readily distinguished from every other by its short, wiry, adpressed hair, and the almost uniform sandy-brown colour of the body; the male is also remarkable for having very powerful incisors, and for having the outer toe much developed, whence results a deep cleft between it and the middle one; the head is also longer and more pointed than in any other species which I have seen.

The Agile Wallaby appears to be abundant on all the low swampy lands of the northern coast of Australia. I have seen many specimens from the Cobourg Peninsula; and it is common both near the settlement of Port Essington and at Raffles' Bay. I have also had others placed at my disposal for the purpose of describing by Mr. Bynoe of H.M.S. the Beagle, which were collected on the shores of Torres Straits. It is stated to be a most agile species, readily eluding the dogs employed in hunting it by its extreme activity in leaping among the high grass; when chased it frequently seeks shelter in the thick beds of mangroves, passing over the muddy flats in such a manner as almost to baffle pursuit.

In size, when full-grown, the male is nearly equal to *H. Ualabatus.*

In some notes by Mr. John M'Gillivray on the animals observed by him at Port Essington, it is stated that a young one, very large in proportion to the size of the mother, was taken from the pouch of a female shot by him at Barrow's Bay, and that it did not differ in its colouring in any respect. He adds that the species is very common at Port Essington, where it frequents the tall grass of the low grounds, especially where the Pandanus-tree abounds, under the shelter of which it generally forms its lair. It is extremely active in its movements, and when pursued by dogs makes for the nearest jungle or mangrove thicket.

I now believe the *Halmaturus Binoë*, described by me as a distinct species in the 10th Part of the " Proceedings of the Zoological Society of London," to be merely the young of this animal, and I have consequently placed that name among its synonyms.

Fur rather short, adpressed, and harsh to the touch; general colour sandy-yellow; the upper surface of the head and body freely pencilled with blackish, the hairs being of this colour at the point; chin, throat and chest dull white; abdomen yellow, the hairs terminated with white; limbs pale sandy-yellow externally and white on their inner side, the arms externally pencilled with blackish; tarsi nearly white, passing into rusty on the toes; lips whitish, and a whitish mark from the lip to beneath the eye, parallel with which is another of a dusky hue; ears white within, externally sandy-yellow at the base and broadly margined with black at the apex, and with a narrow black line along the inner edge; on each side of the rump an oblique whitish line; tail sparingly clothed with nearly white hairs, except at the base, which is like the body; the tip of the tail black.

		Male.	
		feet.	inches.
Length from the nose to the base of the tail		5	3
,, of tail		2	6
,, ,, tarsus and toes, including the nail			10
,, ,, arms and hand, including the nails			9
,, ,, face from the tip of the nose to the base of the ear . . .			6
,, ,, ear			3

The first of the accompanying Plates represents the head of a dead animal of the natural size; the second, reduced figures of both sexes.

HALMATURUS DORSALIS, Gray.

J. Gould and H.C. Richter del. et lith.

Hullmandel & Walton Imp.

HALMATURUS DORSALIS, *Gray*.

Black-striped Wallaby.

HEAD AND FORE PARTS, OF THE SIZE OF LIFE.

THIS is one of the largest species of that section of the *Halmaturi* which comprises the *H. Thetidis, H. Parma*, and their immediate allies,—the old males frequently measuring from four to five feet from the nose to the end of the tail. As far as I am aware, it is confined to the interior of New South Wales; certain it is that I have not seen examples from the sea side of the ranges. Its distinguishing features are the red colouring of its fore quarters, its large ears, long tail, and a distinct stripe of black down the nape and back. Like many other kinds of Kangaroo, the male of this species appears to increase in bulk for several years; and hence, in a scrub frequented by this animal, males of various sizes may be found. The accompanying illustration of the head and forearm was taken from the largest male I have seen. The reduced figures will give an accurate idea of the body-colours, and the annexed letter-press all that is known respecting the species.

HALMATURUS DORSALIS: *Gray*

HALMATURUS DORSALIS, *Gray.*

Black-striped Wallaby.

Halmaturus dorsalis, Gray in Mag. of Nat. Hist., vol. i. new series, p. 583.—Gould, Mon. of Macropodidæ, pl.   .
Macropus (Halmaturus) dorsalis, Waterh. in Jard. Nat. Lib. Marsupialia, p. 230.—Ib. Nat. Hist. of Mamm., vol. i.
p. 152.

THIS fine Wallaby, which is distinguished from all other species by the greater length of its tail, and by the black mark which commences at the occiput and runs down the centre of the back, is an inhabitant of the interior, and is particularly abundant in all the scrubs clothing the sides of the hills that run parallel to the rivers Mokai and Namoi; and, although I cannot positively assert that such is the case, I have reason to believe that it inhabits all similar situations between the above-mentioned localities and the great Murray scrub in South Australia. I have never heard of its having been seen between the ranges and the coast, a circumstance that may be attributed to the brush being of a totally different character, and the vegetation more dense and humid than on the dry stony hills of the interior. Like the other members of the genus, it is strictly gregarious, and is so numerous, that I found no difficulty in procuring as many specimens as I pleased; it was, however, more often shot as an article of food than for any other purpose. Its flesh is excellent, and when the vast continent of Australia becomes more thickly inhabited, it will doubtless be justly esteemed. The natives often resort to the haunts of this species, and commit great havoc among them, both for the sake of their flesh as food, and for their skins as articles of clothing. They have various modes of capturing them, sometimes making use of large nets; at other times they are driven by dogs from side to side of the brush, which affords the hunters abundant opportunities of spearing or killing them with the waddy as they pass the open spots.

It is especially abundant at Brezi, to the northward of the Liverpool Plains, and I also found it extremely numerous in the Brigaloe brush on the Lower Namoi.

The female is distinguished by her smaller size, but in the markings of the two sexes no difference exists.

The full-grown males of this species weigh from twenty to twenty-five pounds.

Fur rather harsh to the touch; general colour brown, with a rusty tinge, produced by each hair being of a rusty-brown in the middle; upper surface and sides of the body freely pencilled with black and white; on the back of the neck, shoulders and outer side of the arms a bright rusty-red hue prevails, and the same hue is observable on the hinder part of the back, outer side of the hind legs (especially near the knee) and sides of the body, but is much paler; chin, throat, and all the under parts of the body white; tail clothed with very short, adpressed, grisly hairs, becoming longer and of a dirty-white on the under side of the apical half; upper surface of the muzzle dusky, with a white line on each side; ears black on the outside, and white internally; a black mark commences near the occiput and proceeds backwards; towards the tail it is broadest, most distinct on the middle of the back, and becomes obliterated as it approaches the tail; on the haunch a transverse white mark; hands and feet black.

	Male.		Female.	
	feet.	inches.	feet.	inches.
Length from the nose to the extremity of the tail	4	7	3	10
„ of tail	2	1	1	9
„ „ tarsus and toes, including the nail		8		7¼
„ „ arm and hand, including the nails		8¼		6¼
„ „ face from the tip of the nose to the base of the ear		5		4¼
„ „ ear		3		2¾

The figures in one of the Plates represent the entire animals, necessarily much reduced, and a head the size of life.

HALMATURUS PARMA, Gould

J.Gould and H.C.Richter del. et lith.

Hullmandel & Walton, Imp.

HALMATURUS PARMA, *Gould*.

Parma Wallaby.

Halmaturus Parma, Gould in Gray's List of Mamm. in Brit. Mus., p. 91.
Macropus (Halmaturus) Parma, Waterh. Nat. Hist. of Mamm., vol. i. p. 149.

THE *Halmaturus Parma* is so very distinct from all the other small *Halmaturi* inhabiting New South Wales, that the aborigines who hunt these animals recognize it immediately by the native term I have selected as a specific appellation; this remark applies more particularly to the natives of Illawarra, in which district I myself saw it in a state of nature. In these extensive brushes it doubtless still exists, as since my return other specimens have been sent to me from thence by the late Mr. Strange. How far its range may extend westwardly towards Port Philip, or eastwardly in the direction of Moreton Bay, I am unable to state.

The following note, by Mr. Waterhouse, may be quoted as confirmatory of my view of the specific value of this animal, the original description of which I intended to publish in the "Proceedings of the Zoological Society of London," but by some inadvertence omitted so to do:—

" The Parma Wallaby, I think, merits the distinction of a species. It is intermediate between the *H. dorsalis* and the *H. Derbianus*, and may be distinguished from either by its deep reddish-brown colour, and the distinct large white patch on the throat and chest; the hairs forming this patch are white to the root, in *H. Derbianus* they are distinctly grey next the skin, and in *H. dorsalis* they are very slightly tinted with grey at the root in the same parts; and this circumstance, combined with the general form and superior size of *H. Parma*, caused me at first sight to think it might be a variety of the latter animal; I soon perceived, however, that it differed much from *H. dorsalis* in the form and size of its incisor teeth, and in the proportion of the tarsus, which is much shorter than in that species."

The following is Mr. Waterhouse's description of my original specimen, which now forms part of the Collection at the British Museum :—

" Fur moderate, both as to length and texture; general colour deep reddish-brown, pencilled with white, and much pencilled with black on the back; on the sides of the body the white is less distinct, and as the black is wanting, or nearly so, the hue is paler; the fur on these parts is of a very deep grey next the skin; on the under parts of the body each hair of the fur has the basal half grey, and the external half whitish, but tinted with rust-colour; on the throat and fore part of the chest, however, the hairs are uniform white; back of the shoulders and fore-legs brownish rust-colour; in some specimens a narrowish longitudinal black mark extends from near the occiput along the back of the neck; in others this mark is not apparent; head ashy-grey, tinted with rufous, and finely tinted with whitish; the pale cheek-mark is indistinct; chin brownish; back of the ears clothed with hairs like those of the head, the few hairs of the inner side are whitish; feet brown, finely pencilled behind with very pale brown; tail sparingly clothed, and excepting at the base the scales very distinct; the small, stiff, scattered hairs of the upper surface are black; quite at the root the tail is clothed with hairs like those of the body; on the under side the hairs are more numerous, and of a dirty white hue."

The accompanying Plate represents the head of the animal of the size of life.

HALMATURUS DERBIANUS, Gray

HALMATURUS DERBIANUS, *Gray.*

Derby's Wallaby.

FACE AND FORE PART OF A MALE AND OF A FEMALE, OF THE NATURAL SIZE.

My figures of this animal were taken from specimens procured on Kangaroo Island, at the entrance of Spencer's Gulf. I mention this particularly, because I have given the name of *Halmaturus Houtmanni* to an animal inhabiting the Abrolhos of Western Australia, which Mr. Waterhouse considers may be only a variety of the present species; but this is a point which time alone can determine. These insular animals are extremely puzzling, and considerable judgment is required in ascertaining their specific value.

The *Halmaturus Derbianus* is somewhat allied to the *H. Parma*; still they are unquestionably distinct. They form, with *H. dorsalis*, a little section of the group quite different from that constituted by *H. Thetidis, H. stigmaticus, H. Billardieri,* and *H. brachyurus.*

The *H. Derbianus* is very numerous in all the thick brushes of the islands on which it has been found. It is bustling and quick in all its actions; and it is only by the aid of dogs that it can be forced from its retreat, or to leave the numerous runs formed by it beneath the underwood in all directions.

For the pleasing life-like representation of this species, much credit is due to Mr. Richter; for nothing could be more faithful.

The reader is referred to the description accompanying the reduced figures given on the next Plate for a full account of this animal.

HALMATURUS DERBIANUS. *Gray.*

HALMATURUS DERBIANUS, *Gray*.

Derby's Wallaby.

Halmaturus Derbianus, Gray in Mag. Nat. Hist., vol. i. new ser. p. 583.—Gould, Mon. of Macropodidæ, pl.
    —Gray, List of Spec. of Mamm. in Coll. Brit. Mus., p. 91.
Macropus (Halmaturus) Derbianus, Waterh. in Jard. Nat. Lib. *Marsupialia*, p. 234, pl. 21.—Ib. Nat. Hist. of
    Mamm., vol. i. p. 154.
Halmaturus Houtmanni, Gould in Proc. of Zool. Soc., pt. xii. p. 31 ?.—Waterh. Nat. Hist. of Mamm., vol. i. p. 156 ?.
——— *Eugenii*, Less. Man. de Mamm., p. 227 ?.
——— *Emiliæ*, Gray, List of Spec. of Mamm. in Coll. Brit. Mus., p. 90 ?.
Bangap, Aborigines of Perth in Western Australia.

ALTHOUGH the name of *Derbianus* is retained for this small species of *Halmaturus*, I am by no means certain that it has any claims to priority; in all probability the older name of *Eugenii* had reference to this animal; Mr. Waterhouse also is of opinion that an animal which I have called *Houtmanni* is merely a variety of the same species. Before me at this moment, while writing the present article, is my type specimen of *Houtmanni* from Wallaby Island, Houtmann's Abrolhos, and two specimens of *Derbianus* from Garden Island lying about five miles off the mouth of Swan River; now the former certainly differs from the latter in being of a darker colour, in having less rufous on the shoulders and rump, and in having stouter legs and feet; notwithstanding I bow to Mr. Waterhouse's opinion, and regard them as local varieties of one and the same species; and I incline to do so the more readily from feeling convinced, after having for a series of years paid considerable attention to these and other nearly allied species, that there is an animal of this family peculiar to the scrubby islands lying off the southern and western coasts of Australia, and one only, and that that one is the species under consideration, whatever its specific name may be. Up to the present time I have never seen examples from the mainland, the brushes of which lying between the mountain ranges and the coast are all tenanted by their own peculiar species, such as *Thetidis*, *Dama*, &c., whilst the Brigaloe brush of the interior has also an animal of this section peculiar to it—the *H. dorsalis*. The *H. Derbianus*, then, inhabits all the islands lying off the west coast, and extends round to those of the south-west as far as Kangaroo Island in Spencer's Gulf, where it is abundant.

Like many others of the small Wallabies, the present species loves to dwell among the densest underwood: hence the almost impenetrable scrub of dwarf *Eucalypti*, which covers nearly the whole of Kangaroo Island, will always afford it a secure asylum, from which in all probability it will never be extirpated,—the vegetation being too green and humid to be burnt, and the land too poor to render it worth the expense of clearing. It is very abundant in the ravines and gullies, through which it makes innumerable runs; and such is the dense nature of the vegetation, that nothing larger than a dog can follow it; still it is taken by men residing on the island in the greatest abundance, both for the sake of its skin and its flesh: they procure it principally by snares, a simple noose placed on the outskirts of the brush; but they also shoot it when it appears on the open glades at night.

Considerable difference exists in this, as well as in the other allied species, in the colour of the hair, which varies very much, not only in the intensity of its hue, but also in being much redder in some specimens than in others.

Fur long and moderately soft; face grizzled grey, reddish and dark brown; on the upper lip a buffy-white mark which extends backwards under the eye, and blends with the general colour of the face; back of the neck, shoulders and arms rufous; a blackish mark commences at the occiput, and continues downwards until it becomes lost in the colouring of the back, which is grizzled black and dull white, caused by the middle portion of each hair being dull white, and the tips black, the base of all the fur being deep-blue grey; rump, base of the tail, hind legs and tarsi grizzled with rufous and black, the former colour predominating; throat, chest and all the under surface buffy white; arms the same as the tarsi, but rather darker; under side of the tail buff.

	Male.	
	feet.	inches.
Length from the nose to the extremity of the tail	3	1
,, of tail	1	1¼
,, ,, tarsus and toes, including the nail		6
,, ,, arm and hand, including the nail		4¼
,, ,, face from the tip of the nose to the base of the ear . . .		4¼
,, ,, ear		2¼

One of the accompanying Plates represents the head and fore quarters of the natural size, the other the entire animal much reduced.

HALMATURUS THETIDIS, F. Cuv et Geoff.

HALMATURUS THETIDIS, *F. Cuv. et Geoff.*

Pademelon Wallaby.

Head and Fore Parts, of the size of nature.

The accompanying life-sized head represents the common Pademelon of the colonists, a shy and timid creature, which bounds away on the least disturbance. It runs in the same brushes with the *H. Ualabatus* and *H. Parma*; and every extensive district of this kind, from Illawarra to the Clarence, was tenanted by great numbers of it at the period of my visit to New South Wales. Its flesh is good, and is frequently eaten by the settlers and the aborigines. Considerable difference occurs in the colouring of this animal,— specimens obtained in one locality having the red hue of the neck predominating over the brown, while in those from another the contrary is the case.

The front figure represents a moderate-sized adult male, of the size of life, while the distant figure of the female is a trifle less. A more lengthened description, and reduced figures of the entire animal, will be found on the next Plate and page.

HALMATURUS TENUIS. _Gould & H.C. Richter_

HALMATURUS THETIDIS, *F. Cuv. & Geoff.*

Pademelon Wallaby.

Halmaturus Thetidis, F. Cuv. et Geoff. Mamm., tab. 56.—F. Cuv. Less. Zool. de M. Bougainville's Journ. de la
    Navig. autour du Monde de la Frigate Thetis, &c., tom. ii. p. 305. pl. 37.—Gould, Mon. of Macro-
    podidæ, pl.
——————— *nuchalis,* Wagn. in Schreb. Saug. Suppl., part 111-112. p. 128 (Waterhouse).
Macropus (Halmaturus) Thetidis, Waterh. Nat. Hist. of Mamm., vol. i. p. 144.
Pademelon of the colonists of New South Wales.

OF the smaller species of Wallaby inhabiting New South Wales, the present is perhaps the one best known to the colonists, inasmuch as it is more abundant than any other. It is strictly a brush animal, and consequently only to be found in such localities. All the brushes I have visited from Illawarra to the Hunter, as well as those of the great range which stretches along parallel with the coast, are equally favoured with its presence; I have also received specimens from Moreton Bay. It is not unfrequently found running in the same locality, and even in company, with the *H. Ualabatus,* although the very humid parts of the forest appear to be less suited to it than to that species.

As an article of food, few animals are so valuable, its flesh being tender and well-flavoured, and more like that of the common Hare than that of any other European animal I can compare it with.

The sexes are precisely alike in colour, but the female is smaller than the male.

The species appears to have been first brought to Europe by the French navigators, who applied to it the inappropriate term of *Thetidis* (after their vessel), which, however, it would not be right to alter. Having seen the original specimen in Paris, which is said to have been brought from Port Jackson, I am satisfied of its identity with my own specimens. I mention this circumstance, particularly as the name of *Thetidis* has been placed as a synonym of *Eugenii,* an animal brought home by Peron, and which I now believe to be identical with *H. Derbyanus.*

The *H. Thetidis* must be classed among the smaller *Halmaturi,* being scarcely so large as *H. Derbyanus* or *H. Billardierii.*

Fur rather soft; general colour deep brown; shoulders, sides and back of the neck rusty-red; ears furnished internally with moderately long dirty-white hairs; upper lip dirty-white; chin and throat white; remainder of the under surface dirty-white; arms greyish; hands brown; tarsi and feet uniform dark brown; tail brownish-grey above and dirty-white beneath; on the sides of the tail the hairs are scanty, and the scales covering the tail are very apparent.

	Male. feet.	inches.
Length from the tip of the nose to the extremity of the tail . . .	3	0
,, of tail	1	0
,, ,, tarsus and toes, including the nail		$5\frac{3}{4}$
,, ,, arm and hand, including the nails		5
,, ,, face from the tip of the nose to the base of the ear . .		$4\frac{1}{4}$
,, ,, ear		2

The accompanying Plates represent the head and shoulders of the size of life, and reduced figures of the entire animal.

HALMATURUS STIGMATICUS, *Gould.*

Gould and H.C.Richter del et lith.   Walter and Cohn, Imp.

HALMATURUS STIGMATICUS, *Gould.*

Branded Wallaby.

HEAD AND FORE QUARTERS, OF THE SIZE OF NATURE.

IT will be seen that this new species has the facial aspect of the *Halmaturus Thetidis* and *H. Billardieri*; and it does, in fact, belong to the same section as those Brush Kangaroos, but, as a species, it is doubtless quite distinct. In size it exceeds both the above-mentioned animals, and differs, in its rich red colouring, from all the other members of the genus. At present only a single specimen has reached this country: its capture was effected under somewhat singular circumstances, and tends to prove the probability that other species of this great group of Marsupials will yet be discovered when the naturalist has an opportunity of exploring the extensive forests of the north-eastern coast of Australia, which at present is impossible, or not to be done without great risk of encountering the treacherous aborigines.

Full details respecting this species will be found in the letter-press accompanying the following Illustration, on a reduced scale, of the entire animal.

HALMATURUS STIGMATICUS, *Gould*

HALMATURUS STIGMATICUS, *Gould*.

Branded Wallaby.

Halmaturus stigmaticus, Gould in Proc. of Zool. Soc., Nov. 13, 1860.

A SINGLE and very fine specimen of this new *Halmaturus* was obtained by Mr. John Macgillivray at Point Cooper, on the north-east coast of Australia, in the month of June 1848; this specimen is now deposited in the British Museum collection. I cannot refer this animal to any described species, but I observe that it is very nearly allied to the *Halmaturus Thetidis*. Now it is well known that this latter animal is strictly an inhabitant of the humid brushes of the south-eastern coast, and that it never leaves them for either the drier hills or the adjacent plains; and the present species may be regarded as its representative on the north-eastern coast, which is, I believe, clothed with brushes of a similar character.

The *Halmaturus Thetidis* and the *H. stigmaticus* are very similar in the smallness of their heads, the comparative shortness of their ears, and their adpressed, short stiff fur; but the latter differs from the former in being of a somewhat larger size and in the more rufous colouring of its fur (particularly of that clothing the legs), and in having a broad brand-like mark of buff on each haunch; similar marks, it is true, exist in some other species of Kangaroo, but in none of them is it so conspicuous as in the animal under consideration; hence the specific name I have assigned to it.

In the 'Voyage of H.M.S. Rattlesnake,' vol. i. p. 92, Mr. Macgillivray says:—"Near this place, while tacking in-shore, a native dog was seen by Lieut. Simpson in chase of a small Kangaroo, which, on being close pressed, plunged into the water and swam out to sea, when it was picked up by the boat, leaving its pursuer standing on a rock, gazing wistfully at its intended prey, until a musket-ball, which went very near its mark, sent it off at a trot. The Kangaroo lived on board for a few days, and proved to constitute quite a new kind, closely allied to *Halmaturus Thetidis*."

Face, sides of the body, outer side of the fore limbs, and the flanks rufous, more or less interspersed with whitish, the tips of the hairs being of that hue, and their middle portion rufous; outer side of the hinder limbs rich rusty red; occiput dark brown, interspersed with silvery-tipped hairs; ears clothed with long black hairs externally, and narrowly fringed with white on the front edge; all the upper surface of the body blackish brown, interspersed with numerous whitish-tipped hairs, gradually blending with the rufous hue of the sides and flanks; down the back of the neck an indistinct line of a darker or blackish hue; across each haunch a broad and conspicuous mark of buff; upper lip, chin, all the under surface of the body, and the inner side of the limbs dirty white; hands and feet dark brown; upper surface of the tail dark brown; on the sides the hairs are less numerous, and the scaly character of the skin becomes conspicuous.

		feet.	inches.
Length from the tip of the nose to the extremity of the tail	. . . . .	3	4
„ of the tail	. . . . .	1	4
„ of tarsus and toes, including the nail			$5\frac{3}{4}$
„ of arm and hand, including the nails	.·		$6\frac{1}{4}$
„ of face from the tip of the nose to the base of the ear	. . . .		$4\frac{3}{4}$
„ of ear			$1\frac{7}{8}$

The accompanying Plates represent the head of the natural size, and a reduced figure of the entire animal.

HALMATURUS BILLARDIERI.

J. Gould and H.C. Richter, del et lith.   Hullmandel & Walton, Imp.

HALMATURUS BILLARDIERI.

Tasmanian Wallaby.

HEAD AND FORE PARTS, OF THE SIZE OF LIFE.

As the Rabbit is to us one of the commonest and most numerous of our native quadrupeds, so is the Tasmanian Wallaby to the colonists of Van Diemen's Land. Exceeding a Hare in size, this useful animal is most numerous in all the scrubby and humid situations of the island. Its physiognomy, which is striking and singular, is well portrayed in the accompanying illustration, while the reduced figures will give a just idea of the entire animal. It will be seen that this species is much darker in colour than most of its allies, and that its coat is longer and more shaggy—a character of fur which is well adapted to its more southern, wetter, and colder climate, while its hue is in unison with that of the herbage amidst which it dwells. The interior of the forest, amid stranded trees and rank vegetation, are the situations in which this animal forms its runs, and from which it is not easily driven; but for these and for all other details respecting the species the reader is referred to the page accompanying the reduced figures.

HALMATURUS BILLARDIERII.

HALMATURUS BILLARDIERI.

Tasmanian Wallaby.

Kangurus Billardierii, Desm. Mamm., Suppl. p. 542.

Macropus (Halmaturus) rufiventer, Ogilby in Proc. of Zool. Soc., part vi. (Feb. 1838) p. 23; and in Ann. of Nat. Hist. for May 1838, vol. i. p. 220.

Halmaturus (Thylogale) Tasmanei, Gray in Ann. of Nat. Hist. for April 1838, vol. i. p. 108.

Macropus (Halmaturus) Billardieri, Waterh. in Jard. Nat. Lib., *Marsupialia*, p. 227.

Halmaturus Billardieri, Gould, Mon. of Macropodidæ, pl. .—Waterh. Nat. Hist. of Mamm., vol. i. p. 159.—Gray, List of Spec. of Mamm. in Coll. Brit. Mus., p. 90.

———— *brachytarsus*, Wagn. Schreb. Saug., Nos. 111, 112. p. 121, November 30, 1842.

Wallaby, Colonists of Van Diemen's Land.

I HAVE but little doubt that the habitat of this Wallaby is limited to Van Diemen's Land and the larger islands in Bass's Straits, in all which localities it is so numerous that the thousands annually destroyed make no apparent diminution of its numbers. In consequence of the more southerly and therefore colder latitude of Van Diemen's Land, the vegetation is there much more dense and humid than on the continent of Australia; indeed the sun never penetrates into many parts of its forests, and accordingly we find this species clothed with a warmer and more sombre-coloured coat. It is consequently of a more hardy nature than any of its congeners, and with care and a slight degree of perseverance it might be easily naturalized in England; indeed I feel confident that if a sufficient number were introduced in a suitable locality, as in some of our forests and large estates of the nobility and gentry, the experiment would be attended with complete success. Independently of the novelty of a species of this singular tribe ranging at liberty in our woods, its flesh could not fail to be highly esteemed for the table. Being one of the best-flavoured of the small Kangaroos, it is very generally eaten in Van Diemen's Land.

The Tasmanian Wallaby may be regarded as strictly gregarious, hundreds generally inhabiting the same localities; the situations which it frequents are gullies, and the more dense and humid parts of the forest, particularly those that are covered with rank high grass, through and under which it forms numerous well-beaten tracks. From these coverts it seldom emerges, and never even approaches the outskirts of the forest except at night: hence it is seldom seen by ordinary observers. It is very easily taken with snares, formed of a noose placed in its run; and thousands are captured in this way, solely for their skins: the sportsman also may readily procure it by stationing himself in some open glade of limited extent, accompanied by two or three small yelping dogs, before which it keeps hopping round and round, and thus affords him an opportunity of shooting it as it passes; for, like the common rabbit, it never quits the locality in which it is bred.

Much diversity of colour is observable in different specimens, some having the throat and under surface deep reddish buff, while others have the same parts much lighter. Its usual weight is from fifteen to twenty pounds, although many are smaller.

This species is readily distinguished from the other small members of the group by its short ears, long, dark-coloured fur, and the rufous or yellow tint of the under surface of the body.

Fur very thick, the hairs blue-grey at the base, buffy brown in the middle, the tips, which are much produced, ending in black; face and all the upper surface very dark brown, approaching to black, particularly on the shoulders and back, where the hairs become much lengthened; arms and tarsi greyish brown; lips, throat, chest and under surface reddish buff; in some specimens these parts are grey tinged with buff; ears dark brown tinged with buff; upper side of the tail dark brown, under side dirty white.

		feet.	inches.
Length from the nose to the extremity of the tail	. . . . . .	3	6
,, of tail.	. . . . .	1	1
,, of tarsus and toes, including the nail	. . . .		6
,, of arm and hand, including the nails	. . . .		5¼
,, of face from the tip of the nose to the base of the ear	. .		4¼
,, of ear	. . . .	. . . .		2¼

The accompanying Plates represent the head of the natural size, and reduced figures of the entire animals.

HALMATURUS BRACHYURUS.

J.Gould and H.C. Richter, del. et lith.

Hullmandel & Walton, Imp.

HALMATURUS BRACHYURUS.

Short-tailed Wallaby.

Kangurus brachyurus, Quoy et Gaim. Voy. de l'Astrolabe, Zoologie, tom. i. p. 114. pl. 19.
Halmaturus (Thylogale) brevicaudatus, Gray, List of Mamm. in Coll. Brit. Mus. p. 90.
Macropus (Halmaturus) brachyurus, Waterh. Nat. Hist. of Mamm. vol. i. p. 162.
Ban-gup, Aborigines around Perth in Western Australia.
Quăk-a, Aborigines of King George's Sound.

BEFORE my visit to Australia, this animal was extremely rare in the collections of Europe; indeed the example in the Paris Museum was the only one then known. The specimen alluded to was said to have been picked up dead at King George's Sound, and there also my specimens were procured. Even now it is still a rare animal, those examples introduced by myself being, so far as I am aware, all that have been transmitted to Europe.

In his notes respecting this species, Mr. Gilbert states that besides meeting with it at King George's Sound, he found it abundant in all the swampy tracts which skirt nearly the whole of Western Australia at a short distance from the sea, and that at Augusta, where its native name, Quăk-a, is the same as at King George's Sound, it inhabits the thickets and is destroyed in great numbers at the close of the season by the natives, who, after firing the bush, place themselves in a clear space and spear them as they attempt to escape from the fire: it is also caught by the settlers with springes placed in their little covered runs beneath the scrub. Mr. Gilbert adds, that he had not heard of its being killed to the eastward of the Darling range.

Mr. Waterhouse has given the relative admeasurements of the Paris specimen, and of an example in the British Museum which had been procured by Mr. Gilbert; the latter is considerably smaller than the former; but I have since received a specimen from the same locality which considerably exceeds both in size, its admeasurements being as follows:—

Length from the nose to the root of the tail . .	1 foot 10 inches.	
„ of the tail	10	„
„ „ tarsus, toes, and nails	4¼	„
„ „ ear	1¼	„

This animal differs from all the other *Halmaturi* in its short bluff head, diminutive ears, and extremely short tail; it is also clothed, especially about the face, with thick, stiff, and wiry hairs; which, combined with the general character of the fur, would lead to the inference that it resorts to more humid and secluded situations than those frequented by the other members of the genus.

The short and rounded ears, which are much hidden by the long fur of the head, are well clothed with hairs, those on the inner side being yellow, while externally they are of the reddish-brown tint which pervades the head and back of the neck, but which is somewhat brighter in the region of the ears; the hairs of the back are grey next the skin, broadly annulated with yellow towards the point, and black at the extremity; the back is also beset with numerous long, interspersed, almost entirely black hairs, which, being most plentiful in the middle of the back, give that part a deeper hue; the hairs of the sides of the body are similar, but the yellow portion is paler and the tips are brownish; on the under surface the hairs are grey next the skin, with a pale yellow external tint; feet deep brown; tail sparingly clothed with small stiff hairs, between which rings of small blackish scales are very perceptible.

Of this rare species I have given two illustrations; one representing the entire animal, much reduced, and the other, the head, tail, and foot, of the natural size.

HALMATURUS BRACHYURUS.

HALMATURUS BRACHYURUS.

Short-tailed Wallaby.

HEAD, ARMS, HIND FOOT, AND TAIL, OF THE SIZE OF NATURE.

THE most remarkable feature in the zoology of Australia is, undoubtedly, the great number of the Kangaroos, and the diversity of their characters, some being conspicuous for their great size, others for their banded or crescentic markings, and others again for their sombre hues and their diminutive sizes. Of that section of the family to which the generic appellation of *Halmaturus* has been assigned, the Short-tailed Wallaby is the smallest. Its nearest ally is the *H. Billardieri*, to which it assimilates not only in the shortness of its ears and the shaggy character of its fur, but in its still more sombre hues, which latter feature indicates that it dwells among grassy and dense herbage, in swampy and humid situations.

The *H. brachyurus* is a native of Western Australia, the *H. Thetidis* of New South Wales, and the *H. Billardieri* of islands of Tasmania and Bass's Straits; and thus we find these little Wallabys distributed along the whole of the south coast, from east to west. The exact localities frequented by these animals will be found in the pages accompanying the entire representations of each of them.

PETROGALE PENICILLATA, *Gray.*

PETROGALE PENICILLATA, *Gray*.

Brush-tailed Rock-Wallaby.

UPPER HALF OF THE ANIMAL, OF THE SIZE OF LIFE.

IT is not a little interesting to observe how varied are the forms of the various species of Kangaroos, and how well each is suited to the physical conditions of that great southern land of our antipodes, Australia,—the plains, the forests, the rocks, and the trees, each being tenanted by members of this extensive family. Of these the Rock-Wallabys constitute a well-defined section, the species of which are active in the extreme among the haunts they affect.

The *P. penicillata*, the Brush-tailed Rock-Wallaby of the colonists, is an inhabitant of New South Wales. A more detailed account of the situations it frequents will be found in the page accompanying the Plate with the reduced figures.

PETROGALE PENICILLATA. Gray

PETROGALE PENICILLATA, *Gray*.

Brush-tailed Rock Wallaby.

Kangurus pencillatus, Gray, in Griff. Anim. Kingd., vol. iii. pl. opp. p. 49.

Macropus penicillatus, Benn. in Proc. Zool. Soc., part iii. p. 1.—Waterh. in Jard. Nat. Lib. Mamm., vol. xi. (Marsupialia) p. 243. pl. 22.—Benn. Cat. of Australian Museum, Sydney, p. 6. no. 27.

Kangurus Pencillatus, Gray, in Griff. Anim. Kingd., vol. v. p. 204.

Petrogale penicillatus, Gray, in Mag. Nat. Hist., vol. i. new ser. p. 583.

Heteropus albogularis, Jourd. Compte rendu de l'Acad. des Sci. Oct. 9, 1837, p. 552 ?, et Ann. des Sci. Nat. 1837, tom. viii. p. 368 ?

Petrogale penicillata, Gould, Mon. of Macropodidæ.—Gray, List of Mamm. in Coll. Brit. Mus., p. 92.

Macropus (Heteropus) penicillatus, Waterh. Nat. Hist. Mamm., vol. i. p. 167. pl. 1.

THE Colony of New South Wales, or the south-eastern portion of Australia, is the native habitat of the *Petrogale penicillata*; it must not, however, be understood that it is universally dispersed over this part of the continent, for the situations it affects and for which its structure is especially adapted are very peculiar, and do not occur in all parts of the colony; those portions of the mountain ranges stretching along the eastern coast from Port Philip to Moreton Bay, the character of which is rocky and precipitous, are among the localities in which it is to be found; hills of a lower elevation than those of the great ranges, and the precipitous stony gullies between the mountains and the sea, are also situations it inhabits; my own specimens were collected in various parts of the Upper Hunter district, both on the Liverpool Range and on the low hills which spur out in a southerly direction. Agile and monkey-like in its actions, few animals are more active among their native rocks; it readily evades the pursuit of the Dingo or native dog (*Canis Dingo*) by leaping from one rocky ledge to another, until, arriving at the edge of the cliff, it is secure from its attacks; it also ascends trees with facility, particularly those the half-prostrate position of which offer it a ready means of ascent; but it more particularly loves to dwell among rocks abounding with deep and cavernous recesses, into which it plunges on the slightest apprehension of danger, when both the natives and its natural enemy the Dingo are generally foiled; at the mouths of these caverns, and for a considerable distance down the hill-sides, regular, hard, well-beaten tracks are formed, which, on the one hand, serve to facilitate the retreat of the animal to its secure asylum, while, on the other, they indicate its proximity. I have used the words "monkey-like" when speaking of its actions; and to show that they appeared as such to others as well as to myself, I may mention that in a note by Capt. Sir Edward W. Parry, R.N., published in the part of the Proceedings of the Zoological Society above referred to, it is stated that "the first intimation received of these animals was that monkeys were to be seen in a particular situation; and the manner in which they jumped about when a number of them were approached left that impression on the mind. They were so wild that it was impossible on the first attempt to obtain a specimen, and one which was wounded escaped into its hole." Sir Edward adds, "As several were seen together on more than one occasion, they appear to be gregarious." It must be regarded rather as a local animal than otherwise, as it is never to be found but in districts similar to those described. Although strictly nocturnal in its habits, individuals may frequently be seen during the day sunning themselves on the face of a rock or on half-prostrate trees. At such times they may be easily crept upon and shot; it was in this way that I procured numerous specimens for my own collection.

Several examples of this species have from time to time lived in the Gardens of the Zoological Society in the Regent's Park, and when placed in a proper inclosure displayed all the actions and attitudes they assume in their native wilds. A fine male now (October 1853) living in the Society's Menagerie formed the model from which Mr. Richter took the correct delineation given on the accompanying Plate; it was an excellent sitter, for it remained perched on the stem of a large tree for hours together. Great diversity of colouring occurs in different individuals, some being much darker than others; again, some have the breast and under surface rich rust-red, while in others the same parts are of a much paler hue, or inclining to buff.

The *Petrogale penicillata* may be regarded as the largest species of the well-defined genus to which it belongs, its entire length from the nose to the end of the tail being three feet ten inches, the tail measuring twenty-three inches ; the arm, hand and nails six inches ; the tarsus, toes and nail six inches and a half ; the face from the tip of the nose to the ear four and a half inches, and the ear two inches. Of its flesh as an article of food I can speak most highly, having frequently partaken of it in the bush and always found it excellent.

The fur is long and rather harsh to the touch ; its general colour is a dusky brown tinged with purple, passing into deep rusty red on the rump, the base of the tail, the hinder part of the thighs and the abdomen ; face dark grey ; along the face from the lip to the ear a dusky white mark ; a narrow dark line runs from the middle of the forehead nearly half way down the back ; shoulders and flanks vinous grey, separated from the general tint of the upper surface by an indistinct line of a lighter tint, scarcely to be distinguished in some specimens ; a narrow white line passes from the throat down the centre of the chest ; ears black, passing into grey at the base, and having in some specimens a band of rufous along the outer edge ; arms and hands, tarsi and feet rusty red, deepening into black on their extremities.

The two sexes when adult are nearly alike in size and similar in colour.

PETROGALE LATERALIS, Gould.

PETROGALE LATERALIS, *Gould.*

Stripe-sided Rock-Wallaby.

UPPER HALF OF THE ANIMAL, OF THE SIZE OF LIFE, AND A REDUCED FIGURE
IN THE DISTANCE.

THIS is the West Australian representative of the *P. penicillata* of New South Wales; in size it is somewhat smaller than that animal, but its markings are more strongly defined. In the colony of Swan River, rocky districts alone are the places of its resort. In their dispositions and general economy the *P. lateralis* and *P. penicillata* are very similar. I always observed that the furry coat of the former is thicker or more dense than that of the latter; this difference, however, cannot be depicted, but is readily seen when skins of the two animals are seen side by side.

The reader is referred to the page accompanying the succeeding Plate for a more detailed account of this species.

PETROGALE LATERALIS. _Gould._

PETROGALE LATERALIS, *Gould.*

Striped-sided Rock Wallaby.

Petrogale lateralis, Gould, Mon. of Macropodidæ, pl.
Macropus (Heteropus) lateralis, Waterh. Nat. Hist. of Mamm., vol. i. p. 172.
Moö-roo-rong, Aborigines of the Perth and Toodyay districts of Western Australia.

THIS conspicuously marked species is very abundant in all the rocky districts of Swan River, particularly the Toodyay, and I have little doubt that the whole of the line of coast of Western Australia will hereafter be found to be inhabited by it, wherever the character of the country is suitable. Independently of the difference in its markings and the more woolly texture of its fur, it is a much more diminutive animal than the *P. penicillata*; the crania of the two animals also exhibit sufficient differences to satisfy the most sceptical mind of their being specifically distinct; in their disposition and economy, however, but little variation is found to exist.

Mr. Gilbert states that "the *Petrogale lateralis* is only to be met with in the rocky parts of the interior intersected with caverns. It is a remarkably shy and wary animal, feeding only at night in little open patches of grass, and never, from all that I have been able to observe, going more than two or three hundred yards from its rocky retreats. When alarmed, it leaps most extraordinary distances from rock to rock and point to point with the utmost rapidity. When running along a level surface, its tail is very much curved upwards like that of a greyhound, and the best way to procure specimens is to walk over the rocks without shoes, and station yourself within gunshot distance of the principal entrance to their caverns, when, on making their appearance in the middle of the day for the purpose of sunning themselves, they are easily shot."

Fur shorter and much softer than that of *P. penicillata*; general colour reddish-brown, passing into silvery-grey on the neck and shoulders; basal half of the tail brownish-grey, the remainder black, with a brush at the end; face greyish-brown; a distinct white mark from the tip to the base of the ear; a black mark between the ears, extending in a distinct narrow line half-way down the back; ears dark brown, becoming of a light sandy colour at the base; a deep rich brown mark extends from behind the shoulders, down the back of the arm, along the flanks and down the inside of the thigh; this mark is separated from the general colour of the back by a very distinct stripe of white; chin, throat, chest and abdomen sandy-red; under sides of the neck grey; arms light sandy-red, passing into black on the hands; tarsi reddish brown, passing into blackish-brown on the toes.

	Male.	
	feet.	inches.
Length from the tip of the nose to the extremity of the tail	3	5
„ of the tail	1	5
„ „ tarsus and toes, including the nail		5¼
„ „ arms and hands, including the nails		4¾
„ „ face from the tip of the nose to the base of the ear		4
„ „ ear		1⅞

The first Plate represents the head of the size of life; the second, reduced figures of the entire animal.

PETROGALE XANTHOPUS, _Gray._

PETROGALE XANTHOPUS, *Gould.*

Yellow-footed Rock-Wallaby.

A SINGLE FIGURE OF THE ENTIRE ANIMAL, OF THE SIZE OF LIFE.

Two reasons have induced me to give double figures of this animal—one to show its peculiar brushy tail and richly coloured ears, of the size of life, while, in the scenery accompanying the reduced figures, I have endeavoured to portray the kind of country inhabited by this new and very fine species. It was one of the last discoveries made by one who sacrificed his life in the pursuit of natural history on the east coast of Australia; and it would have been well if the name of Frederick Strange had been associated with the species. Dr. Gray has, however, seized upon a good specific character in the name of *xanthopus*, which will for ever serve to distinguish this fine species of the genus *Petrogale*.

PETROGALE XANTHOPUS, Gray.

PETROGALE XANTHOPUS, *Gray*.

Yellow-footed Rock Wallaby.

Petrogale xanthopus, Gray in Proc. of Zool. Soc., Nov. 14, 1854.

ALL that is known respecting this fine animal is, that two examples, a male and a female, were collected on Flinders' Range in South Australia, and sent to this country by Mr. Strange, and that they were subsequently purchased by Dr. Gray for the British Museum collection, wherein they are now deposited.

The *Petrogale xanthopus* is a typical example of the genus to which it belongs, and may be regarded as one of the finest species of the form yet discovered. Its large size and rich colouring render it very conspicuous, while the buffy hue of the ears and legs at once distinguishes it from the whole of its congeners.

The habits, actions and economy of the *Petrogale xanthopus* are doubtless as similar to those of the other members of the genus as it is like them in form, but on these points nothing is at present known.

Fur long, soft, and yielding to the touch; face, head, and all the upper surface vinous brown-grey, becoming greyest on the rump; a narrow line of dark rich brown extends from the crown of the head down the centre of the back; on each cheek a distinct mark of white; eyelashes full, prominent, and brownish-black; behind each arm a large patch of reddish-brown, separated from the general tint of the upper surface by a streak of buffy-white; ears ochre-yellow, becoming lighter at the base, fringed internally with white, and tipped externally with brown; front of the arms bright buff; hands rich dark brown; outer side of the legs light ochreous-brown, fading into white on the inner side, and passing into the rich dark brown of the toes; throat and under surface white; tail ochreous-brown, irregularly barred with a darker tint, and ending in a conspicuous tuft which is rich brown above and ochreous below.

Of so fine a species I have considered it desirable to give two illustrations,—the entire animal, necessarily much reduced, and a foreshortened figure of the size of life. Nor must I omit to call attention to the interest which would attach to the introduction of living examples to our menageries, and to the acquisition of additional examples for our museums.

PETROGALE INORNATA, Gould.

J Gould and H.C. Richter, del. et lith.   Hullmandel & Walton, Imp.

PETROGALE INORNATA, *Gould.*

Unadorned Rock-Wallaby.

UPPER HALF OF THE ANIMAL, OF THE SIZE OF LIFE.

THE northern as well as the southern portions of the Australian continent are evidently tenanted by members of that section of the Kangaroos to which the present species belongs; for although a single example only has yet reached me from the north coast, it is sufficient to show that such is the case.

The *P. inornata* is a true *Petrogale*, and, like the *P. concinna*, merely differs from its congeners in the total absence of stripes or markings on its sides.

This plain-coloured but rare species is one I would recommend to the notice of the naturalists and explorers who may visit the north coast of Australia, where it was discovered by Mr. Bynoe.

A very reduced figure of the entire animal will be found on the next Plate.

PETROGALE INORNATA, Gould.

PETROGALE INORNATA, *Gould.*

Unadorned Rock Wallaby.

Petrogale inornata, Gould in Proc. of Zool. Soc., part x. p. 5.—Ib. Mon. of Macropodidæ, pl. .—Gray, List of
Spec. of Mamm. in Coll. Brit. Mus., p. 92.
Macropus (Heteropus) inornatus, Waterh. Nat. Hist. of Mamm., vol. i. p. 175.

THIS new species, for which I am indebted to the kindness of B. Bynoe, Esq., of H.M.S. Beagle, differs
from all the other members of the genus in the unusual uniformity of its colouring. Mr. Bynoe collected
it on the north coast of Australia. In size it is about equal to the *P. lateralis* of the western coast, to
which, as also to *P. penicillata*, it is very nearly allied, but differs from them both in being destitute of
any markings on the sides, in the absence of any dark colouring behind the ears, and in the light colouring
of the arms and tarsi.

During the interval of nearly twenty years which has elapsed since I first characterized this animal, no
additional examples of this, or any other mammal of the rarely visited part of Australia it inhabits, have
reached this country; but when the north coast of Australia shall have been thrown open to the settler,
it will doubtless be found that the *Petrogale inornata* is as abundant in the rocky districts of that part
of the country as the *P. penicillata* is in the brushes of New South Wales.

General colour of the upper parts sandy grey, grizzled over the shoulders, and becoming much lighter
on the flanks; an indistinct line of a lighter hue along the face under the eye; a dusky red patch behind
the elbow; under surface sandy white, inclining to rufous on the lower part of the abdomen; arms and
tarsi sandy grey, passing into dark brown at the extreme tips of the toes; basal half of the tail sandy brown,
the remainder black, the former colour extending along the sides of the tail for some distance towards
the tip; ears sandy grey, bordered by a very narrow line of dark brown on their inner edge; a dark patch
at the occiput, passing into a dark line down the forehead.

	Female.	
	feet.	inches.
Length from the tip of the nose to the extremity of the tail	3	2
„ of tail	1	3¼
„ of tarsus and toes, including the nail		5¼
„ of arm and hand, including the nails		5
„ of face, from the tip of the nose to the base of the ear		4¼
„ of ears		1⅞

The accompanying Plates represent the head of the natural size, and a reduced figure of the entire animal.

PETROGALE BRACHYOTIS, Gould.

J. Gould and H.C. Richter, del et lith.   Hullmandel & Walton, Imp.

PETROGALE BRACHYOTIS, *Gould.*

Short-eared Rock-Wallaby.

Macropus (Petrogale) brachyotis, Gould in Proc. of Zool. Soc., part viii. p. 127.—Ib. Mon. of Macropodidæ, pl. .
———— *brachiotis,* Waterh. in Jard. Nat. Lib. *Marsupialia,* p. 247.
———— *(Heteropus) brachiotis,* Waterh. Nat. Hist. of Mamm., vol. i. p. 176.
Petrogale brachyotis, Gray, List of Spec. of Mamm. in Coll. Brit. Mus., p. 92.

THE discovery of this species of Rock Kangaroo is due to the researches of His Excellency Sir George Grey, the present (1859) Governor of the Cape of Good Hope, who procured it on the North-west coast of Australia, near Hanover Bay, on the 29th of December, 1837. The two specimens, a male and a female, then obtained, which appear to be fully adult, are in the British Museum, and are at present unique. Sir George Grey states that the animal " is excessively wild and shy in its habits, frequenting in the day-time the highest and most inaccessible rocks, and only coming down to the valleys to feed early in the morning and late in the evening. When disturbed in the day-time it bounds among the roughest and most precipitous rocks, apparently with the greatest facility, and is so watchful and wary that it is by no means easy to get a shot at it. How it can support the excessive heat of the sand rocks amongst which it always lies is to me truly astonishing, the temperature there during the hottest part of the day being frequently 136°. I have never seen this animal on the low land or the plains, and I consequently believe it to be entirely an inhabitant of the mountains."

No other species of Rock Kangaroo has yet been discovered with such short and scanty hair as the *Petrogale brachyotis,* which scantiness of covering may be due to the great heat of the latitudes it inhabits, and the peculiar localities to which it resorts—hard craggy rocks exposed to the burning sun. In confirmation of this being a genuine species, Mr. Waterhouse remarks :—

" The Short-eared Rock Kangaroo is readily distinguished from the *penicillatus* and *lateralis* by the absence of the black band on the sides of the body, the only remains of this dark hue being confined to a patch immediately behind the base of the fore leg ; its general colour is paler, and the fur is much shorter ; the tail is less bushy ; its bulk is moreover inferior, and the proportionately small size of the ears is an important distinguishing character ; in its smaller size and in the reddish hue of the upper parts of the body, it approaches to *concinnus* ; but besides other differences, that animal does not possess any dark mark or spot on the sides of the body."

Fur short and rather close to the body ; general colour of the upper surface greyish brown, suffused with rust-colour ; under surface dirty yellowish white ; head pale brown, with a dirty white mark on each side ; cheeks almost white ; ears pale internally, dusky externally ; a rusty black patch on the body, immediately behind the base of the fore leg ; fore feet brown ; nails of the toes very short and scarcely projecting beyond the fleshy portion, which is extremely rough beneath ; tail moderately bushy, coloured at the base like the body, but the apical third dusky black.

	Male.	
	feet.	inches.
Length from the nose to the extremity of the tail	3	0
„ of tail	1	3
„ „ tarsus and toes, including the nail	0	5
„ „ arms and hands, including the nails	0	3¼
„ „ face from the tip of the nose to the base of the ear . . .	0	3¾
„ „ ear	0	1¼

The larger figure is about the size of life.

PETROGALE CONCINNA, Gould

J.Gould and H.C.Richter, del. et lith.

Hullmandel & Walton, Imp.

PETROGALE CONCINNA, *Gould.*

Little Rock Wallaby.

Petrogale concinna, Gould in Proc. of Zool. Soc., part x. p. 57.—Gray, List of Mamm. in Coll. Brit. Mus., p. 92.
Macropus (Heteropus) concinnus, Waterh. Nat. Hist. of Mamm., vol. i. p. 177.

MANY parts of Australia are even yet almost unknown both to travellers and naturalists; particularly the countries bordering its northern and north-western coasts—not to mention the distant interior; and in all these parts, numerous new species of quadrupeds, birds, and other classes of natural history are in my opinion to be discovered. The north-west coast has, it is true, been visited by the officers of H.M. Surveying Ship Beagle, and such species as fell in their way have been collected by them, but their official duties prevented them from giving that attention to the subject that could be desired. Nearly all they did collect proved to be new to science. The present interesting Little Rock Wallaby may be cited as a case in point, having been one of the specimens thus procured and brought home by Lieut. Emery, R.N. The single specimen obtained by this gentleman, and which is now in the British Museum, is fully adult, and is remarkable for its brilliant colouring and diminutive size. Mr. Waterhouse remark, that "it may be readily distinguished from its congeners, not only by its small size and bright. colouring, but by the absence of any black spot behind the base of the fore-leg."

Fur moderately long and somewhat soft to the touch; general colour bright rusty-red; head palish ash-colour, slightly suffused with rust-colour, which tint is most conspicuous above the eyes; cheeks rusty-white, with an indistinct greyish-brown mark extending forwards from the front of the eye; ears very pale brown externally, and lined with a few white hairs internally; fur on the back grey next the skin, and this tint at the root of each hair is followed by brilliant rusty-red, then a broad space which is white, and lastly, the tip is deep rusty-brown; on the under parts of the body the fur is grey next the skin, and yellowish-white on the visible portion; fore-legs rusty-white; hands brownish-white; hind-legs pale rust-colour externally; tarsi brownish-white, slightly pencilled with brown; on the back of the neck an indistinct trace of a mesial darker mark; tail clothed at the base with fur like that of the body; beyond this the hairs are of a harsher nature, at first about half an inch in length, and on the apical third about an inch and a half, of a brownish-white, tipped with black.

The accompanying Plate represents the animal about the natural size.

DENDROLAGUS URSINUS, *Müll.*

J. Gould and H.C. Richter, del. et lith.   Hullmandel & Walton, Imp.

DENDROLAGUS URSINUS, *Müll.*

Black Tree-Kangaroo.

Dendrolagus ursinus, Müll. Zoogd. van den Indischen Archipel., part iv. pl. 19; pl. 22. fig. 1, head ; pl. 23. figs. 1–3, and pl. 24. fig. 1, skull ; figs. 2 & 3, bones of hind-leg.—Gould, Mon. of Macropodidæ, pl.   .—
Waterh. Nat. Hist. Mamm., vol. i. p. 185.—Gray, List of Mamm. in Coll. Brit. Mus., p. 87.

As an evidence of how little we know of the productions of New Guinea, and of the infrequency of our intercourse with that country, I may state, that, although twelve years have elapsed since the publication of the second part of my "Monograph of the Macropodidæ, or Family of Kangaroos," I have not been able to obtain any information respecting the history and habits of this singular animal beyond the meagre account there given, the substance of which I here repeat.

Both the *Dendrolagus ursinus* and the *D. inustus* are natives of New Guinea, where they inhabit the trees, and feed upon the bark and leaves of the smaller branches, fruits and berries. They were discovered in Triton Bay by Dr. M. S. Müller, who states that they also frequent the interior of the country: in all probability they are generally dispersed over the forests of that *terra incognita*. What a field for enterprise here presents itself to the notice of the scientific explorer !

The specimen from which my former illustration was taken, and which was then in the Royal Museum at Leyden, now forms part of the fine collection at the British Museum ; the half-figure, of the size of life, on the accompanying Plate was also taken from the same example.

The following accurate description of this animal is transcribed from Mr. Waterhouse's "Natural History of the Mammalia," a work of great scientific value, which it is to be regretted has been discontinued for want of a due appreciation of its merits on the part of the public :—

" This animal has received the specific name of *ursinus*, no doubt on account of a certain superficial resemblance it has to a small Bear, arising in a great measure from the nature of its fur, which differs much from that of the ordinary Kangaroos, not only in being harsh and glossy, but in being composed of one kind of hair only ; it would appear that that kind of hair which forms the chief clothing in the ordinary Kangaroos is here entirely, or almost entirely, wanting ; and that the hairs representing the longer interspersed hairs in the fur of those animals, here forms the entire coat. With all the essential characters of the true Kangaroos, we find, in these tree-climbing animals, the limbs modified for their different mode of life : the long hind-legs of the Kangaroo proper are replaced by comparatively short legs, and the fore-legs are but little inferior in size to the posterior limbs ; the strong fore-feet are armed with stout and long claws, compressed and much curved, and fitted for clinging to the inequalities of the bark of the trees. The enormously long tail no doubt helps to balance the animal whilst on the branches of the lofty trees which it ascends in quest of food.

" On the upper parts and sides of the body, as well as the limbs (excepting at the base internally), the fur is black and glossy, and very nearly uniform to the skin, an indistinct brownish hue being only observable at the roots of the hairs ; tail densely clothed throughout, and black, but tinted with brown at the roots of the hairs ; ears densely clothed with very long hairs which completely conceal them ; the hairs springing from the top of the ears are brown, but the rest are black ; the head in front of the ears and the whole of the under parts of the body are brown, but varying in intensity in parts, being darker round the eye and on the muzzle, and yellowish on the cheeks ; the belly is also yellowish, whilst the chest assumes a deeper hue ; the muffle appears as if naked, but has in fact very minute hairs scattered over it."

DENDROLAGUS INUSTUS, *Mull.*

DENDROLAGUS INUSTUS, *Müll.*

Brown Tree-Kangaroo.

Dendrolagus inustus, Müll. Zoogd. van den Indischen Archipel., part iv. pl. 20 ; pl. 22. fig. 2, head ; pl. 23.
   figs. 4–6, and pl. 24. fig. 4, skull ; figs. 5 & 6, bones of hind-leg.—Gould, Mon. of Macropodidæ,
   pl.   .—Waterh. Nat. Hist. of Mamm., vol. i. p. 188.

SINCE the appearance of the second part of my " Monograph of the Macropodidæ, or Family of Kangaroos," in which I published a reduced figure of this animal, taken from a preserved specimen in the
Royal Museum at Leyden, a living example has been transmitted to the Gardens of the Zoological Society
of London, and lived there for some years.   In disposition it appeared to be more slothful than the terrestrial Kangaroos, as it spent the greater part of the day on the large branch of the tree placed in the cage
in which it was kept, and there it would sit for hours together in a moping, sleepy attitude, with its great
brush tail coiled round the front of its body ; at other times it was somewhat more active, and would then
sit erect, with the tail hanging down nearly straight, much after the manner of the Monkeys.

The *Dendrolagus inustus* is a native of New Guinea, where it was discovered by Dr. W. S. Müller in
Triton Bay.   The description of the habits of the animal, so long promised by this gentleman, has not yet,
I believe, appeared ; at least I am unable to find it in any of the great works on the Dutch possessions in
the Indian Archipelago, to which I have access.

The following note, respecting the living specimen above mentioned, occurs in Mr. Mitchell's " Popular
Guide to the Gardens of the Zoological Society of London," p. 58 :—

" The Tree-Kangaroo (*Dendrolagus inustus*) has only in one instance been brought alive to Europe.   This
specimen was presented to the Society by Lieut.-Col. Butterworth, C.B., Governor of Singapore.   The
beautiful modification of structure in the extremities, by which it is enabled to ascend the straightest palm-
trees, presents a most instructive contrast, when compared with the same organs in the Kangaroos, which
bound in leaps of twenty feet along the ground."

The accompanying drawing, which represents half of the animal of the size of life, was made by
Mr. H. C. Richter, from the living example in the Society's Gardens.   The entire figure, from the Leyden
specimen, is much reduced.

The following is Mr. Waterhouse's careful description of this species, which I transcribe rather than give
one of my own, as the animal mentioned was somewhat out of condition when it died :—

" This species is about the same size as *D. ursinus,* from which it differs not only in being of a brown
colour, but in having the muzzle and tarsi rather more elongated, and the ears less densely clothed with
fur : the hairs of the back do not so distinctly radiate from a point, rather behind the shoulders, as in
D. ursinus ; over the shoulders, however, the hairs are directed outwards, and on the back part of the neck
they are directed forwards, but are semi-erect, and those of the head are directed backwards.   The fur is
rather less harsh than in *D. ursinus* ; its general hue is deepish brown on the upper parts of the body,
but here each hair is brown at the base, shaded into brownish-black externally, whilst at the point they are
of a very pale brown inclining to white ; on the under parts of the body, the exposed portions of the hairs
are white, or very nearly so, but in the middle they are of a very pale brown, at the base still paler, and
nearly white in some parts ; the sides of the head are pale brown, and the upper surface dusky-brown ;
the muzzle is clothed with very short hairs ; the ears tolerably well clothed with longish hairs, brown on the
inner side, and dusky on the outer ; the limbs are brownish-white, but the hairs on these parts are brown
at the root ; the hinder part of the haunches and the under surface of the base of the tail are whitish ; the
tail is well clothed with longish harsh hairs, partly brownish-white and partly pale brown, the general hue
being paler than that of the body ; all the feet are dusky-brown, pencilled with whitish on the hinder
parts."

DORCOPSIS BRUNI.

J. Gould and H. C. Richter del. & lith.   Hullmandel & Walton, Imp.

DORCOPSIS BRUNI.

Filander.

Filander, Le Brun, Voy. par Muscovie, en Perse, et aux Ind. Orient., tom. i. p. 347. f. 213. 1718.—Ib. Edition of 1725, vol. v. p. 45. pl. at p. 43.

Didelphis Brunii, Schreb. Säug., tom. iii. p. 551. pl. 153.—Gmel. Edit. of Linn. Syst. Nat., tom. i. p. 109.—Shaw, Gen. Zool., vol. i. p. 480.—Quoy et Gaim. Voy. de l'Astrolabe, Zool., p. 116. pl. 20.

———— *Asiatica*, Pall. Act. Acad. Sci. Petrop. 1777, pt. 2. p. 229. tab. 9. figs. 4, 5.

Javan Opossum, Penn. Hist. of Quad., p. 305.

Halmaturus Brunii, Ill. Prod. Syst. Mamm. et Av., p. 80.

Macropus veterum, Less. Man. de Mamm., p. 227.

———— *Brunii*, Fischer, Syn. Mamm., p. 283.

Hypsiprymnus Bruni, Müll. Zoogd. der Indesch. Archipel., pt. 4. pl. 21; head, pl. 22. fig. 3; skull, pl. 23. figs. 7 and 8, and pl. 24. fig. 7; bones of hind leg, pl. 24. figs. 8, 9.

Halmaturus Asiaticus, Gray, List of Mamm. in Coll. Brit. Mus., p. 91.

Macropus Brunii, Waterh. Nat. Hist. of Mamm., vol. i. p. 180.

Dorcopsis Brunii, Müll. Verh. Zool. Mamm., p. 131. pl. 21.—Sclater in Journ. of Proc. of Linn. Soc., Zoology, vol. ii. p. 154.

BEING desirous of rendering my account of the Kangaroos as perfect as possible, I have considered it advisable to figure and describe in this work the species of that group of animals inhabiting New Guinea, in addition to those found in Australia and Van Diemen's Land. Independently of the two species of *Dendrolagus*, this contiguous island presents us with another animal belonging to the same family, which is rendered especially remarkable from the circumstance of its being the earliest known species of this singular group of quadrupeds; its discovery dating as far back as 1711, long before the geographical limits of Australia had been ascertained, or its productions become known to us. But, although so long a time—nearly one hundred and fifty years—has elapsed since its discovery, little or nothing is known of its habits and economy, and specimens are still rarely to be found in our own museums or those of the Continent. In his work on "The Natural History of the Mammalia," Mr. Waterhouse states:—

"This singular animal is the first of the Kangaroo family with which naturalists became acquainted, being imperfectly described, but better figured, as early as the year 1711, by Le Brun; its characters were subsequently more carefully pointed out by Pallas, and it is upon the accounts of these two authors that all the various descriptions and notices in systematic works, chiefly under the specific names of *Filander* and *Brunii*, have been founded until a comparatively recent period. Several specimens of the Filander were seen, in a state of captivity, at Batavia by Le Brun; these, however, must have been transported from New Guinea, whence it has since been procured during the French expedition of the Astrolabe, and still more recently by the naturalists sent out by the Dutch Government, to whom we are indebted for many important additions to our knowledge of the natural-history productions of the islands of the Indian Archipelago. One of the specimens of this last-mentioned expedition is now in the British Museum, and enables me to give an original description."

The following quaint note is a translation of the passage referring to this animal in Le Brun's "Voyage par Muscovie, en Perse, et aux Indes Orientales," published in 1725:—

"Being at the country-house of my general in Batavia, I there saw a certain animal which is called Filander, and which is something very singular. There were several of them which ran about with the rabbits, and had their burrows under a little hill surrounded by a balustrade. This animal, which I have represented in the plate, has the hind-legs much longer than those of the front, and is nearly of the size and texture of hair of a large Hare; it has a pointed tail, and the head approaches that of a Fox; but the most singular thing about it is, that it has an opening under the belly in the form of a bag, in which the young enter and go out again even when they are tolerably large: one may often see the head and the neck out of this bag; but when the mother runs, they do not appear to keep at the bottom of the bag, because she jerks strongly in running."

The following is Mr. Waterhouse's description of the specimen in the British Museum, which was formerly in my own possession, I having received it from the late M. Temminck, of Leyden:—

"The Filander, like the Tree-Kangaroos, has the fur radiating from a point rather behind the shoulders, and the hair on the neck directed forwards as in those animals. The fur is remarkably short, rather soft, and has very little gloss; on the crown of the head the hairs have their points directed inwards and backwards, and there meeting the hairs of the neck, which have the points directed forwards, a small tuft is formed at their point of junction on the back of the head. The general tint of the animal is brown, slightly inclining to greyish brown on the back; the sides of the body are of a somewhat brighter colour, being slightly tinted with yellowish; the whole of the under parts, as well as the fore-legs and feet, are of a dirty yellowish white; the hind-legs are of the same tint externally as the sides of the body, but paler on their inner sides; the tarsi are of a uniform pale brown. The ears are rather small, rounded at the tip, and clothed externally with short and almost velvet-like black hairs; internally the hairs are few in number and of a greyish hue; the tail is well clothed with short and soft hairs, brown on the upper surface, and brownish white beneath; on the sides of the tail the hairs, instead of pointing backwards as usual, are directed upwards; the tip of the tail is almost destitute of hair (apparently worn off by friction), and exhibits the scales very distinctly. The head is of a pale brown colour, and the muffle is naked.

		inches.	lines.
" Length from the tip of the nose to the root of the tail	. . . .	29	3
,, of tail .	. . . .	18	3
,, ,, tarsus .	. . . .	6	0
,, from nose to ear	. . . .	5	1
,, of ear .	. . . .	1	10
,, ,, fore-arm, from elbow to ends of fingers .	. . . . . .	7	6."

The Plate represents the fore part of the animal of the natural size; and complete figures in the distance, which are necessarily much reduced.

ONYCHOGALEA UNGUIFER, *Gould.*

J. Gould and H. C. Richter, del et lith.

Hullmandel & Walton, Imp.

ONYCHOGALEA UNGUIFERA, *Gould.*

Nail-tailed Kangaroo.

HEAD, ARMS, AND TIP OF THE TAIL, IN TWO POSITIONS, OF THE SIZE OF NATURE.

A MORE singular Kangaroo than the present does not exist among the known species of this great family of animals, its whole contour being characterized by a degree of elegance seldom seen among the Mammalia; but its most peculiar feature is the well-developed but somewhat flat nail at the end of the tail. It is said that the Lion has this organ terminating in a spiny hook; and here we find an analogous feature among the Kangaroos. From the discovery of Australia to the present time (March 1863), the single specimen of this animal in the British Museum is the only one that has been sent to Europe. It will be seen by the accompanying illustration, that the appendage which renders this species so remarkable is covered and protected by a well-developed tuft of lengthened hairs. Reduced figures of the entire animal will be found on the succeeding Plate, and a more lengthened description on the accompanying page.

ONYCHOGALEA UNGUIFER, Gould

ONYCHOGALEA UNGUIFER, *Gould.*

Nail-tailed Kangaroo.

Macropus unguifer, Gould in Proc. of Zool. Soc., part viii. p. 93.—Ib. Mon. of Macropodidæ, pl.   .—Gray, List of Spec. of Mamm. in Coll. Brit. Mus., p. 88.—Waterh. Nat. Hist. of Mamm., vol. i. p. 75.

THIS very elegant little Kangaroo, of which I have only seen a single example, was liberally placed in my hands, for the purpose of being described and figured, by Mr. Bynoe, of Her Majesty's Ship the Beagle, who had obtained it on the north-west coast during the present expedition of that vessel, whose captains and other officers, not only in this, but in her former voyage, have so largely extended our knowledge of the zoological productions of the little-known countries they have visited in the course of their explorations.

This animal peculiarly attracts our attention by the circumstance of its possessing a character not found in any other known member of its family, namely, a broad flattened nail much resembling that of the finger, situated at the extremity of the tail, but which is not ordinarily observable, from its being hidden in the tuft of long black hairs clothing the apical portion of that organ. It is true that a somewhat similar character exists in the *Onychogalea frænata*, but in that species it is merely rudimentary.

The foregoing remarks were published in my Monograph of the Kangaroos, and although nearly twenty years have since elapsed, no additional information has been obtained, nor has any other specimen than the original one in the British Museum been procured; it consequently still remains unique. It may be as well, however, to add Mr. Waterhouse's opinion respecting this animal, since it tends to establish a species dependent upon the skin of a single specimen; though no doubt could, I should suppose, be entertained by any one on this point, since the extreme elegance of the animal, both in shape and colour, and its lengthened tail terminated with an extraordinary-shaped nail, serve to distinguish it from every other known species.

" The muffle in *O. unguifer*," says Mr. Waterhouse, " is covered with hair, with the exception of a very narrow margin next the nostril-openings : the foremost of the three incisor teeth on either side of the upper jaw is distinctly the broadest, the other two are very nearly equal in width ; the hindermost has a strong oblique external fold ; these teeth are small compared with the incisors of most Kangaroos. A canine tooth is present, but it is very small. The tarsi are very long and slender ; the ends of the nails of the double inner toe terminate 2¼ inches short of the end of the nail of the great central toe, and the tip of the nail of the outer toe is 1¼ inch short of the same point ; the nails of the two larger toes are long, narrow, and much compressed above. The nails of the finger are rather short and broad."

Fur very short and moderately soft ; general colour buffy yellow, extending on to the outer side of the legs and the base of the tail, and gradually passing into the all but pure white of the head, ears, legs and under surface ; on each side of the body just before the knee a pale rusty patch ; a brownish mark commences about the middle of the back, runs backward over the rump, and extends to about four inches along the upper surface of the tail ; arms and tarsi cream-white ; an indistinct yellowish-white mark, curving upwards, crosses the thigh at the base ; middle portion of the tail brownish, the tip being clothed with a long black tuft, in the centre of which is a thinnish black nail half an inch in length and a quarter of an inch in breadth, convex above and concave beneath, considerably resembling the nail of the human finger.

	feet.	inches.
Length from the nose to the extremity of the tail	4	4
„ of tail	2	2
„ „ tarsus and toes, including the nail		7¼
„ „ arm and hand, including the nails		5
„ „ face from the tip of the nose to the base of the ear		4¼
„ „ ear		2½

One of the accompanying Plates represents the head and fore part of the body, and two views of the extremity of the tail, all of the natural size ; the other Plate, reduced figures of the entire animal.

ONYCHOGALEA FRÆNATA: Gould.

J. Gould and H. C. Richter del. et lith.

Hullmandel et Walton Imp.

ONYCHOGALEA FRÆNATA, *Gould*.

Bridled Nail-tailed Kangaroo.

Macropus frænatus, Gould in Proc. of Zool. Soc., Part VIII. p. 92.—Ib. Mon. of Macropodidæ, pl.   .—Waterh.
Nat. Hist. of Mamm., vol. i. p. 77.
Onychogalea frænata, Gray, List of Mamm. in Brit. Mus., p. 88.

THE large ears, full eyes, delicate limbs and lengthened tail of the *Onychogalea frænata*, conjoined with the soft grey fur of its face and body, beautifully relieved by diverging lines of black and white, render it one of the most graceful objects that can be conceived. In its disposition it is timid, peaceful and shy in the extreme, and the faculty of hearing, as indicated by the great development of its ears, being remarkably acute, it is exceedingly difficult to procure. It is a native of the south-eastern portions of Australia, and the locality nearest to the colony of New South Wales in which I observed it was Brezi, on the river Mokai, whence it extended into the interior as far as I had an opportunity of proceeding; Mr. Gilbert subsequently discovered that it was common in the thick patches of scrub which are dispersed over all parts of the Darling Downs. It inhabits all the low mountain ranges, the elevation of which varies from one to six hundred feet, and which are of a sterile character—hot, dry, stony, and thickly covered with shrub-like stunted trees. These situations are also the abode of the *Halmaturus dorsalis*, with which I sometimes found the *Onychogalea frænata* associating; but it differs from that species, which is strictly an inhabitant of the brushes, in frequenting the more open parts, and occasionally even the plains. When started from its seat, which is formed like that of a hare, beneath the shelter of a tuft of grass or a small bush, it bounds away with remarkable fleetness, generally giving the best dogs a sharp run, and frequently making its escape by gaining the thick part of the brush, or the hole of a decayed tree; one of those I procured, on being sharply pressed, mounted the inside of a tree, to an opening nearly fifteen feet from the ground; whence it leaped down before the dogs and succeeded in reaching the hollow of a fallen trunk, from which it was finally taken by the hand.

In the neighbourhood of Brezi the natives hunt this species with dogs, and often kill it with spears, bommerengs and other weapons; at Gundermein on the Lower Namoi I found myself among a tribe of natives who succeed in capturing them with nets, which, although rudely constructed, are very well adapted to the purpose. On being made acquainted with my object, they were easily induced to accompany me to a "Brigaloe brush," in which the present species and the *Halmaturus dorsalis* were very abundant. Arriving at the skirts of the brush, the oldest men of the tribe separated from the rest, each two taking a net about twenty-five yards long by three and a half feet wide, with which they proceeded to those parts where the runs of the animals were most frequent, while the rest of the natives entered the brush on the opposite side, and with loud shouts and yells drove the Kangaroos towards the nets; by this means in a single afternoon they obtained for me as many specimens as I desired.

Its food consists of grass and various kinds of herbage, and its flesh, like that of the other small Kangaroos, is excellent, and when procurable was eaten by me in preference to other meat.

Some diversity of colour is found to occur in examples from different localities; those obtained by Mr. Gilbert on the Darling Downs being of a much browner hue than those I procured on the Mokai and Namoi.

Fur short and soft, general tint grey, being finely penciled with black and white; under surface of the body and inner side of the limbs white; on the cheeks a white mark, beneath which is a dusky line; ears clothed externally with grey hairs, edged with black at the apex and lined internally with white hairs;

muzzle blackish in front of the eye; from the occiput two conspicuous white marks run backward, and diverging, pass one on each side over the shoulder, and recurve at a short distance behind the insertion of the fore-leg; the space between these lines is black on the occiput, and brownish black on the back of the neck; sides of the neck suffused with pale ochreous yellow; tarsi and arms nearly white; hands and toes dusky, but most of the hairs round the nails of the former white; tail coloured like the body at the base, but black along the upper surface of the apical third, and at the point where the hairs being longer than elsewhere, hide a small horny tubercle with which the tail is terminated; under surface of the tail dirty yellowish white.

The female is not distinguished by any difference in marking; the stripes are quite as intense as in the male, and are even apparent in the fœtus.

Considerable variation occurs in the weight of individuals, and particularly in the size and weight of the sexes, fully adult males weighing from ten to twelve pounds, while the females do not exceed four or six.

The following are the admeasurements of the largest specimens I have seen; ordinarily they are about one-fifth smaller :—

	Male.		Female.	
	feet.	inches.	feet.	inches.
Length from the nose to the root of the tail	3	7¼	2	8½
,, of the tail	1	7	1	3
,, ,, tarsus and toes, including the nails	0	6¼	0	5
,, ,, arm and hand, including the nails	0	5	0	4
,, ,, face from the tip of the nose to the base of the ear . .	0	3¾	0	3¼
,, ,, ear	0	3½	0	3¼

The Plate represents a male about three-fourths of the natural size.

ONYCHOGALEA LUNATA: *Gould.*

ONYCHOGALEA LUNATA, *Gould.*

Lunated Nail-tailed Kangaroo.

Macropus lunatus, Gould in Proc. of Zool. Soc., Part VIII. p. 93.—Gray, List of Mamm. in Brit. Mus., p. 88.—
Waterh. Nat. Hist. of Mamm., vol. i. p. 79.

WHEN writing upon the Birds of Australia I frequently had occasion to allude to the very remarkable manner in which different species of the same form represent each other on opposite parts of the continent; and that a similar law of representation exists among the Mammals, is evidenced by the present and preceding species, which, although most nearly allied, inhabit portions of the country as widely apart as if seas had flowed between, as at some distant period was probably the case. We have no evidence that they approach each other in the interior of the country, as neither of them have yet been discovered within the limits of the intervening colony of South Australia; consequently they must be regarded as beautiful representatives of each other in the respective countries they inhabit.

Although assimilating in form and markings to the *Onychogalea frœnata*, the present species is certainly less ornamental, and is also much smaller in all its dimensions. The habits and economy of the two species are very similar; both exhibit a remarkable degree of shyness and timidity, and seek safety in flight upon the slightest alarm. I had no opportunity of observing it in a state of nature myself, but Mr. Gilbert's notes inform me that "the *Waurong*," by which name it is known to the natives, "is found in the gum forests of the interior of Western Australia, where there are patches of thick scrub and dense thickets, in the open glades intervening between which it is occasionally seen sunning itself, but at the slightest alarm immediately betakes itself to the shelter of the thick scrub; the dogs sometimes succeed in driving it out to the open spots, when, like the Kangaroo rats, it runs to the nearest hollow log, and is then easily captured. I remarked, that when sitting quietly cleaning itself, there was a constant twitching of the tail in an upward direction; an action which I have never seen performed by any other Kangaroo. I was not sufficiently near to ascertain whether this motion of the tail had any connection with the claw or nail at its extremity, but I think it not improbable. The *Waurong* makes no nest, but forms a hollow in the soft ground beneath a thick brush in which it lies during the heat of the day."

Fur soft and of moderate length; general tint ashy grey, finely pencilled with dusky and yellowish white; back of the neck and shoulders vinous rust-colour; a short distance behind the base of the fore-leg a distinct curved white mark; under surface of the body pale grey, the hairs tipped with dirty white; on the sides of the body a faint rusty tint, more distinct in some specimens than in others; around the eye a ring of pale rust-colour, and the muzzle suffused with the same tint; ear clothed with long white hairs within, and externally with very minute dusky hairs finely freckled with yellowish white; on the hinder half the hairs are longer and almost white, at the apex a delicate fringe of blackish hairs; fore-feet in some specimens brown, in others dirty white; tarsi chiefly dirty white, but the sides of the toes suffused with pale brown; tail clothed for the most part with short adpressed hairs, having a general greyish tint; on the upper surface the hairs are somewhat lengthened, and on the apical portion they form a slight crest which is usually blackish; at the tip of the tail is a small conical horny appendage like a nail, of about an eighth of an inch in length.

	Male.	
	ft.	in.
Length from the nose to the root of the tail	1	8
„ of the tail	0	12
„ „ tarsus and toes including the nails	0	$4\frac{3}{4}$
„ „ arm and hand including the nails	0	$3\frac{1}{4}$
„ „ face from the tip of the nose to the base of the ear	0	3
„ „ ear	0	$2\frac{1}{4}$

It is to be regretted that this as well as all other Kangaroos lose the delicate tints of their colouring on exposure to light; so much so in the present instance, that Museum and recent specimens could scarcely be considered as identical.

The Plate represents the two sexes rather under the natural size.

LAGORCHESTES FASCIATUS.

LAGORCHESTES FASCIATUS.

Banded Hare-Kangaroo.

Kangarus fasciatus, Péron et Lesueur, Voy. aux Terr. Aust., tom. i. p. 114. pl. 27.—Desm. Mamm., part i. p. 274.
Halmaturus elegans, Cuv. Règne Anim., tom. i. p. 187.
Bettongia fasciata, Gould, Mon. of Macropodidæ, pl.
Lagorchestes albipilis, Gould, Ann. and Mag. of Nat. Hist., vol. x. 1842, p. 2.
Macropus (Lagorchestes) fasciatus, Waterh. Nat. Hist. of Mamm., vol. i. p. 87.

I BELIEVE that this beautiful species is noticed for the first time in the Voyage of the celebrated Dampier, but MM. Péron and Lesueur are undoubtedly entitled to the credit of making it known to science. It was during their voyage to the " Terres Australes," and while exploring the western coasts of Australia, that they met with it on Dirk Hartogs and the neighbouring islands, where it was found among the impenetrable low thickets formed of a species of *Mimosa*; " from these bushes," says MM. Péron and Lesueur, " it cuts away the lower branches and spines so as to form galleries communicating one with another, in which it takes refuge in time of danger. The females bring forth but one young one at a time. Although abundant on the islands, none were to be found on the main land. These little Kangaroos, like all feeble animals which have neither the power of attack nor of defence, are extremely timid. The slightest noise caused them to take flight to the thick brushwood in which their galleries are constructed, and where it is impossible to pursue them; hence, although very common, they are difficult to procure." Although the above-mentioned naturalists were unable to discover the animal on the main land, the researches of more recent travellers, aided by the facilities afforded by the colonization of the country, have shown that it is not only abundant there, but enjoys a most extensive range. Mr. Gilbert found it far in the interior of the Swan River Colony, and Mr. Eyre, one of the most indefatigable of Australian explorers, states that he frequently observed it in the Murray Scrub of South Australia; here then we have a range of many thousand square miles of country as the known habitat of this beautiful species, and we may consequently infer, that every intermediate district between Southern and Western Australia favourable to its habits is tenanted by it. Mr. Gilbert states that it is called *Marnine* by the natives of the interior of Western Australia, and is only to be found in densely thick scrubs, on flats and on the edges of swamps, where the small brush *Melaleuca* grows so thickly, that it is almost impossible for a man to force his way through; its runs being under this, the animal escapes even the quick eye of a native. The only possible means of obtaining it is by having a number of natives to clear the spot, and two or three with dogs and guns to watch for it.

During an excursion into the interior Mr. Gilbert was so fortunate as to cross one of its haunts, but so dense was the vegetation, that after three days of severe toil, he was only able to secure a single specimen; he adds, that it appears to run in company with the *Damas*, which being more numerous were continually presenting themselves and disappointing him, the vegetation being much too thick to distinguish the one from the other until after they had been shot. The natives are in the habit of burning these thickets at intervals of three years, and by this means destroy very great numbers; and this, in fact, appears to be the only plan they could very well adopt for capturing both the *Marnine* and the *Dama*, for the mere treading down an open space, as is done at King George's Sound, will not answer here: the specimen he procured was a male, and weighed three pounds and three-quarters.

From the period of MM. Péron and Lesueur's Voyage in 1800—1804 until 1842, when M. Priess, a German naturalist, visited Western Australia, no example of this little Kangaroo was sent to Europe. The specimens from which the description published by the former gentlemen was taken, still form part of the collection of the " Jardin des Plantes " at Paris, but from long exposure to a powerful light, and their muzzles having in the course of time become denuded of hair, their appearance is so much altered, that I was induced to believe that the animal brought home by M. Priess was both generically and specifically distinct; and therefore, while, from their apparently naked muzzles, I placed the Paris animals in the genus *Bettongia*, I referred the recently received specimen to the genus *Lagorchestes*, and gave it the name of *L. albipilis*; this error has been corrected by Mr. Waterhouse, who, after a careful examination and comparison of specimens, satisfied himself that they were identical,—an opinion in which I now entirely coincide.

The flesh is said to resemble that of the rabbit, but has a slight aromatic flavour, arising probably from the nature of the plants on which they feed, nearly all of which are fragrant.

When MM. Péron and Lesueur visited the islands, all the females carried young in their pouch, and the devotedness with which they sought to save their offspring was truly admirable; although wounded they flew with the young in the pouch, and never left them until, overcome with fatigue and loss of blood, they could no longer carry them; they then stopped, and squatting themselves on their hind-legs, helped the young to get out of the pouch by means of the fore-feet, and sought to place them in a situation favourable for retreat.

Mr. Gilbert states that it makes no nest, but when on the plains squats precisely like a hare.

The sexes are very similar, and may be thus described :—

The fur is very long and soft; its general colour greyish, variegated with black, white, and rusty red, the latter colour being most conspicuous round the eyes; on the back are numerous narrow, transverse black bands; these are somewhat irregular and not well-defined; the spaces between the bands are partly of a rusty red, and partly whitish; the white joins the dark band, and is gradually shaded into the rusty red, to be followed by the next dark band; over the whole of the upper surface, sides and cheeks are numerous very long interspersed hairs, which have the exposed portion white, but, like the ordinary fur, are nearly black at the root; under surface dirty white, with a considerable admixture of grey; the ears are clothed with lengthened white hairs internally, and externally with short hairs finely freckled with brownish black and white; fore-feet dirty rust-red; tarsi pale rusty red penciled with blackish; sides of both tarsi and toes pale brown; tail tolerably well covered with short adpressed brownish grey hairs; on the under side the hairs are somewhat longer and of a brownish white; on the upper surface is a narrow blackish streak, and on the apical third the hairs are lengthened, and form a small dark crest at the point; they are sometimes an inch in length.

The Plate represents an adult male about the size of life.

LAGORCHESTES LEPOROÏDES, Gould.

LAGORCHESTES LEPOROÏDES, *Gould*.

Hare Kangaroo.

Macropus Leporides, Gould in Proc. of Zool. Soc., part viii. p. 93.—Waterh. in Jard. Nat. Lib. *Marsupialia*, p. 204.
Lagorchestes Leporoïdes, Gould, Mon. of Macropodidæ, pl.   .—Gray, List of Spec. of Mamm. in Coll. Brit.
Mus., p. 95.
Macropus (Lagorchestes) Leporoides, Waterh. Nat. Hist. of Mamm., vol. i. p. 82.

THE name of Hare Kangaroo has been given to this species, as much from the similarity of its size and the colour and texture of its fur, as from its habits assimilating in many particulars to those of that animal. I usually found it solitary, and sitting close in a well-formed seat under the shelter of a tuft of grass on the open plains. For a short distance its fleetness is beyond that of all others of its group that I have had an opportunity of coursing; its powers of leaping are also equally extraordinary, in proof of which I may mention an incident connected with the chase of the animal which occurred to myself. While out on the plains in South Australia I started a Hare Kangaroo before two fleet dogs; after running to the distance of a quarter of a mile, it suddenly doubled and came back upon me, the dogs following close at its heels; I stood perfectly still until the animal had arrived within twenty feet before it observed me, when to my astonishment, instead of branching off to the right or to the left, it bounded clear over my head, and on descending to the ground I was enabled to make a successful shot, by which it was procured.

Considerable diversity of colour is observable in different specimens, some being much redder than others; but the sexes are scarcely distinguishable in size.

I have but little doubt that this animal enjoys a wide range over the interior of New South Wales; it certainly inhabits the Liverpool Plains as well as those in the neighbourhood of the Namoi and the Gwydyr, from all of which localities I have received numerous examples; it is equally certain that it is found on the grassy plains of South Australia, for I not only found it there myself, but specimens have since been sent to me from thence by the late Mr. Strange. Now as the character of all these districts is very similar, it is probable that the Hare Kangaroo is equally abundant in the intermediate countries as it is in those above mentioned; as yet, I have never seen specimens in collections from the Northern or Western portions of Australia.

The following is Mr. Waterhouse's description of the character of the fur and the colouring of this animal, which being taken from my own specimen, and more minute than that given by myself in the "Proceedings of the Zoological Society," is here transcribed.

"Fur long and soft; on the upper parts of the body variegated with black, rust-colour and rusty white, the white most conspicuous and the rust-colour but little seen; the back of the neck and shoulders, and a considerable space round each eye tinted with palish rust-colour, sometimes inclining to buffy yellow; sides of the body and haunches suffused with rust-colour; under parts greyish white tinted with rust-colour, but nearly pure white between the hind legs; fore legs with a more or less strongly marked black patch, at the base externally (or behind the elbow), but the hairs on this part are pencilled with white; fore arm and hand with short brown hairs, pencilled with very pale brown; on the middle of the tibia is a dusky patch; tarsi impure palish rust-colour finely freckled with brown; toes brownish; tail clothed throughout with small adpressed hairs, which are partly black and partly white; beneath brownish white;" nails of the hind feet very long, pointed, and jet-black.

The figure is about the natural size; if at all different, it is a trifle less.

LAGORCHESTES HIRSUTUS: Gould.

LAGORCHESTES HIRSUTUS, *Gould.*

Rufous Hare Kangaroo.

Lagorchestes hirsutus, Gould in Proc. of Zool. Soc. 1844, p. 32.
Macropus (Lagorchestes) hirsutus, Waterh. Nat. Hist. of Mamm., vol. i. p. 92.

ALL the examples I have seen of this species, some of which are at the British Museum, and the remainder in my own collection, have been procured in Western Australia, whence they were sent to this country by Mr. Gilbert; judging from the size of *Lagorchestes fasciatus*, I should suppose that the present animal would weigh about four or six pounds, the weight of a moderate-sized hare. The lengthened shaggy reddish hairs, which are abundantly distributed over the lower part of the back, and particularly near the base of the tail, at once distinguish it from all the other members of the genus. The only note transmitted by Mr. Gilbert, respecting the habits of the species, is as follows :—

"It has a hairy muzzle : in its habits it assimilates in an equal degree to those of the *Bettongiæ* and the *Lagorchesti.* It constructs a burrow, open at both ends, with a seat at the side of the entrance, from which it plunges into the burrow the instant it is alarmed. It feeds on the open country adjacent to the thickets, where there is a low thick scrub about two feet high : when running, and particularly when hunted, it utters a singular note, resembling the syllable *ting* rather quickly repeated. Some slight difference is found to exist in specimens from various localities, which I presume must be regarded as due to the difference of situation, and nothing more." He adds, that it is called *Woô-rup* by the Aborigines of the interior of Western Australia, who appear to give the name of *Môr-da* to the animal during the period of immaturity; at all events, the young example sent by him with that name attached to it, is undoubtedly the young of the present species. Both the adult and the young were procured in the Walyemara district.

Mr. Waterhouse having given a very accurate description of this animal from the specimens in the British Museum, I take the liberty of transcribing it :—

"The fur is long and moderately soft; the upper parts of the body grey, much tinted with rufous brown and freely pencilled with white; the sides of the body, rump, hind- and fore-legs are of a bright rust-red, deepest on the hinder and palest on the fore-legs; the throat, chest and mesial line of the belly rusty white; crown of the head grey; a broad space around the eye is of a bright, but palish rust-red, which tint extends on to the muzzle; a whitish line on the upper lip runs back past the angle of the mouth; ear clothed internally with somewhat lengthened white hairs, externally they are pencilled with rusty yellow and dusky, the former being, however, the prevailing tint; the hinder half is almost entirely clothed with small white hairs; the fore-feet are clothed with glistening yellowish white hairs; the tarsus is almost entirely of a pale rusty red, but is of a rusty white towards the hinder part, and the toes are obscurely suffused with brownish rust-red; the tail is clothed throughout with short, stiff, adpressed hairs, scarcely hiding the scaly skin; they are finely pencilled with black and rust-red at the base of the tail, but on the upper surface they assume an uniform brownish black tint, which is continued to the point; on the under surface they are of a dirty pale rust-red, and towards the apex is a naked scaly space of about an inch in length; the fur of the back is nearly black next the skin, but a considerable portion of each hair is of a brownish rust-red; near the point the hairs are broadly annulated with white, and at the point they are dusky or black; on the belly the fur is ashy grey next the skin."

The figure is rather less than the natural size.

LAGORCHESTES CONSPICILLATA, *Gould.*

LAGORCHESTES CONSPICILLATA, *Gould.*

Spectacled Hare Kangaroo.

Lagorchestes conspicillatus, Gould in Proc. of Zool. Soc., part ix. p. 82. — Ib. Mon. of Macropodidæ, pl. —
Gray, List of Spec. of Mamm. in Coll. Brit. Mus., p. 95.
Macropus (Lagorchestes) conspicillatus, Waterh. Nat. Hist. of Mamm., vol. i. p. 85.

I HAVE again to offer my thanks to the Officers of H.M.S. Beagle for subjects they have contributed to my illustrations of Australian zoology, and especially for the loan of two fine specimens of this highly interesting *Lagorchestes,* the second species yet discovered of this beautiful form. It is to Capt. Wickham and Mr. Bynoe that science is indebted for its discovery. It was procured on Barrow Island, which lies off the north-western coast of Australia, about thirty miles from the main land. The two specimens collected by those gentlemen are fortunately male and female, and hence the subject is rendered so much the more complete. The specimen sent me by Captain Wickham has, by his desire, been presented to the national collection at the British Museum, and his example will, I feel assured, be followed by my esteemed friend Mr. Bynoe, as no exertion should be spared to render that collection, already so fine, as complete as possible.

This species is rather less in size than the *Lagorchestes Leporöides,* from which it is distinguished by its fur being more dense and harsh to the touch, by the extreme blackness of the basal part of the hair, by the shortness of its ears, by the want of the black patch at the base of the arm, and by the red colouring around the eyes being of a more brilliant rusty hue than in that animal.

The sexes are alike in colour and size.

The above was published in my 'Monograph of the Macropodidæ' nearly twenty years ago; and, as Barrow Island has not been visited since, we have not received any additional examples of this very distinct species of *Lagorchestes.* Both the specimens above alluded to are in the British Museum.

Fur very long, dense, and rather soft to the touch; on the back it is of a black colour next the skin, yellowish white towards the apex, shaded into deeper yellow still nearer to the point, and black at the point; on the lower part of the back the portion of each hair which is yellow on the back is replaced by white, and there is an oblique white mark on each side of the rump; fur on the side of the body deep grey next the skin, brownish yellow in the middle, followed by black, then whitish, and at the point black; on the under surface of the body the fur is ash-coloured next the skin, and white externally, excepting on the sides of the belly, where it is of a rusty yellow hue externally; the hair on the upper surface of the head is black, freely pencilled with yellowish white; a broad space round the eye is covered with bright rusty red hairs, and this hue, though less bright, is extended backwards beneath the ear; lips and chin dirty white; throat white; ears internally clothed with whitish hairs, and externally with dirty white hairs on the apical portion, but towards the base there is an admixture of black; fore- and hind-legs and feet pale, the hairs being dirty white at the point and brown next the skin; tail slender, and, being but sparingly clothed with short dirty white bristly hairs, permits the scaly character of the skin to be seen; on the under surface it is more densely clothed, and the hairs are longer and of a dirty yellowish hue.

		Male.	
		feet.	inches.
Length from the nose to the extremity of the tail	. . . .	2	$8\frac{1}{4}$
„ of tail		1	$1\frac{1}{2}$
„ of tarsus and toes, including the nail			$5\frac{1}{4}$
„ of arm and hand, including the nails			3
„ of face from the tip of the nose to the base of the ear . .			$3\frac{1}{4}$
„ of ear			$1\frac{1}{4}$

The accompanying Plates represent the head of the natural size, and the entire animal reduced.

LAGORCHESTES LEICHARDTI, Gould.

LAGORCHESTES LEICHARDTI, *Gould.*

Leichardt's Hare Kangaroo.

Two specimens—one adult, the other immature—of this beautiful species of *Lagorchestes* have been transmitted to me by direction of the Council of the Australian Museum at Sydney, New South Wales, for the purpose of being described and figured in the present work. The specimens in question formed part of the Mammalia collected by Dr. Leichardt during his extensive overland journey from Moreton Bay to Port Essington; unfortunately no information has been furnished me respecting them; perhaps, indeed, none was obtained. I am, therefore, unable to state the precise locality in which they were procured; but as I find no mention of them in the late Mr. Gilbert's Journal, we may infer that they were not obtained until after his lamented death, and that the country between the Gulf of Carpentaria and Port Essington is the natural habitat of the species. I have named it *Leichardti*, from a desire to assist in perpetuating the name of the intrepid traveller who has done so much in the exploration of Australia, and whose life it is to be feared has fallen a sacrifice to his zeal for discovery in the previously untraversed portions of that strange country.

The only species with which the *L. Leichardti* could be confounded is the *L. conspicillatus*, but on comparison it will be found to differ from that animal in the richly contrasted colouring of its crisp and wiry fur, in the whiteness of its rump and tail, in the brighter rusty hue of the space surrounding the eye, in the chestnut colour of the basal portion of the fur, and in its smaller ears.

I cannot conclude without offering my thanks to the Council of the Australian Museum for their kindness and liberality in permitting this rare and interesting animal to be sent to Europe, the discovery of which adds so much to the interest of " The Mammals of Australia."

Face grizzled grey and brown, passing into rufous between and on the ears, which are margined with white; around the eye a conspicuous oval patch of lively ferruginous red; hairs of the cheeks stiff and bristly; all the upper surface mottled rufous black and white, the base of the fur being chestnut, passing into black about the middle, then into white, and lastly into dark rufous at the tip; on the rump and base of the tail these colours give place to greyish white, intermingled with black; all the under surface greyish white; at the insertion of the hinder limbs two curved marks of grey; hands and toes washed with buff; nails black; tail greyish white.

Total length from the nose to the extremity of the tail two feet four inches; of the tail thirteen inches; of the tarsus and toes including the nail five inches and three-quarters; of the arm and hand including the nail three inches; of the face from the tip of the nose to the base of the ear two inches and five-eighths; of the ear one inch and an eighth.

The figure is rather under the size of life.

BETTONGIA PENICILLATA, *Gray.*

BETTONGIA PENICILLATA, *Gray*.

Jerboa Kangaroo.

Bettongia penicillata, Gray, in Mag. of Nat. Hist. new ser. vol. i. p. 584.—Waterh. Nat. Lib. Marsupialia, p. 183.—
    Gould, Mon. of Macropodidæ.—Gray, List of Mamm. in Brit. Mus., p. 93.
Hypsiprymnus murinus, Ogilby, in Proc. of Zool. Soc., Part vi. p. 63.
———— *setosus*, Waterh. Cat. of Mamm. in Mus. Zool. Soc., p. 65.
———— (*Bettongia*) *penicillatus*, Waterh. Nat. Hist. of Mamm., vol. i. p. 212. pl. 9.

The eastern parts of Australia, particularly the districts on the interior side of the ranges of New South Wales, constitute the true habitat of the species figured on the accompanying Plate. I observed it to be very abundant on the Liverpool Plains, and on the banks of the river Namoi, from its source to its junction with the Gwydyr; but between the ranges and the coast I did not meet with it. I do not, however, assert that it is not an inhabitant of those districts also; but, if it be, it is certain that it is far less abundant there than on the other side of the ranges. I have never seen an example from South Australia; its place in that part of the country appearing to be supplied by its near ally the *Bettongia Ogilbyi*, a species dispersed in abundance from thence to the western limits of the country, or the colony of Swan River. Mr. Waterhouse is inclined to believe that these eastern and western animals (*B. penicillata* and *B. Ogilbyi*) are merely varieties of one and the same species; and, while I admit the feasibility of this opinion, the markings and colouring of the two animals are so different, that, in a work on the Mammals of Australia, I cannot do otherwise than figure both of them, leaving their specific value to be ascertained by future zoologists, should no opportunity for fully investigating the subject occur to myself: it is just one of those cases in which a careful examination of a great number of specimens and skeletons from both localities is required to determine so dubious a question, and such materials are not at present accessible.

Like the other members of the genus, this species constructs a thick grassy nest, which is placed in a hollow scratched on the ground for its reception, so that when completed it is only level with the surrounding grass, which it so closely resembles, that without a careful survey it may be passed unnoticed: the site chosen for the nest is the foot of a bush or any large tuft of grass; during the day it is generally tenanted by one, and sometimes by a pair of these little creatures, which lying coiled in the centre are perfectly concealed from view; there being no apparent outlet, it would seem that after they have crept in they drag the grass completely over the entrance, when, as I have before stated, the whole is so like the surrounding herbage that it is scarcely perceptible. The natives, however, rarely pass without detecting its presence, and almost invariably kill the sleeping inmates, by dashing their tomahawk or heavy clubs at it. The most curious circumstance connected with the history of the Jerboa Kangaroo is the mode in which it collects the grasses for its nest: these, as may be seen in the accompanying Plate, are carried with its tail, which is strongly prehensile, and, as may be easily imagined, their appearance when leaping towards their nests with their tails loaded with grasses is exceedingly grotesque and amusing: this curious feat is even exhibited in a state of confinement, a pair in the Menagerie of the late Earl of Derby having evinced the same natural habits, by frequently loading their tails with the hay of their nests, and carrying it round the cage in which they were kept. The usual resorts of the Jerboa Kangaroo are low grassy hills and dry ridges, thinly intersected with trees and bushes; and although not strictly gregarious, numbers may be found in the same locality. It is a nocturnal animal, lying curled up in the shape of a ball during the day, and sallying forth as night approaches in quest of food, which consists of grasses and roots, the latter being procured by scratching and burrowing, for which its fore-claws are admirably adapted, and its vicinity is frequently indicated by the little excavations it has made. When startled from its nest, it bounds with amazing rapidity, and always seeks the shelter of a hollow tree, or a small hole in a rock, etc.

Fur moderately long, and not very soft to the touch; general colour brown; the hairs on the upper surface grey at the base, pencilled with rusty white near the tip, and black at the point; under surface dirty white; internal surface of the ear yellow; feet very pale brown; tail brown above and pale brown beneath, the apical quarter clothed with brownish-black hairs, which are longer than those of the other parts of the tail, and form a kind of tuft.

	Male.		Female.	
	feet.	inches.	feet.	inches.
Length from the nose to the extremity of the tail	2	6	2	0
„ of tail	1	1		$11\frac{1}{4}$
„ of tarsus and toes, including the nail		5		4
„ of arm and hand, including the nails		$3\frac{1}{4}$		$2\frac{3}{4}$
„ of face from the tip of the nose to the base of the ear . . .		$3\frac{1}{4}$		3
„ of ear		$1\frac{1}{4}$		1

The figures are of the natural size.

BETTONGIA OGILBYI, Gould.

J. Gould and H. C. Richter del. et lith.

Hullmandel & Walton, Imp.

BETTONGIA OGILBYI, *Gould.*

Ogilby's Jerboa Kangaroo.

Bettongia Ogilbyi, Gould, MSS.
Hypsiprymnus Ogilbyi, Waterh. Nat. Lib. Marsupialia, p. 185.
Bettongia Ogilbii, Gray, List of Mamm. in Brit. Mus., p. 93.
Hypsiprymnus (Bettongia) Ogilbyi, Waterh. Nat. Hist. of Mamm., vol. i. p. 214.
Bettongia Gouldi, Gray ?

THE *Bettongia Ogilbyi* is as abundantly distributed over Western Australia as the *B. penicillata* is over New South Wales, but while the latter appears to be confined to the country within the ranges, the former inhabits the districts near the coast. Besides specimens from Swan River, I have received others from the late Mr. Harvey, procured at Port Lincoln, and I have also seen examples in collections formed in the neighbourhood of Adelaide ; those from the last-mentioned locality have the rufous tints of the tail and tarsi somewhat less highly coloured, but in other respects specimens from these distant countries are perfectly similar.

The *B. Ogilbyi* always appeared to me to have a longer head and proportionately longer ears and somewhat more slender tarsi than *B. penicillata* ; these, however, are only slight differences ; but the darker colouring of the body, the rusty red hue of the base and sides of the tail, and the rufous colouring of the feet, are characters always observable in the western animal, and constitute a style of colouring never seen in any example of the eastern species, or *B. penicillata* of New South Wales.

In Mr. Waterhouse's remarks on my specimen of *B. Ogilbyi*, published in the volume of the ' Naturalist's Library,' on the Marsupialia, he says, " This species is very closely allied to *B. penicillata*, but its tarsi are proportionately rather longer and more slender, and differ in being of a deeper hue ; the ears are longer, and the apical half of the tail is black both above and below. In *B. penicillata* the black hair is confined to the upper surface of the tail ; on the under part, lengthened, brown, adpressed hairs extend to the tip ; this under part is, moreover, much more densely clothed than in the present species, in which the hairs are not sufficiently numerous to hide the scales : this does not arise from the wearing away of the hair, as is often the case ; for the under side of the tail is better covered than the sides." In his more recent work, ' A Natural History of the Mammalia,' Mr. Waterhouse is rather doubtful as to the distinctness of the southern and western animals, and remarks, " All that can be said is, that the specimens of the tufted-tailed *Bettongiæ*, from the western and southern districts, are generally somewhat darker in the colouring of the feet and tail than those from New South Wales ; but it is certainly, in some cases, difficult to distinguish these, which I can but regard as local varieties, by a difference of colouring."

" This species," says Mr. Gilbert in his notes on the ' Mammals of Western Australia,' " appears to be equally abundant in all parts of the colony, but to evince a preference, perhaps, for the white-gum forests. It makes a nest of dried sticks or strong coarse grass, under the shelter of the overhanging grasses of the *Xanthorrhœa*, or a bunch of dried grasses or sticks ; the entrance being on one side and lengthened out so as to form a tube or porch. When driven from the nest it generally resorts to a hollow tree or stump ; if this is not to be found, it makes a long circuit before returning to the nest. This animal is one of the favourite articles of food of the natives, who are very quick in detecting the nest, and generally capture the little inmate by throwing a spear through the nest and transfixing it to the ground, or by placing the foot upon and crushing it to death. It is almost invariably found in pairs, and, like the true *Macropi*, the female throws the young from the pouch when pursued."

Fur dense, the under fur very abundant, soft, long and woolly, general colour brown, obscurely washed with yellow on the sides of the face and body ; under surface of the body dirty yellowish white ; ears clothed with yellow hairs ; hind feet brown, darkest on the sides, especially of the toes ; fore-feet paler brown ; tail well-clothed, a very small space at the base covered with fur, like that of the body ; beyond this and extending to about the middle of the tail the hairs are of a rusty hue on the upper side, and very pale brown on the under ; the apical half of the tail is clothed with black hairs, which vary from rather more than half to three quarters of an inch in length ; those nearest the tip are the longest : on the sides of the tail the hairs are comparatively short, and excepting at the tip they are of a deep brown colour ; on the under side of the apical half of the tail the hairs are longer than on the sides, and of a black colour ; the ordinary hairs of the back are rather broadly annulated with pale rusty yellow, sometimes rusty white, and at the point they are blackish brown ; the longer interspersed hairs are black ; the fur both on the upper and under parts of the body is grey at the base.

The figures are of the natural size.

BETTONGIA CUNICULUS.

BETTONGIA CUNICULUS.

Tasmanian Jerboa Kangaroo.

Hypsiprymnus cuniculus, Ogilb. in Proc. of Zool. Soc., part xi. p. 43.—Waterh. in Jard. Nat. Lib., Marsupialia, p. 186.

Bettongia setosa, Gray in Mag. Nat. Hist., new. ser. vol. i. p. 584.—Ib. List of Mamm. in Brit. Mus. Coll., p. 93.

———— *cuniculus*, Gould, Mon. of Macropodidæ, pl.   .

Hypsiprymuus (Bettongia) cuniculus, Waterh. Nat. Hist. of Mamm., vol. i. p. 200.—Gunn, Proc. of Roy. Soc. of
      Van Diemen's Land, vol. ii. p. 86.

Forest Kangaroo-Rat of the colonists of Van Diemen's Land.

FEW of the indigenous quadrupeds of Van Diemen's Land are better known than the present, which may be said to be universally dispersed over that island, wherever localities occur favourable to its habits and mode of life ; these are grassy plains and the stony ridges of the outskirts of the forest, precisely the reverse of the situations affected by the *Hypsiprymnus apicalis*, which resorts to the low and swampy districts covered with green and dense vegetation.

The *Bettongia cuniculus* is altogether a larger and more robust animal than either *B. Grayi* or *B. penicillata*. From the former it differs in having a rather more lengthened face, and from the latter in the tip of the tail being white. Mr. Waterhouse states that it is not only distinguished from the latter by its size, but that the proportions of the crania of the two animals differ very considerably. I believe the species to be strictly confined to Van Diemen's Land, as I have never received examples from any part of the continent. The only outward difference in the sexes consists in the smaller size of the female. It makes a thick and warm grassy nest in a slight depression under the shelter of a bush or large tuft of grass, and feeds on bulbous and other roots, which it readily scratches up with the powerful claws of its fore feet.

Mr. Richter has handed me the following notes made by him while engaged in drawing the animal in the Gardens of the Zoological Society in the Regent's Park :—

" The *Bettongia cuniculus* collected together a large mass of straw, &c. with its fore feet, threw it backward between its hind legs, curled the tail around it and hopped about with it in this position for several hours during the night. Both *B. cuniculus* and *B. Ogilbyi* have the power of elevating the duplex toe of the hind foot to scratch their ears, &c. In fighting, the teeth and fore legs are but little used ; their chief attack being made by throwing themselves on one side and lashing out, with great velocity and strength, with their hind legs. In confinement they are very partial to bread and milk sweetened with sugar. They are very tame, seldom biting or showing anger on being handled. When angry they emit a succession of short hisses. The two species seem very inimical to each other. They drink a great quantity of water, as much as two or three ounces at a time, by lapping with the tongue. They invariably sleep with the tail brought between the hind legs and curled round the head, which is depressed to the ground. If given plenty of clean hay they cover themselves completely with it, forming a sort of bower or nest."

Fur rather long and not very soft ; general colour brownish grey, pencilled with white ; feet brownish white ; tail well clothed with pale brown hairs, gradually passing into dark brown near the extremity and tipped with pure white ; margin of the ears slightly tinged with yellowish ; under surface of the body dirty white ; fur both of the upper and under surface grey at the base.

The figures are of the natural size.

BETTONGIA GRAII, Gould.

BETTONGIA GRAII, *Gould.*

Gray's Jerboa Kangaroo.

Hypsiprymnus Graii, Gould in Proc. of Zool. Soc., part viii. p. 178.—Waterh. in Jard. Nat. Lib. Mamm., vol. xi.
    (*Marsupialia*) p. 190.
Bettongia Grayii, Gray, List of Mamm. in Coll. Brit. Mus., p. 93.
Hypsiprymnus (Bettongia) Graii, Waterh. Nat. Hist. of Mamm., vol. i. p. 203.
——————— *Lesueuri*, Quoy et Gaim. Voy. de la Coquille ?
Boör-dee, Aborigines of the mountain districts of Western Australia.

———————

I first described this species in the "Proceedings of the Zoological Society of London" for 1840, from Swan River specimens, and remarked that it differed from its near ally, the *Bettongia rufescens*, in being of an ashy-brown colour above, and in having the hairs which clothe the back of the ears of the same colour as those of the head. During the years which have elapsed between 1840 and the time at which I am now writing (1855), many other specimens have come under my notice, the examination of which has confirmed my views as to its specific value : although in some of its characters it approximates to *B. rufescens*, its most near ally is the species found in Van Diemen's Land, and figured under the name of *B. cuniculus* ; it differs, however, from that animal in its more bluff head and in its shorter hind feet. Mr. Waterhouse remarks also, that although the many specimens which have come under his notice exhibited considerable variation in their colouring, and sometimes approximated very closely to other species, yet, with the assistance of the skull, he found no difficulty in distinguishing them.

I have received examples of this animal from various parts of the south-western coasts of Australia, and it appears to be equally abundant in the plains around Adelaide as in those in the neighbourhood of Perth in Western Australia. My drawing was taken from living examples in the Menagerie of the Zoological Society, and I mention this because the positions may appear somewhat singular, but they are correct representations of those the animals assumed at the time. Mr. Gilbert, who had many opportunities of observing the *Bettongia Graii* in Western Australia, states that :—

"It is truly gregarious, many dwelling together in extensively ramified burrows with several entrances, before which the excavated earth is formed into large mounds ; the openings are not, as usual, mere round holes, but are dug out in the form of tunnels with perpendicular sides, as correct as if dug with a spade. These burrows are usually constructed in a bank sloping down to a brook or river, and are very numerous along both banks of the river Avon. I made several attempts to dig them out, but failed in every instance in consequence of the depth, six or eight feet, and sometimes even more, at which the burrows are constructed, and of their running one into the other in endless confusion. The Boöt-dee is exclusively a nocturnal feeder, and, by quietly watching near the entrances to the burrows at sunset, may be shot in considerable numbers either when they emerge or while feeding in the immediate vicinity. It is one of the most destructive animals to the garden of the settler that occurs in Western Australia, almost every kind of vegetable being attacked by it, but especially peas and beans ; and I know of no species of its size which makes so loud a thumping noise while hopping along the ground on being alarmed ; besides making this noise with its feet, it also utters, when first started, a most singular succession of sounds, which I find it impossible to describe. Many of the specimens brought in by the natives were much discoloured, either by their dirty cloaks, or the clayey soil in which they had been captured. A remarkable circumstance connected with this animal is, that it is extremely difficult to meet with specimens which are not more or less denuded of the fur of the back, and I have often shot examples almost destitute of fur on any part of the body ; whether this is the result of disease or some accidental circumstance I am unable to say, but the skins of several I examined certainly presented a very similar appearance to that of dogs afflicted with mange.

"The Boöt-dee is confined to the interior, and, besides burrowing as above described, sometimes dwells among the rocks like the *Petrogalæ*."

Fur of the upper and under surface grey at the base ; hairs of the under surface dirty-white externally ; those of the back dirty-white, inclining to ash-colour near the apex, and tipped with brownish-black ; on the sides of the head and body a very faint wash of yellow ; ears sparingly clothed, internally with small yellowish hairs, externally with fur like that of the head ; feet, greyish-brown in Western Australian specimens, and dark brown, inclining to chestnut, in those from South Australia. A similar difference occurs in the colouring of the tail ; there is also an absence of white hairs near the tip of South Australian specimens ; nose and other denuded parts flesh-colour.

The figures are about the size of life.

BETTONGIA RUFESCENS, Grey.

BETTONGIA RUFESCENS, *Gray*.

Rufous Jerboa Kangaroo.

Bettongia rufescens, Gray, Mag. Nat. Hist. 1837, vol. i. p. 584.—Gould, Mon. of Macropodidæ, pl.   .—Gray, List
of Mamm. in Coll. Brit. Mus., p. 94.

Hypsiprymnus rufescens, Waterh. in Jard. Nat. Lib. Mamm., vol. xi. (Marsupialia) p. 188.—Ib. Nat. Hist. of
Mamm., vol. i. p. 196.

———— *melanotis*, Ogilby in Proc. of Zool. Soc., part vi. p. 62.

————

THERE will be but little difficulty in distinguishing this species from every other member of the genus *Bettongia* yet discovered. It is the largest and most powerful of its tribe, and this remark applies particularly to its strong hind feet and legs : the hair with which it is clothed is also more harsh and bristly than that of its allies ; again, the back part of the ears is nearly black, and the back and upper surface generally are strongly suffused with chestnut-brown, with which the stiff silvery-white interspersed hairs present a strong contrast. The south-eastern portion of the continent is its true habitat ; and it is almost universally dispersed over New South Wales, both on the sea and interior side of the mountain ranges. I found it very abundant on the stony sterile ridges bordering the grassy flats of the Upper Hunter, and in all similar situations. It constructs a warm nest in which it lies coiled up during the day, the nests being placed under the shelter of a fallen tree or some scrubby bush : it sometimes sits in a form like the Hare Kangaroo, but never sits out on the open plains like that species : on being startled, it runs for a short distance with remarkable rapidity ; but, from its invariably seeking shelter in the hollow logs, easily falls a prey to the natives, who hunt it for food. In size it fully equals that of a full-grown rabbit : its food consists of roots and grasses. There is no material difference in the colouring of the sexes ; but in size the female is somewhat smaller than the male.

Fur harsh and wiry ; general colour grizzled-grey and rufous, the latter hue predominating on the back ; ears black externally and buffy-white internally ; under surface greyish-white, slightly tinged with buff ; tail strongly prehensile, covered with short wiry grizzled-grey hairs, becoming whiter towards the tip, where they are much lengthened ; under side of the tail, throughout its whole length, dirty-white ; hands grey ; nails white ; tarsi and feet greyish.

The figures are about the size of life.

BETTONGIA CAMPESTRIS.

BETTONGIA CAMPESTRIS, *Gould.*

Plain-loving Jerboa-Kangaroo.

Bettongia campestris, Gould in Proc. of Zool. Soc. Part xi. p. 81.
Hypsiprymnus (Bettongia) campestris, Waterh. Nat. Hist. Mamm., vol. i. p. 221.

It will be readily seen, on glancing at the accompanying Plate, that the *Bettongia campestris* cannot be confounded with any other species ; its bluff head, the yellow colouring of its sides, and the peculiarly rigid texture of its fur being characters not combined in any of its congeners.

The stony and sandy plains of the interior of South Australia partially clothed with scrub are its native habitat, and I have not yet seen specimens of it from the other colonies either to the east or to the westward.

As confirmatory of its specific value I quote from Mr. Waterhouse, who says :—

"This is a very distinct species, remarkable for its short and broad head, and its general pale yellowish colouring.

"The hairs of the back are grey at the root, yellow in the middle, then blackish, followed by a long yellow white space, and black tip; on the chest and belly they are pale grey at the base and yellowish externally, but on the lower part of the abdomen the grey is wanting ; the upper lip is white ; the muffle is naked ; tarsi rusty white ; the tail is sparingly clothed with small pale hairs on the upper surface and sides ; on the under part the hairs are more dense, harsher, and of a brownish white colour ; the sides of the body and the outer surface of the hind legs are of a more distinct yellowish hue than the other parts."

The figures are of the natural size.

Gould and H.C.Richter. del. et lith.

Hullmandel & Walton Imp.

HYPSIPRYMNUS MURINUS.

HYPSIPRYMNUS MURINUS.

New South Wales Rat-Kangaroo.

Poto-Roo or *Kangaroo Rat*, White's Journ., p. 286. pl.
Macropus minor, Shaw, Gen. Zool., vol. i. part 11. p. 513. pl. 116.
Hypsiprymnus murinus, Ill. Prod. Syst. Mamm., p. 79.
Potoroüs murinus, Desm. Dict. d'Hist. Nat., tom. xxvii. p. 79, 80.—Ib. Mammalogie, p. 271.
Hypsiprymnus setosus, Ogilb. in Proc. of Comm. of Sci. and Corr. of Zool. Soc., part 1. p. 149.
———— *Peron*, Quoy et Gaim. Zool. de l'Uranie, p. 64.
———— *myosurus*, Ogilb. in Proc. of Zool. Soc., part 6. p. 62?
———— (*Potoroüs*) *murinus*, Waterh. Nat. Hist. of Mamm., vol. i. p. 225.
Bettong of the Aborigines of New South Wales.

THE hot and dry climate of the Australian continent appears not to be so well adapted for the members of the genus *Hypsiprymnus* as the more humid atmosphere of Van Diemen's Land, and hence it is only in the swampy and damp parts of the brushes of New South Wales that the *H. Murinus* is to be found in any abundance. The district of Illawarra, Botany Bay, the low scrubs bordering the rivers Hunter, Manning, and Clarence, are the principal localities in which it may be successfully sought for.

The *Hypsiprymnus murinus* is one of the very oldest known species of the Australian quadrupeds, and this will, in some measure, account for the long list of synonyms assigned to it, and the diversity of opinion entertained by zoologists regarding its identity. Mr. Waterhouse, who has carefully investigated the subject, has cleared up these difficulties so successfully, that, my own opinion coinciding with his, it will be as well, perhaps, to transcribe the entire passage :—

" The present species was first described by Hunter under the name Potoroo, or Kangaroo-Rat, in the Appendix to " White's Journal," and from the description and somewhat rude figure there given, it would have been difficult to determine to which of the numerous species of Rat-Kangaroo since discovered, the Potoroo of White should be referred, were it not that the skull of that animal is still preserved in the Museum of the Royal College of Surgeons. By the aid of that skull we are enabled clearly to identify the Potoroo of " White's Journal " (upon which Shaw founded his *Macropus minor*) with the *Hypsiprymnus murinus* of Pander and D'Alton, and with the *H. Peron* of Quoy and Gaimard, founded upon a skull contained in the Paris Museum, of which Professor Owen has been so kind as to lend me a drawing.

" Mr. Ogilby states that the animal to which he has given the name of *H. setosus*, is known in the colony of New South Wales by the name ' *Bettong* '; and this remark no doubt has reference to the Rat-Kangaroo, so labelled in the collection of the Linnæan Society, which specimens not only agree with Mr. Ogilby's description, but also with the animal I identify with the *Macropus minor* of Shaw."

The following note was made by Mr. Richter from living examples in the menagerie of the Zoological Society :—

" Though these *Hypsiprymni* stand as much on the hind legs as the *Bettongiæ*, they run in an entirely different manner, using the fore as well as the hind legs in a sort of gallop. They also never attempt to kick with their hind legs. They are very gentle and inoffensive in their manners, and much more stupid than the *Bettongiæ*. They feed like pigs, very seldom using the fore hands to convey their food to their mouths, and seemed to be very partial to boiled rice. They are very pert in their attitudes, sitting up and wriggling the tail laterally; express disapprobation by a slight hiss or sharp expiration; are very quick in their movements, and equally lively during day or night."

Fur long, loose, slightly glossy, and on the upper surface of the body of a dusky brown, a tint produced by the visible portion of the longer hairs being black, and the shorter fur of a pale yellow hue; fur of the under surface deep grey next the skin and of a dirty yellowish white on the surface; ears short and rounded, clothed internally with dirty white hairs; feet brown.

The figures represent both sexes of the size of life.

HYPSIPRYMNUS APICALIS.

J. Gould and H. C. Richter, del. et lith.   Hullmandel & Walton, Imp.

HYPSIPRYMNUS APICALIS, *Gould.*

Tasmanian Rat-Kangaroo.

EVER since my visit to Australia I have been induced to consider the animal here figured, which is a native of Van Diemen's Land, to be distinct from the species known as the *Potoroo* or Kangaroo-Rat of "White's Journal," (*Potoroüs murinus*, Desm.), which inhabits New South Wales; it has not been an examination of dried skins which has induced this opinion, but abundant opportunities for observing the animal in a state of nature. Mr. Waterhouse, although he has made them identical, evidently had some doubt on the subject, since, when figuring the skull of the Van Diemen's Land animal, in comparison with that of the one from New South Wales, he places a note of interrogation after the name he has given to the former. I must admit that they are very closely allied, at the same time I find peculiar and well-marked characters by which they may each be distinguished from the other. The Tasmanian animal is always nearly a third larger in size, and has the tip of its tail white, a feature I have never seen in any other of the three species inhabiting the continent of Australia.

The *Hypsiprymnus apicalis* is very generally, I may say universally, dispersed over Van Diemen's Land; and I seldom failed to find it in low damp situations clothed with dense herbage: during the daytime it lies coiled up in its nest among the herbage in a depression of the ground; a very little noise near its retreat is, however, sufficient to disturb its repose, and cause it to dart away with rabbit-like rapidity to a place of security: it can seldom be induced to break covert into the open space, and if sharply pressed, invariably takes to the shelter of a large tree or stone, which everywhere abound; its food consists of roots, herbage, and the bark and leaves of trees. I must not omit to remark, that in no instance have I known dogs to partake of the flesh of this species either raw or dressed; while that of the members of the genus *Bettongia* is seldom refused. Mr. Richter has made so correct a drawing of this animal from life, and has so well represented in the reduced figures two of the positions frequently assumed by it, that a glance at the Plate will give more information on this point than any description.

The fur is long and of a dark hue; on the upper parts of the body it is of a dusky brown, a general tint produced by the admixture of brown and pale brownish yellow, the visible portion of the longer and coarser hairs being black, and that of the shorter fur of a pale yellow hue; the under surface of the body is of a dirty yellowish white or pale buff tint, with the fur of those parts as well as that of the back of a deepish grey colour next the skin; the ears are clothed internally with dirty white hairs, and externally with hairs of the same colour as the rest of the head; the feet are brown; the tail is of a darker hue than the body, and is tipped for about an inch with pure white; the muzzle is not only naked in front, but a narrow naked space continues upward towards the forehead.

The front figure represents the animal of the natural size.

HYPSIPRYMNUS GILBERTI, *Gould.*

Gilbert's Rat-Kangaroo.

Hypsiprymnus Gilberti, Gould in Proc. of Zool. Soc., part ix. p. 14.—Ib. Mon. of Macropodidæ, pl. .—Gray,
List. of Mamm. in Coll. Brit. Mus., p. 94.
——————— *micropus*, Waterh. in Jard. Nat. Lib., Marsupialia, p. 180.
——————— (*Potoroüs*) *Gilbertii*, Waterh. Nat. Hist. of Mamm., vol. i. p. 229.
Ngil-gyte, Aborigines of King George's Sound.

In its outward appearance this little animal closely resembles the *Hypsiprymnus murinus*, but on a comparison of the skulls of the two species a marked difference is observable, that of the present having the nasal bone more produced or swollen out at the sides; the tarsi and tail also are shorter, and the general colour is of a deeper hue in Gilbert's than in the *Hyp. murinus*. These *Hypsiprymni* are evidently analogues of each other, the former being found only on the western coast, while the other is confined to the eastern portions of Australia.

The animal here represented was procured at King George's Sound, where it is called *Ngil-gyte* by the Aborigines. In dedicating it to the late Mr. Gilbert, who proceeded with me to Australia to assist in the objects of my expedition, I embraced with pleasure the opportunity afforded me of expressing my sense of the great zeal and assiduity he displayed in the objects of his mission; and as science is indebted to him for the knowledge of this and several other interesting discoveries, I trust that, however objectionable it may be to name species after individuals, in this instance it will not be deemed inappropriate.

The above remarks were published in the first Part of my " Monograph of the Macropodidæ or Family of Kangaroos," soon after which Mr. Gilbert made a second journey to the interior of Western Australia, and while there, transmitted to me the following additional information respecting this species :—

" This little animal may be said to be the constant companion of *Halmaturus brachyurus*, as they are always found together amidst the dense thickets and rank vegetation bordering swamps and running streams. The natives capture it by breaking down a long, narrow passage in the thicket, in which a number of them remain stationed, while others, particularly old men and women, walk through the thicket, and by beating the bushes and making a yelling noise, drive the affrighted animals before them into the cleared space, where they are immediately speared by those on the watch: in this way a tribe of natives will often kill an immense number of both species in a few hours. I have not heard of the *Hypsiprymnus Gilberti* being found in any other part of the colony than King George's Sound."

General colour of all the upper surface mingled grey, brown and black, produced by the base of the hairs being grey, the middle portion brown and black; centre and lower part of the back washed with reddish brown; a blackish line commences at the nose and blends into the general colour on the forehead; all the under surface greyish white; hands greyish brown; feet blackish brown; tail black, very thinly clothed with short hairs.

The figures are of the natural size.

HYPSIPRYMNUS PLATYOPS.

HYPSIPRYMNUS PLATYOPS, *Gould.*

Broad-faced Rat-Kangaroo.

Hypsiprymnus platyops, Gould in Proc. of Zool. Soc., part xii. p. 103.

——————— (*Potoroüs*) *platyops*, Waterh. Nat. Hist. Mamm., vol. i. p. 231.

This species is the least of the family of Kangaroos yet discovered; and is so rare that an adult male in my own collection and another in that of the British Museum, both procured by Mr. Gilbert in Western Australia, one in the Walyema Swamps, near Northam in the interior, and the other at King George's Sound, are all the examples that have yet been seen. When compared with *Hypsiprymnus Gilberti* and its allies, the present species will be found to differ from the whole of them in several particulars; the more important of which are its smaller size and the great breadth of its zygomatic arches, which, together with the brevity of its nose, give to the facial aspect of the animal a very bluff appearance, not unlike that of the young Wombat.

At the time Mr. Waterhouse wrote the first volume of his "Natural History of the Mammalia," the specimen from which he took his description was the only one that had reached this country; the adult male I have since received differs in no material respect, and I therefore transcribe his remarks and description verbatim.

"This," says Mr. Waterhouse, "is a small and very distinct species, readily distinguished from *Hyps. minor* and *H. Gilberti* by its having the tip of the muzzle naked in front only; while in the two species first named the naked part of the muzzle is extended somewhat on the upper surface; the zygomatic arches (so far as may be judged from the skull enclosed in the skin) must be thrown boldly out from the cranium, and thus give the breadth to the face which suggested the specific name."

"The hairs constituting the fur, are, on the back, grey at the root, then yellowish brown, and this is followed by a long space in each hair which is white, and this again is succeeded by black, that being the colour of the tips of the hairs; the white portion, showing conspicuously, gives the upper parts of the body the appearance of being distinctly pencilled with that hue; on the under parts of the body each hair is pale grey at the root, and dusky white externally; the feet are dirty white, indistinctly grizzled with brownish; this latter tint being most distinct on the sides of the toes: the ears are short and rounded, externally clothed with longish hairs, which are partly brown and partly white, and internally with hairs which are of a dirty white."

The figures are of the natural size.

THE

MAMMALS OF AUSTRALIA.

BY

JOHN GOULD, F.R.S.,

F.L.S., F.Z.S., M.E.S., F.R.GEOG.S., M.RAY S.: HON. MEMB. OF THE ROYAL ACADEMY OF SCIENCES OF TURIN; OF THE ROYAL
ZOOL. SOC. OF IRELAND; OF THE PENZANCE NAT. HIST. SOC.; OF THE WORCESTER NAT. HIST. SOC.; OF THE
NORTHUMBERLAND, DURHAM, AND NEWCASTLE NAT. HIST. SOC.; OF THE NAT. HIST. SOC.
OF DARMSTADT; OF THE TASMANIAN SOC. OF VAN DIEMEN'S LAND; OF THE NAT.
HIST. SOC. OF STRASBOURG; OF THE NAT. HIST. SOC. OF IPSWICH; AND
CORR. MEMB. SOC. OF NAT. HIST. OF WÜRTEMBERG.

IN THREE VOLUMES.

VOL. III.

LONDON:

PRINTED BY TAYLOR AND FRANCIS, RED LION COURT, FLEET STREET.

PUBLISHED BY THE AUTHOR, 26 CHARLOTTE STREET, BEDFORD SQUARE.

1863.

LIST OF PLATES.

VOL. III.

Hapalotis albipes	White-footed Hapalotis	1
——— apicalis	White-tipped Hapalotis	2
——— hemileucura	Elsey's Hapalotis	3
——— hirsutus	Long-haired Hapalotis	4
——— penicillata	Pencil-tailed Hapalotis	5
——— conditor	Building Hapalotis	6
——— murinus	Murine Hapalotis	7
——— longicaudata	Long-tailed Hapalotis	8
——— Mitchellii	Mitchell's Hapalotis	9
——— cervinus	Fawn-coloured Hapalotis	10
Mus fuscipes	Dusky-footed Rat	11
——— vellerosus	Tawny Rat	12
——— longipilis	Long-haired Rat	13
——— cervinipes	Buff-footed Rat	14
——— assimilis	Allied Rat	15
——— manicatus	White-footed Rat	16
——— sordidus	Sordid Rat	17
——— lineolatus	Plain Rat	18
——— Gouldi	White-footed Mouse	19
——— nanus	Little Rat	20
——— albocinereus	Greyish-white Mouse	21
——— Novæ Hollandiæ	New Holland Field-Mouse	22
——— delicatulus	Delicate-coloured Mouse	23
Hydromys chrysogaster	Golden-bellied Beaver-Rat	24
——— fulvolavatus	Fulvous Beaver-Rat	25
——— leucogaster	White-bellied Beaver-Rat	26
——— fuliginosus	Sooty Beaver-Rat	27
Pteropus poliocephalus	Grey-headed Vampire	28
——— conspicillatus	Spectacled Vampire	29
——— funereus	Funereal Vampire	30
Molossus Australis	Australian Molossus	31
Taphozous Australis	Australian Taphozous	32
Rhinolophus megaphyllus	Great-leaved Horse-shoe Bat	33
——— cervinus	Fawn-coloured Bat	34
——— aurantius	Orange Horse-shoe Bat	35
Nyctophilus Geoffroyi*	Geoffroy's Nyctophilus	36
——— Geoffroyi	Geoffroy's Nyctophilus	37
——— unicolor	Tasmanian Nyctophilus	38
——— Timoriensis	Western Nyctophilus	39
Scotophilus Gouldi	Gould's Bat	40
——— morio	Chocolate Bat	41
——— microdon	Small-footed Bat	42
——— picatus	Pied Scotophilus	43
——— nigrogriseus	Blackish-grey Scotophilus	44
——— Greyi	Grey's Scotophilus	45
——— pumilus	Little Bat	46
Vespertilio macropus	Great-footed Bat	47
——— Tasmaniensis	Tasmanian Bat	48
Arctocephalus lobatus	Cowled Seal	49
Stenorhynchus leptonyx	Sea Leopard	50
Canis dingo	The Dingo	51, 52

HAPALOTIS ALBIPES, Licht.

Hullmandel & Walton, Imp.

Drawn and on Stone by J. Gould & H.C. Richter.

HAPALOTIS ALBIPES, *Licht.*

White-footed Hapalotis.

Hapalotis albipes, Licht. Darst. der Saugth., tab. 29.—Gray, Ann. Nat. Hist., vol. ii. p. 308.—*Id.*, List of Mamm.
    in Coll. Brit. Mus., p. 115.—Gould in Proc. Zool. Soc. 1851, p. 126.
Conilurus Constructor, Ogilb. Trans. Linn. Soc., vol. xviii. p. 126.
Bar-roo, Aborigines of the Darling Downs, New South Wales.
The Rabbit Rat of the Colonists, Benn. Cat. of Australian Museum, Sydney, p. 6. no. 30.

THE native habitat of the *Hapalotis albipes* is the south-eastern portions of Australia generally; it is dispersed over all parts of New South Wales, Port Philip and South Australia, but is nowhere very abundant. The South Australian specimens and those of New South Wales assimilate very closely, while those from the Darling Downs district are rather browner in the colouring of the fur and have shorter hind feet. Although I regard this latter animal from the table lands as only a local variety, it may at some future time prove to be distinct.

Judging from my own observations I should say that the *Hapalotis albipes* is strictly nocturnal in its habits, for it sleeps during the day in the hollow limbs of prostrate trees, or such hollow branches of the large *Eucalypti* as are near the ground, in which situations it may be found curled up in a warm nest of dried leaves; more than once have I, after detecting the animal in its retreat, sawn off the hollow limb and secured it without injury. In a note with specimens from Darling Downs in New South Wales, Mr. Gilbert states that " it is generally found inhabiting hollow logs or holes in standing trees."

The following note respecting this species was sent to me by my kind friend His Excellency Sir George Grey, now Governor of New Zealand, during his Governorship of the Colony of South Australia :—

" This animal lives among the trees. The specimen I send you, a female, had three young ones attached to its teats when it was caught; the mother has no pouch, but the young attach themselves with the same or even greater tenacity than is observable in the young of the *Marsupiata*. While life remained in the mother they remained attached to her teats by their mouths, and grasped her body with their claws, thereby causing her to present the appearance of a Marsupial minus the pouch. On pulling the young from the teats of the dead mother, they seized hold of my glove with the mouth and held on so strongly that it was difficult to disengage them."

I had frequent opportunities of observing this animal in a state of nature during my rambles in the interior of Australia, and Mr. Gilbert was equally fortunate during his short sojourn in New South Wales. I mention this, because certain habits and nest-making propensities have been referred to this animal by Sir Thomas Mitchell, W. Ogilby, Esq., and others, which belong not to this species, but to the *Hapalotis conditor*, a fact which is fully established by the drawings, specimens and notes of that species made on the spot and communicated to me by Captain Sturt.

Fur long, close and soft; head, ears, upper surface, flanks and outer surface of the limbs grey at the base and ashy brown on the surface, interspersed with numerous fine black-tipped hairs; whiskers and a narrow line around the eye black; under surface of the body, inner surface of the limbs, hands and upper surface of the feet white; upper surface of the tail dark brown; sides and under surface white.

The figures are of the size of life.

HAPALOTIS ALBIPALIS, Gould

HAPALOTIS APICALIS, *Gould.*

White-tipped Hapalotis.

Hapalotis apicalis, Gould, in Proc of Zool. Soc., 1851, p. 126.

THIS new species is about the size of, and similar in colour to, *H. albipes,* but it differs in having larger ears, much more delicately formed feet, the tail nearly destitute of the long brushy hairs towards the tip, and smaller eyes.

I possess a single example only of this species; it was procured by Mr. Strange in South Australia. There is an animal in spirits in the British Museum, presented by R. C. Gunn, Esq., from Van Diemen's Land, which accords very closely with it in the colouring of the fur, and in the rat-like form of the tail; it is, however, of much smaller size, and in all probability will prove to be a new species.

Face and sides of the neck blue-grey; upper part of the head, space between the ears, the ears and upper parts of the body pale brown, interspersed with numerous fine black hairs; under surface white; flanks mingled grey and buffy white; fore feet white, with an oblique mark of dark brown separating the white from the greyish brown of the upper surface; hinder tarsi and feet white; basal three-fourths of the tail brown, apical fourth thinly clothed with white hairs.

The figures are the size of life.

HAPALOTIS HEMILEUCURA, Grey

HAPALOTIS HEMILEUCURA, *Gray*.

Elsey's Hapalotis.

Hapalotis hemileucura, Gray in Proc. of Zool. Soc., part xxv. p. 243.

Iᴛ is with a degree of mixed pleasure and regret that I bring before the notice of the scientific world this new species of *Hapalotis*. It was brought home by that young and intelligent naturalist, the late Mr. J. R. Elsey, Surgeon to the expedition conducted by A. C. Gregory, Esq., from the north-western coast of Australia to Moreton Bay : all who like myself had an opportunity of becoming acquainted with the amiable qualities of this gentleman, cannot but regret the loss the science of natural history has sustained by his premature decease. On the part of Dr. Gray, I brought this animal before the Meeting of the Zoological Society held on the 24th of November, 1857, and gave it the name of *hemileucura*, a term suggested by the parti-colouring of the tail. Only a single specimen was procured, and this is now in the British Museum. I am unable to state the precise locality in which it was obtained, but believe it was about midway between the Gulf of Carpentaria and Moreton Bay.

The *Hapalotis hemileucurus* is a harsh wiry-furred animal, nearly allied to, but considerably larger than, the *H. melanura*, from which it also differs in having the apical half of the tail white.

Head, all the upper surface and flanks very light sandy brown, with numerous, but thinly placed, fine, long black hairs ; under surface buffy white, with even lighter feet and fore-arms ; tail brown, deepening into black about the middle, beyond which the apical portion is white, the white hairs being prolonged into a small tuft at the tip.

	inches.
Length from the nose to the base of the tail	8
,, of the tail	6¼
,, ,, fore-arm	1¼
,, ,, tarsus and toes	1¼

The figures are of the natural size.

HAPALOTIS ? HIRSUTUS, Gould

HAPALOTIS HIRSUTUS, *Gould.*

Long-haired Hapalotis.

Mus hirsutus, Gould in Proc. of Zool. Soc., part x. p. 12.—Ib. Ann. & Mag. Nat. Hist., vol. x. p. 405.
Hapalotis hirsutus, Gould in Proc. of Zool. Soc., part xix. p. 127.

THE discovery of this rare Australian animal is due to the researches of the late Mr. Gilbert, who obtained a single specimen during his sojourn at Port Essington on the Cobourg Peninsula in 1840 ; since that period a second example from the same locality has been sent to this country, and, as well as the former, deposited in the British Museum. It will be seen, by the synonyms above given, that I at first regarded this animal as a true *Mus*, and that I subsequently assigned it a place in the genus *Hapalotis*. I am, however, by no means satisfied that this is its right situation, and think it possible that, when a sufficient number of specimens have been received to justify the formation of a correct opinion upon the subject, it may be found desirable to constitute it the type of a new genus.

The following is a copy of my original description of the animal, published in the 10th Part of the Proceedings of the Zoological Society of London :—

" Fur coarse and shaggy ; on the upper parts of the body the shorter hairs are of a yellowish-brown colour, but the longer interspersed hairs, being numerous and of a black colour, give a deep general tint to those parts ; the under parts of the body are of a rusty-yellow colour, tinted with brownish on the neck and chest, and having a more decided rust-colour on the abdomen ; tail well clothed with lengthened hairs, especially on the apical half, where the scales are hidden by them ; those at the point of the tail measure upwards of an inch in length ; on this part they have a rusty hue, but on the remaining portions they are black."

The Plate represents the animal of the natural size.

HAPALOTIS PENICILLATA, *Gould.*

J. Gould and H.C. Richter, del. et lith.

Hullmandel & Walton, Imp.

HAPALOTIS PENICILLATA, *Gould.*

Pencil-tailed Hapalotis.

Mus penicillatus, Gould in Proc. of Zool. Soc., Part X. p. 12.—Ann. and Mag. Nat. Hist., vol. x. p. 405.—List
of Mamm. in Brit. Mus., p. 109.
Hapalotis melanura, List of Mamm. in Brit. Mus., p. 115. ?

THIS animal was procured by Mr. Gilbert during his short sojourn at Port Essington on the Cobourg
Peninsula in Northern Australia, in which part of the country it was also obtained by Mr. MacGillivray and
sent by him to the late Earl of Derby; specimens from the same country are also contained in the collection
at the British Museum.

It is in every respect a true *Hapalotis*, and may be readily distinguished from the other members of the
genus by the blackness of its tail, the hairs of which are much lengthened; and by the rigid, almost spiny,
nature of the hairs clothing the back. Its habits would seem to be somewhat singular, inasmuch as it is
frequently found among the swamps on the sea-shore; I have no evidence, however, that it is not also
found in the interior of the country. I find the following note respecting it among the papers of the
late Mr. Gilbert:—

"This little animal is only seen on the beach where there are large *Casuarina* trees, in the dead hollow
branches of which it forms a nest of fine dry grass, and retires during the day; in the evening it leaves its
retreat and proceeds to the beach, where it may be seen running along at the edge of the surf as it rolls up
and recedes, apparently feeding upon any animal matter washed up by the waves."

The fur of the upper surface is greyish brown grizzled with buff, with a rusty tint on the region of the
occiput and back of the neck; around the angle of the mouth, the chin, throat, and all the under parts of
the body, as well as the feet and inner side of the legs, are white with a faint yellow tint or cream-coloured,
and the hair of these parts is of a uniform tint to the roots except on the chest, where they are grey next the
skin: the tail is sparingly clothed at the base with minute bristly hairs; but about the middle the hairs
become of a black colour and longer, and towards the apex attain a considerable length, measuring at and
near the tip half an inch or more: the ears are sparingly clothed with minute hairs.

The figures represent the two sexes of the size of life.

HAPALOTIS? CONDITOR: *Grald.*

HAPALOTIS CONDITOR, *Gould.*

Building Hapalotis.

Mus conditor, Gould in Sturt's Narr. of Exp. to Central Australia, vol. i. pl. in p. 120 ; vol. ii. App. p. 7.

FOR a knowledge of this curious little animal, we are indebted to the researches of Captain Sturt, who, during his recent expedition into the central portion of Australia, found it inhabiting the brushes of the Darling; there is little doubt that it had been previously met with by Major Mitchell, who, in the second volume of his " Three Expeditions into the Interior of Eastern Australia," page 263, when speaking of the specimens collected during the journey, mentions, among others, " the flat-tailed rat from the scrubs of the Darling, where it builds an enormous nest of branches and boughs, so interlaced as to be proof against any attacks of the native dog;" but as the specimen he procured appears never to have been described, the credit of its first introduction to science is due to the first-mentioned traveller.

In its general form and dentition it is very nearly allied to the members of the genus *Mus*, but its lengthened and broad hind-feet, large ears, and its habit of constructing a nest, are characters which in an equal degree ally it to the *Hapaloti*, with which, upon a closer examination of its structure, I am induced now to associate it.

Captain Sturt states that it " inhabits the brushes of the Darling, but was not found beyond latitude 30°. It builds a nest of small sticks varying in length from three to eight inches, and in thickness from that of a quill to that of the thumb, arranged in a most systematic manner so as to form a compact cone like a bee-hive, about four feet in diameter and three feet high; those at the foundation are so disposed as to form a compact flooring, and the entire fabric is so firm as almost to defy destruction except by fire. The animal, which is like an ordinary rat, only that it has longer ears and the hind-feet are disproportioned to the fore-feet, lives in communities, and traverses the mound by means of passages leading into the apartments in the centre. One of these nests or mounds had five holes or entrances at the base, nearly equidistant from each other, with passages leading from them to a hole in the ground beneath, in which I am led to conclude they had their store. There were two nests of grass in the centre, with passages running up to them diagonally from the bottom; the nests were close together, but in separate compartments, with passages communicating from the one to the other."

Fur soft and silky to the touch; general colour greyish brown, becoming of a darker hue down the centre of the head and back, in consequence of the tips of the hairs being dark brown; under surface pale buff, the whole of the fur dark slate-grey at the base; a slight wash of rufous between the ears; whiskers very long, exceedingly fine, and of a blackish brown hue; fore-feet brown, hinder feet pale brown; tail brown above, paler beneath.

The figures are of the natural size.

HAPALOTIS MITCHELII , Gould

Gould and H.C.Richter del et lith

Hullmandel & Walton Imp

HAPALOTIS MURINUS, *Gould.*

Murine Hapalotis.

Hapalotis murinus, Gould, in Proc. of Zool. Soc., Part xiii. p. 78, and 1851, p. 127.

THE large size of its ears, the peculiar softness of its fur, and the whiteness and length of the hinder feet of this animal induced me some years ago to characterize it under the generic name of *Hapalotis* rather than that of *Mus*, and I still adhere to the opinion I then formed, that it must not be associated with the true Rats. The original specimen from which my description was taken was procured by Mr. Gilbert on the plains bordering the rivers Namoi and Gwydyr, where the natives informed him it was very abundant. Mr. Strange subsequently sent me examples from the neighbourhood of Lake Albert in South Australia, which, although of somewhat larger size, are, I believe, identical. He states that he found them in families on the edge of the small dry salt-water lagoons of the plains, and that they did not appear to go far from their habitations.

Fur remarkably soft and delicate, and of a slate-grey tint next the skin; on the upper surface and the sides of the body the exposed portions of the hairs are of a delicate ochreous yellow, with a considerable admixture of black, the points of the hairs being of that colour; ears tolerably well clothed with small hairs of a white hue, excepting on the fore part of the outer surface, where they assume a dusky greyish tint; under surface buffy white; tail moderately clothed with hairs, but not so thickly as to hide the scales; on the upper surface some of these hairs are white and others blackish, on the sides and under surface of the tail they are pure white; whiskers black at the root, greyish at the point; hands and feet buffy white.

The figures represent the two sexes of the size of life.

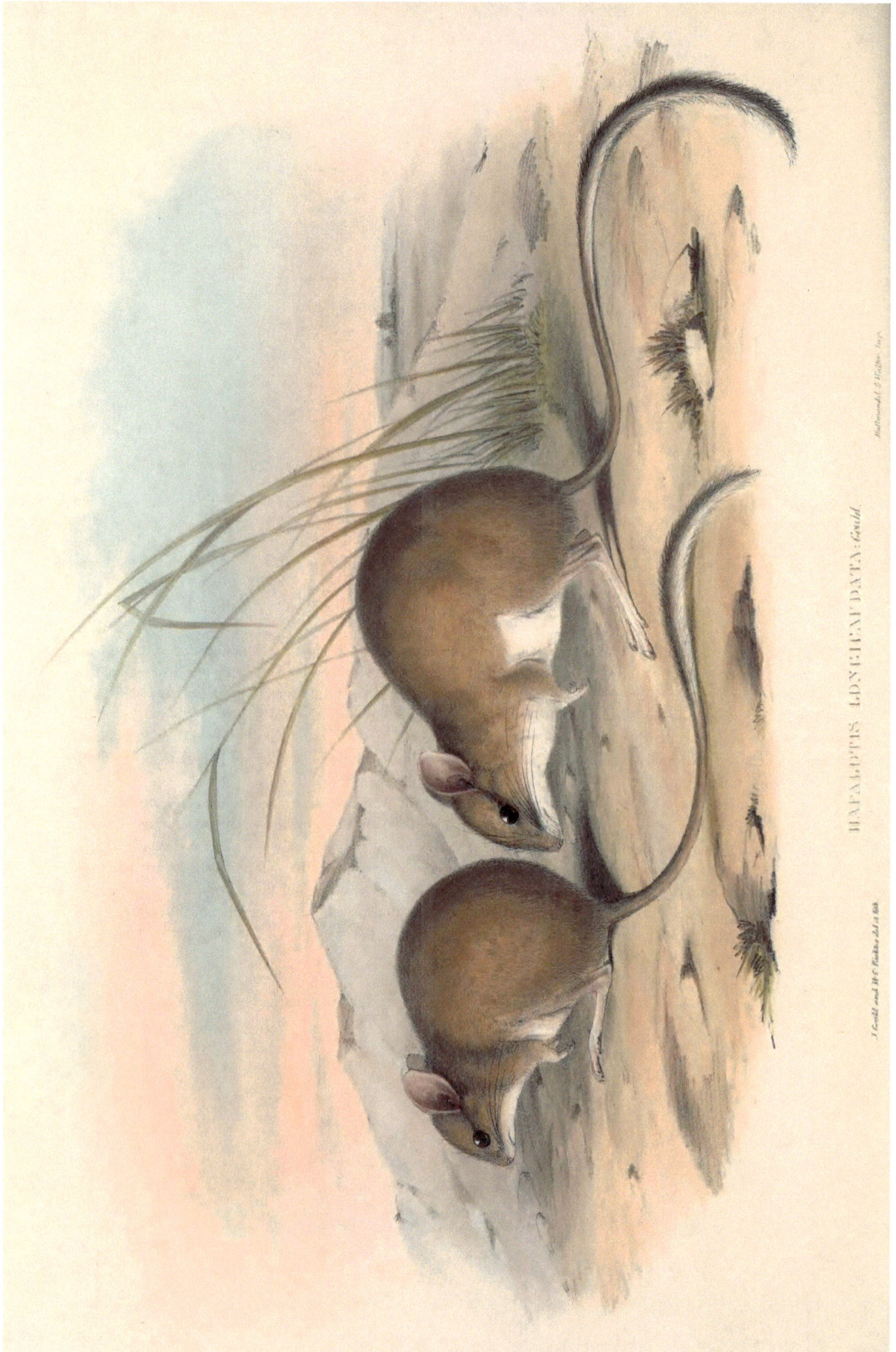

HAPALOTIS LONGICAUDATA, Gould

J. Gould and H. C. Richter del. et lith.

Hullmandel & Walton Imp.

HAPALOTIS LONGICAUDATA, *Gould.*

Long-tailed Hapalotis.

Hapalotis longicaudata, Gould in Proc. of Zool. Soc., Part XII. p. 104.
Kor-tung and *Gool-a-wa*, Aborigines in the neighbourhood of Moore's River in Western Australia.

ALTHOUGH very similar in form and style of colouring to the *Hapalotis Mitchellii*, the larger size and the greater length of its tail are characters by which the present animal may be distinguished from that species.

The interior of Western Australia is the only locality in which the *Hapalotis longicaudata* has been procured. The individuals forming the subject of the accompanying Plate were obtained in the vicinity of Moore's River, and now form part of the collection at the British Museum. They were sent to me by Mr. Gilbert, whose notes relative to the present species may not prove uninteresting :—

" This species differs considerably in its habits from the *Djyr-dow-in* (*Hapalotis Mitchellii*), for while that animal burrows in sandy districts, the favourite haunt of the present species is a stiff and clayey soil. It is also very partial to the mounds thrown up by the *Boordee's* (*Bettongia Grayii*) and the *Dal-goitch* (*Peragalea lagotis*). It is less destructive to the sacks and bags of the store-rooms, but, like the *H. Mitchellii*, is extremely fond of raisins."

All the upper surface and the outside of the limbs pale sandy, interspersed on the head and over the back with numerous fine black hairs, which becoming longer on the lower part of the back and rump, give that part a dark or brown hue; ears naked and of a dark brown; sides of the muzzle, all the under surface and the inner surface of the limbs white; tail clothed with short dark brown hairs at the base, with lengthened black hairs tipped with white on the apical half of its length, the extreme tip being white; tarsi white; whiskers very long, fine and black; the fur is close, very soft, and of a dark slaty grey both on the upper and under surface.

The figures represent a male and a female of the natural size.

HAPALOTIS MITCHELLII.

HAPALOTIS MITCHELLII.

Mitchell's Hapalotis.

Dipus Mitchelli, Ogilby in Linn. Trans., vol. xviii. p. 129.—Mitch. Trav., vol. ii. p. 144. pl. 29.—Gould in Proc.
    of Zool. Soc., Part VIII. p. 151.
Hapalotis Mitchellii, Gray, App. to Grey's Trav., vol. ii. p. 404.
————— *Gouldii*, Gray, App. to Grey's Trav., vol. ii. pp. 404 and 413.
Djyr-doib-in, Aborigines around Perth, and
Màt-tee-getch, Aborigines in the neighbourhood of Moore's River, Western Australia.

———————————

THE animal here represented was originally described by Mr. Ogilby under the name of *Dipus Mitchellii*,
from a drawing by Major Sir Thomas L. Mitchell of a specimen obtained by him on the banks of
the river Murray in South Australia, and now deposited in the Museum at Sydney; since that period
specimens have been sent to the Zoological Society of London by the late Mr. J. B. Harvey from South
Australia, and to myself by Mr. Gilbert from Western Australia, all of which appear to be identical with
the animal discovered by Sir Thomas Mitchell; at least the specimens from Southern and Western
Australia have been found on comparison to be precisely similar, and Mr. Gilbert informs me that on
examining the Major's specimen in the Sydney Museum, he could perceive no specific difference between
it and those transmitted by himself from Western Australia. That they are identical there can be little
doubt, when we take into consideration that Sir Thomas Mitchell's specimen was procured at no great
distance from the locality in which Mr. Harvey obtained his.

The range of this species is very extensive, and it is probable that the greater portion of the interior
of the country will hereafter be found to be inhabited by it.

The only information received respecting the habits of this animal is, that in Western Australia it burrows in the ground; taking up its abode on the sides of grassy hills tolerably well-clothed with small trees
growing in a light soil. It occasionally makes its way into the stores of the settlers, and commits depredations on the provisions, particularly sugar and raisins, of which it is exceedingly fond.

The sexes in size and colour offer no material difference.

All the upper surface and the outside of the limbs very pale sandy, interspersed over the head and back
with fine black hairs, which becoming numerous and longer on the lower part of the back and rump, give
that part a black or brown hue; ears naked and of a dark brown; sides of the face, all the under surface,
inner side of the limbs and feet greyish white; down the centre of the throat and chest a broad patch of
pure silky white; upper surface of the tail dark brown, under surface white, the hairs becoming much
lengthened on the upper surface at the tip; whiskers very long, fine and black; the fur is close, very soft,
and of a slaty grey at the base, both on the upper and under surface.

The accompanying Plate represents the animal in three positions, and of the natural size.

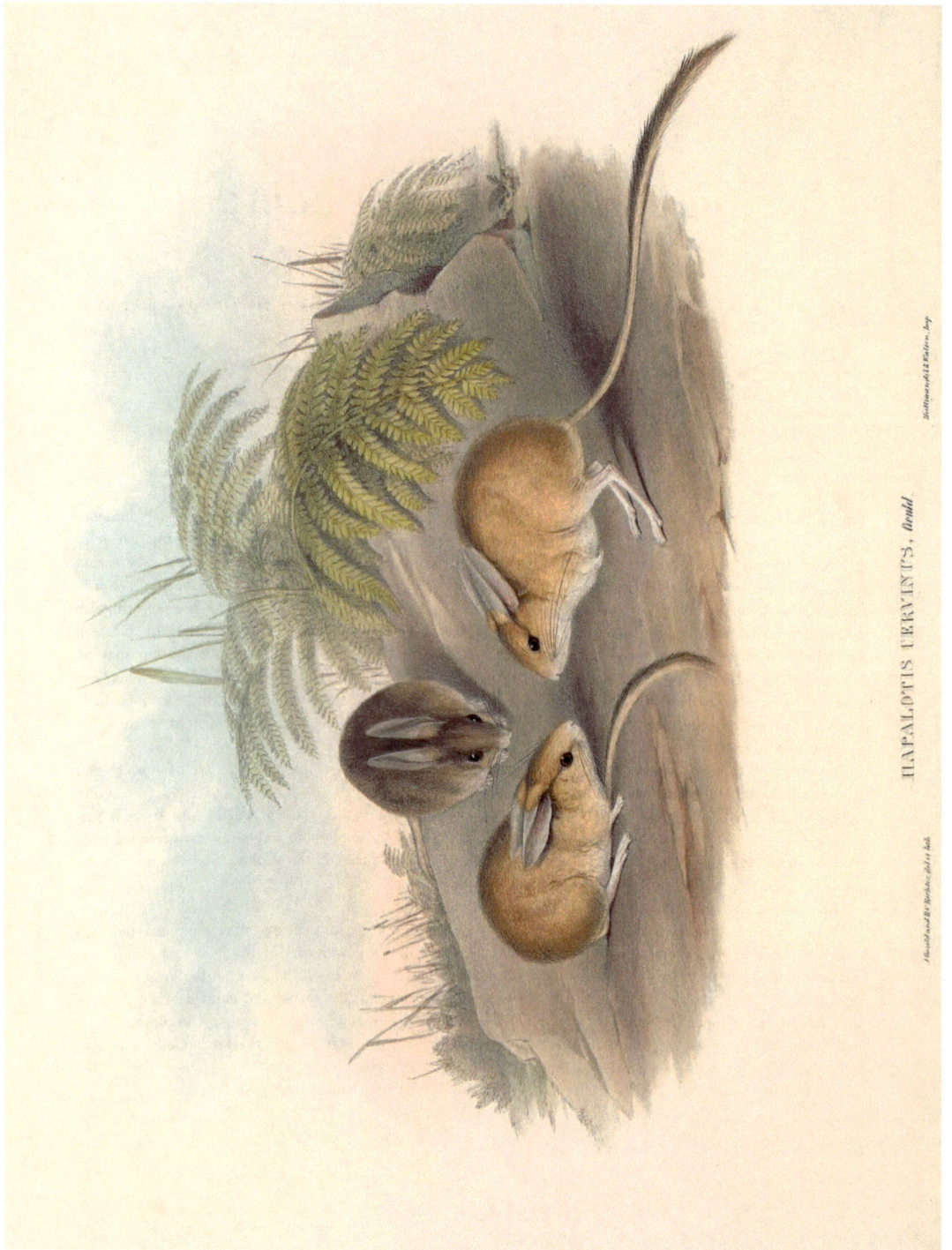

HAPALOTIS CERVINUS, *Gould.*

HAPALOTIS CERVINUS, *Gould.*

Fawn-coloured Hapalotis.

Hapalotis cervinus, Gould, in Proc. of Zool. Soc. 1851, p. 127.

IF the Great Red Kangaroo may be extolled as the finest of the Kangaroos, it must be conceded that the present animal is the most graceful and elegant of the Jerboa-like rodents to which the generic term of *Hapalotis* has been applied. For its discovery and introduction to this country we are indebted to the researches of Captain Sturt, who has thus afforded another instance of the anxiety with which this intrepid traveller seeks to promote the cause of science, not only in his immediate vocation as a soldier and explorer, but in the department of zoology, a department never neglected by him whenever he has had opportunities of adding to its stores. It was during the most hazardous of his journeys towards the centre of Australia, that Captain Sturt first met with this pretty species.

"On the 20th," says Captain Sturt, " we found ourselves in lat. 29° 6', and halted on one of those clear patches on which the rain-water lodges, but it had dried up, and there was only a little for our use in a small gutter not far distant. Whilst we were here encamped, a little Jerboa was chased by the dogs into a hole close by the drays, which, with four others, we succeeded in capturing by digging for them. This beautiful little animal burrows in the ground like a mouse, but their habitations have several passages leading straight, like the radii of a circle, to a common centre, to which a shaft is sunk from above, so that there is a complete circulation of air along the whole. We fed our little captives on oats, on which they thrived and became exceedingly tame. They generally huddled together in a corner of their box; but when darting from one side to the other, they hopped on their hind legs, which, like those of the Kangaroo, are much longer than the fore ones, and held the tail perfectly straight and horizontal. At this date they were a novelty to us, but we subsequently saw great numbers of them, and ascertained that the natives frequented the sandy ridges in order to procure them for food. Those we succeeded in capturing were, I am sorry to say, lost from neglect. This species feeds on tender shoots of plants, and must live for many months without water, the situation in which it is found precluding the possibility of its obtaining any for lengthened intervals."

The whole of the head, upper surface and sides of the body of the most delicate fawn-colour, interspersed with numerous fine black hairs on the head and back; whiskers greyish black; nose and under surface white; tail pale brown, lighter beneath; ears very large, somewhat pointed, and nearly destitute of hairs.

The figures are of the natural size; the darker-coloured figure representing a variety sometimes met with.

MUS FUSCIPES, Waterh.

MUS FUSCIPES, *Waterh.*

Dusky-footed Rat.

Mus fuscipes, Waterh. in Darwin's Zool. of the Voy. of H.M.S. Beagle, Mammalia, p. 66. pl. 25.—Cat. of Mamm.
   in Brit. Mus., p. 111.
—— *lutreola*, Gray, App. to Grey's Journ. of Two Exp. of Disc. in Australia, vol. ii. p. 409.

THIS species of Rat is distributed in abundance over the whole of the southern portion of Australia ; but I have no evidence that its range extends to the north coast. Specimens from Swan River in Western Australia, the swamps and thick brushes of New South Wales, the intermediate colony of South Australia, and the islands in Bass's Straits, differ in no respect from each other. Its favourite haunts are low and humid situations where long grass and herbage abound, and the banks of freshwater brooks and lagoons

Although belonging to a different genus, it presents in its aquatic habits and in many of its actions a striking resemblance to the common Water Vole (*Arvicola amphibius*) of Europe ; like that animal, it swims with the greatest ease, and may be constantly seen crossing and recrossing the small brooks and water-holes so abundant in the localities it frequents. It is rather less than the *Mus Rattus* in size, but is of a stouter form, and is moreover remarkable for the great length and softness of its fur, and the brown colour of its feet.

The general tint of the upper surface and the sides of the head and body is blackish brown, with an admixture of grey ; of the under surface greyish white ; the feet are brown, the hairs being greyish at the tip ; the tail is black, and but sparingly clothed with short bristly hairs ; the ears are rather sparingly clothed with hairs, which are for the most part of a brownish grey colour ; the ordinary fur of the back is about three-quarters of an inch in length and very soft, of a deep grey colour broadly annulated with brownish yellow near, and blackish at, the tip ; the longer black hairs measure upwards of an inch and a quarter in length ; the incisor teeth are orange-coloured.

The figures in the accompanying Plate represent the animal correctly both in size and colour.

MUS VELLEROSUS, *Gray*

MUS VELLEROSUS, *Gray.*

Tawny Rat.

Mus vellerosus, Gray in Proc. of Zool. Soc., part xv. p. 5.

In the fifteenth part of the " Proceedings of the Zoological Society of London," above referred to, will be found the description of a species of Rat, sent from South Australia by His Excellency Governor Grey. This supposed species received from Dr. Gray the name of *Mus vellerosus* : I say supposed species, because I believe it to be a *lusus*, either of the *Mus fuscipes*, or some nearly allied species ; still, although entertaining this opinion, I have considered it necessary to give an accurate figure of the animal in the present work, and I must leave it to future zoologists to ascertain if it be or be not a true species. It differs from the *Mus fuscipes* not only in its tawny colouring, but in the great length of its furry coat, all the hairs of which are of an equal length, or nearly so ; it is also very different from the *Mus longipilis*, with which indeed I am convinced it has no relationship whatever. Only a single example has yet reached this country, and it is on this that Dr. Gray has founded the species, accompanied with the following remark :—

" This rat has the dentition and somewhat the general appearance of *Mus fuscipes*, Waterh., but the skull and animal are considerably larger, and the fur is very much longer and paler."

Fur long and rather soft to the touch ; general colour reddish brown, varied with whitish interspersed hairs, becoming paler on the sides and still paler beneath, the base of the fur being bluish grey ; feet and tail brown.

The animal is figured on the accompanying Plate of the natural size.

MUS LONGIPILIS, *Gould.*

MUS LONGIPILIS, *Gould.*

Long-haired Rat.

I AM indebted to the Directors of the Australian Museum at Sydney for permission to figure this remarkable species of Rat, and for the loan of the unique specimen from which my drawing was taken. In size it approximates very closely to the Common Rat of Europe (*Mus Rattus*), but is at once distinguished from that species by the light buffy hue of its fur, and by the great length of the numerous black hairs interspersed along the back, which latter feature has suggested the specific name of *longipilis*.

In the brief notes kindly transmitted to me by Mr. William Sheridan Wall, that gentleman informs me that it was killed by his late brother, Mr. Thomas Wall, during his expedition to the Victoria River, on a desert which abounded with these animals. " In the absence of vegetation, it was interesting to ascertain, if possible, their means of existence. The stomachs of several were examined with this view, and all were found to contain a fleshy mass, leading to the supposition that they preyed upon each other, for no other animal was found to inhabit the locality." This mode of feeding was doubtless only temporary, probably caused by the entire absence, at the time, of the seeds and other vegetable substances suitable to its economy. It is to be regretted that more examples of this new species were not procured, especially as the one I have figured must be returned to the Australian Museum ; examples of so curious a Rat would be very desirable accessions to our national and other collections.

Fur very long, hairy and somewhat harsh to the touch, of a greyish brown at the base, and tawny buff on the surface, numerously interspersed, especially along the back, with very long, fine, black hairs ; under surface of the body buffy grey ; feet flesh-colour, sparingly clothed with silvery white hairs ; tail thinly beset with fine, stiff, black hairs, between which the usual scaly appearance is perceptible.

Total length, from the tip of the nose to the end of the tail, $13\frac{1}{4}$ inches ; of the tail, $5\frac{3}{4}$; of the nose to the ear, $1\frac{1}{4}$; of the ear, $\frac{3}{4}$; of the tarsi, $1\frac{6}{16}$ inch.

The figures are of the natural size.

MUS CERVINIPES, *Gould.*

MUS CERVINIPES, *Gould.*

Buff-footed Rat.

Mus cervinipes, Gould in Proc. of Zool. Soc., 1852.

THE species of Rat figured on the accompanying Plate, which is rather widely dispersed over the eastern coast of New South Wales, possesses characters which distinguish it from all the known members of the genus inhabiting that country; its short, soft, adpressed, furry coat, destitute of any lengthened hairs along the back and sides of the body, is one of the characters alluded to, the nearly uniform rufous colouring of its upper surface is another, and its slender, hairless, reticulated tail forms a third. The eastern brushes generally from the River Hunter to Moreton Bay are known to be inhabited by it; but how far its range may extend to the northward is as yet unascertained. Among the numerous specimens sent to me by Mr. Strange, several are labelled with the localities in which they were killed,—viz. Stradbrook Island, Moreton Bay, where it is called *Corrill* by the natives,—Richmond River, where the Aborigines term it *Cunduoo*, —and the plains bordering the upper parts of the River Brisbane.

The specific name has been suggested by the fawn-like colouring of its broad tarsi and feet.

Head, all the upper surface and flanks sandy brown, the base of the fur being dark slate-grey; tarsi and feet fawn-colour; under surface mottled buffy white and grey, the base of the fur being grey, and the extremity buffy white; tail purplish flesh-colour.

In some specimens the buffy white hue predominates and becomes conspicuous on the throat and breast.

In the young animal the upper surface is bluish grey and the under surface greyish white.

The figures in the accompanying Plate represent an adult of each sex and three very young individuals, all of the natural size.

MUS ASSIMILIS, *Gould.*

MUS ASSIMILIS, *Gould.*

Allied Rat.

Mus assimilis, Gould in Proc. of Zool. Soc., part xxv. p. 241.
Moor̆-deet, Aborigines of King George's Sound.

THE Allied Rat is somewhat numerous in New South Wales. The two specimens from which the characters of the species were taken for the " Proceedings of the Zoological Society," above quoted, were procured by the late Mr. Strange on the banks of the Clarence. I have three other specimens collected by Mr. Gilbert at King George's Sound, which differ only in being about a fifth smaller in all their admeasurements: it is just possible that it will hereafter be found that these latter animals are distinct from the former, but at present I regard them as identical, and if such be the case, the range of the species extends along the whole southern seaboard of the continent from east to west.

The *Mus assimilis* is about the same size as the *Mus decumanus* of Europe, and has a very similar aspect ; its hair, however, is more soft and silky, and its incisor teeth very long and narrow.

Face, all the upper surface and sides light brown, very finely pencilled with black ; under surface greyish buff ; the base of the fur all over the body dark slaty grey ; whiskers black ; tail nearly destitute of hairs ; all the feet clothed with very fine silvery-white hairs, giving those organs a very delicate appearance.

	inches.
Length from the nose to the base of the tail	$7\frac{1}{4}$
„ of the tail	6
„ „ fore-arm	1
„ „ tarsus and toes	$1\frac{1}{4}$

The figures are of the natural size.

MUS MANICATUS, Gould

MUS MANICATUS, *Gould.*

White-footed Rat.

Mus manicatus, Gould in Proc. of Zool. Soc., Part xxv. p. 242.

THE *Mus manicatus* is a remarkable species of Rat, of nearly the same colour and size, and of a similarly delicate structure, as the well-known Black Rat of the British Islands (*Mus Rattus*), but from which it differs in having the tip of the nose, the front part of the lips, a longitudinal stripe on the breast, and the fore- and hind-feet white, which latter peculiarity suggested the specific appellation of *manicatus* or " gloved."

The only specimen I have yet seen of this animal was procured at Port Essington, on the north coast of Australia, and was subsequently presented to me by J. B. Turner, Esq.

Head, ears, and all the upper surface black, gradually passing into the deep grey of the under surface; nose, fore part of the lips, stripe down the centre of the throat and chest, fore- and hind-feet white; whiskers deep black; tail denuded of hairs.

		inches.
Length from nose to base of tail	. . . .	7
„ of tail	. . . .	5
„ „ fore-arm	. . . .	1¼
„ „ tarsi and toes	. . . .	1⅜

The figures are of the natural size.

MUS SORDIDUS, *Gould.*

MUS SORDIDUS, *Gould.*

Sordid Rat.

Mus sordidus, Gould in Proc. of Zool. Soc., part xxv. p. 242.
Dil-pea of the Aborigines of New South Wales.

———————

Very fine examples of this robust and compact Rat were procured by the late Mr. Gilbert on the Darling Downs in New South Wales. At present these specimens are in my own collection, but when this work is completed, they will form part of the rich stores of natural history at the British Museum. Mr. Gilbert states that it is common on the plains, and is occasionally found on the banks of creeks, and adds, that it mostly feeds on the roots of stunted shrubs.

The *Mus sordidus* is nearly equal in size to the common Water Vole of England (*Arvicola amphibius*), but it is rather smaller than the *Mus fuscipes* of Australia. It is in every respect a true *Mus* : its incisor teeth, when compared with those of *M. assimilis,* are broad and less elongated, its hair also is coarser and more wiry. Its colouring is as follows :—

Head, all the upper surface, and flanks clothed with a mixture of black and brown hairs, the former hue prevailing along the centre of the back, and both nearly equal in amount on the flanks ; whiskers black ; under surface greyish buff ; hind-feet silvery grey ; fore-feet greyish brown ; tail thinly clothed with extremely fine black hairs.

	inches.
Length from the nose to the base of the tail	$6\frac{3}{4}$
„ of the tail	5
„ „ the fore-arm	$\frac{3}{4}$
„ „ the hind-leg and toes	$1\frac{1}{4}$

The name of *sordidus* has been assigned to this animal from the dark colouring of its upper surface. The figures are of the natural size.

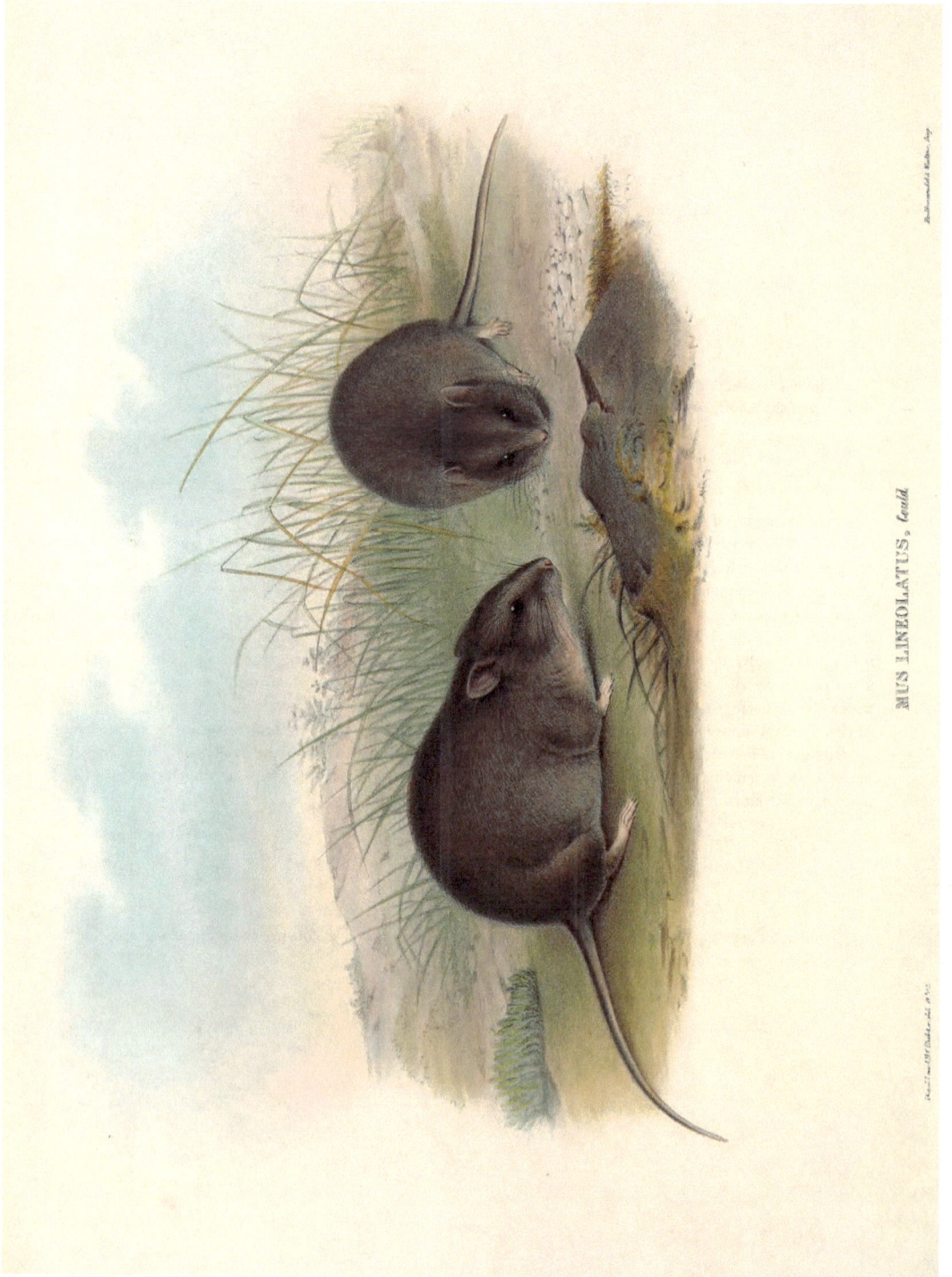

MUS LINEOLATUS, Gould

MUS LINEOLATUS, *Gould.*

Plain Rat.

Mus lineolatus, Gould in Proc. of Zool. Soc., part xiii. p. 77.

THIS species of *Mus* was discovered by Mr. Gilbert on the Darling Downs, where it appears to be abundant. In size it is just intermediate between a Rat and a Mouse, taking our own well-known animals for comparison. Two fine examples are now in my own collection, but will hereafter be added to the stores of the National Museum, where they will be at all times accessible to the mammalogist who may wish to investigate this intricate group of animals ;—I say intricate, because so great a sameness of colouring prevails among the species, that it is exceedingly difficult to distinguish one from another ; indeed it can scarcely be effected without reference to the specimens themselves ; for, although the utmost care is always taken to secure the accuracy of my illustrations, the minute characters which distinguish them cannot be rendered sufficiently apparent in a drawing.

The dark colouring of the upper surface of the well-clothed tail, contrasted with the light hue of its under portion, are the points which distinguish this species.

Mr. Gilbert states that it is called *Yar-lie* by the natives of the Darling Downs ; that it is common in all the open parts of the grassy plains, and that he believes it is confined to the interior of the country.

The fur of this animal is long and very soft ; on the back the hairs are of a deep slate-grey, but with the exposed portion of a dirty yellowish hue, the points however being black ; long interspersed black pointed hairs are abundant on the back, and give a deep general tint to that part ; on the sides of the body the prevailing tint is greyish-yellow, and the under parts are grey-white, faintly suffused with yellowish ; the hairs on these parts are however of a deepish grey, excepting at the points ; the hairs of the moustaches are rather small and black ; the eye is encircled with black ; the ears are of moderate size and well covered with minute hairs ; those on the outer side are black, excepting on the hinder part, where they assume a greyish-white tint, like those on the inner side of the ear. The feet are rather small and white ; the forefeet are however greyish at the wrist, and the tarsi are indistinctly suffused with yellowish. The tail is about equal in length to the head and body taken together, well clothed with smallish hairs, which do not however perfectly hide the scales ; those on the upper surface are chiefly brownish-black, but slightly pencilled with whitish in parts ; on the sides and under part they are white.

The Plate represents the animal of the natural size.

MUS GOULDI, *Waterh*.

MUS GOULDI, *Waterh.*

White-footed Mouse.

Mus Gouldii, Waterh. Zool. of Voy. of Beagle, Mamm., p.   . pl. 32. fig. 18, teeth.—Gray, List of Mamm. in Coll. Brit. Mus., p. 111.

—— *Greyii,* Gray in Grey's Journ. of Discoveries in Australia, App. vol. ii. p. 410.

Kurn-dyne, Aborigines of the neighbourhood of Moore's River, in the interior of Western Australia.

THE *Mus Gouldi* is a very distinct and well-marked species, of a size intermediate between that of a Rat and a common Mouse, and may be at all times distinguished by its lengthened, slender, and, white hind feet. It evinces a preference for the plains and sand-hills of the interior, and, as I have seen specimens from the Liverpool Plains, from South Australia, and from the neighbourhood of Moore's River, in Western Australia, appears to range across the southern part of the continent from east to west. The original example from which Mr. Waterhouse took his description was probably from Mr. Coxen's collection, made either on the Upper Hunter or on the interior side of the Liverpool range. Two others transmitted by Mr. Strange were said to have been found between the River Courong and Lake Albert, " and to make their burrows under bushes." Mr. Gilbert states that in Western Australia the animal inhabits the sides of grassy hills where the soil is loose; that its burrows, which are constructed about six inches below the surface, are often of great extent, and that it is generally found in small families of from four to eight in number, inhabiting the same burrow, and even the same nest of dried soft grasses.

Fur soft; general hue buffy-brown, interspersed on the head, upper surface and sides, but particularly on the back, with numerous somewhat longer black hairs; under surface pale buffy-white, washed with a deeper tint of buff on the cheeks and lower portion of the sides; whiskers black; hands and feet white; tail brown above, paler beneath.

The figures are of the natural size.

MUS NANUS, Gould.

MUS NANUS, *Gould.*

Little Rat.

Mus nanus, Gould in Proc. of Zool. Soc., part xxv. p. 242.
Jib-beetch, Aborigines of Moore's River in Western Australia.

THE *Mus nanus* is a very diminutive Rat, with coarse hair and a somewhat short tail; it is even smaller in size than the *Mus Gouldi* and *M. gracilicaudus,* but is more nearly allied to the latter than to any other. Three or four specimens, all of the same size, are contained in the collection at the British Museum, and there are others in the Derby Museum at Liverpool; some of these were collected by Mr. Gilbert on the banks of Moore's River, and the others on the Victoria plains in Western Australia.

Head, all the upper surface, flanks, outer sides of the limbs, and hairs clothing the tail brown, with numerous interspersed fine black hairs; under surface greyish white, becoming much lighter, and forming a conspicuous patch immediately beneath the tail; whiskers black; feet light brown; base of the whole of the fur bluish grey.

		inches.
Length from the nose to the base of the tail	. . . .	4
„ of the tail	. . . .	$3\frac{1}{4}$
„ „ fore-arm	. . . .	$\frac{1}{8}$
„ „ tarsus and toes	. . . .	$\frac{3}{4}$

The figures are of the natural size.

MUS ALBOCINEREUS: Gould

MUS ALBOCINEREUS, *Gould.*

Greyish-white Mouse.

Mus albocinereus, Gould in Proc. of Zool. Soc., Part XIII. p. 78.
Noô-jee, Aborigines of Perth, Western Australia.
Jûp-pert, Aborigines of Moore's River in the interior of Western Australia.

As yet we have only seen this pretty little Mouse from Western Australia, where it inhabits the sandy districts bordering the sea-shore, particularly those at the back of the sand-hills near the beach a few miles to the northward of Fremantle; in such situations it forms burrows nearly three feet beneath the surface, with two or more openings, one of which is apparently used for no other purpose than that of bringing out the sand, when it becomes necessary to extend the burrow, and this hole is in general nearly filled up by the sand rolling down from the heap. I regret to say that the above meagre account is all that is yet known respecting it; at present its range appears to be very restricted, but future research will doubtless prove that it extends over all parts of Western Australia, wherever suitable localities occur. The remarkable similarity of the colouring of many animals to that of the soil they inhabit, has often been noticed, and the present is another instance of this curious law, which doubtless tends much to enable these little defenceless animals to elude the attacks of their natural enemies; for no two objects so dissimilar in character can be more alike in hue than are the fur of the *Mus albocinereus* and the sandy districts of Western Australia.

This Mouse is rather larger than the Common Mouse of Europe (*Mus musculus*), and its body is considerably stouter in proportion; the head is large; the ears moderate, or perhaps they may be described as rather small; the tail is nearly equal to the head and body in length; the tarsi are very slender: the fur is very long and very soft, and its general hue is pale ashy grey; on the hinder part of the back is a slight brownish tint, produced by a very fine and indistinct pencilling of dusky or pale greyish yellow; the lower part of the sides of the body and the whole of the under parts are white, but not quite pure, having a faint greyish hue; the head is grey-white, pencilled with black; the sides of the muzzle white; the ears are well-clothed with minute greyish white hairs; the feet are white, and if we except some scattered blackish hairs on the upper surface, the tail is also white.

The figures are of the natural size.

MUS NOVÆ-HOLLANDIÆ, *Waterh.*

MUS NOVÆ-HOLLANDIÆ, *Waterh.*

New Holland Field Mouse.

Mus Novæ-Hollandiæ, Waterh. in Proc. of Zool. Soc., part x. p. 146.—Gray, List of Mamm. in Coll. Brit. Mus.,
    p. 112.

It is very generally believed that all, or nearly all, the Mammals of Australia are marsupial; but this is
not the case; one order at least—the Rodentia—being as fairly represented in that country as in any
other. Both rats and mice are in abundance, but they are specifically distinct from those of the northern
hemisphere.

The *Mus Novæ-Hollandiæ* inhabits the plains and stony ridges of New South Wales, both in the districts
between the mountain-ranges and the sea and in those of the interior. Mr. Waterhouse took his description
of this species from an example collected at Yarrundi, on the Upper Hunter, and I have now before me
additional specimens from the same district, and others collected on the banks of the river Gwydir, where
they were procured by Mr. Gilbert. The animal described by Mr. Waterhouse was, I believe, somewhat
immature; his measurements, therefore, will not answer for the adult, which is represented on the
accompanying Plate of the natural size. I usually found this species among stones, or under flat slabs of
bark, left by the aborigines at their encampments; but Mr. Gilbert states that, while travelling among the
high grass in the neighbourhood of the Gwydir, he constantly started it from out of the fissures in the dry
ground.

Mr. Waterhouse states, that it approaches most nearly to the *Mus sylvaticus* in form and colouring, but
that the tail is considerably shorter than in that animal; he remarks that it also approaches that species in
the form of the skull, but has the nasal portion shorter; the molar teeth are of the same structure, but
apparently rather larger in proportion.

The fur is rather long and very soft; on the upper parts the hairs are of a deep grey, tipped with
brownish-yellow; on the belly the hairs are of a paler grey next the skin, and white externally; the tarsi
are rather long and slender; the tail is white beneath and dusky above.

The figures are of the size of life.

MUS DELICATULUS, Gould.

J. Gould and H. C. Richter del. et lith.   Hullmandel & Walton Imp.

MUS DELICATULUS, *Gould.*

Delicate-coloured Mouse.

Mus delicatulus, Gould in Proc. of Zool. Soc., part x. p. 13.—Ann. and Mag. Nat. Hist., vol. x. p. 406.—Gray,
List of Mamm. in Coll. Brit. Mus., p. 112.
Mo-lyne-be, Aborigines of Port Essington.

THE contour and general colouring of this, the smallest and most beautiful species of *Mus* yet discovered in the great country of Australia, strongly remind one of the pretty little harvest mouse, *Mus messorius,* of our own islands. It is a native of Port Essington, where it was discovered by the late Mr. Gilbert, and all we know respecting it is comprised in the following brief notice of it in his Journal:—

"I only met with this species on one occasion, on the Native Companion plains near Point Smith, at the entrance of the harbour, when I found four in a hole which ran along a few inches below the surface for about five feet in a zigzag manner, and terminated in a circular space, wherein was a nest of fine dried grass, in which I captured them."

Two specimens of this little animal are in the collection at the British Museum. Mr. Gray states that I had attached the MS. name of *albirostris* to them; but that appellation not having been published, the term *delicatulus,* under which the animal was certainly described in the "Proceedings of the Zoological Society," is the one retained.

The fur is soft and short; that on the upper parts of the body is of a pale yellow-brown; the sides are of a delicate yellow tint; and the lower part of the sides of the muzzle, chin, throat, under surface and feet are pure white; on the throat and along the mesial line of the abdomen, the hairs are of a uniform colour to the base; ears small; feet delicate; tail slender, and nearly as long as the head and body.

The figures are of the natural size.

HYDROMYS CHRYSOGASTER Geoff.

HYDROMYS CHRYSOGASTER, *Geoff.*

Golden-bellied Beaver-Rat.

Hydromys chrysogaster, Geoff. Ann. Mus., tom. vi. p. 81. tab. 36. fig. A.—Gray, List of Mamm. in Coll. Brit. Mus.,
p. 121.

THE first specimens of Mammalia transmitted to Europe from Australia after the discovery of that country were from New South Wales and Van Diemen's Land; and among others the present animal attracted, at a very early period, the attention of the French naturalists, one of whom—the celebrated Geoffroy St. Hilaire—assigned to it the name of *Hydromys chrysogaster*, a name very suitable to the animal from the localities above mentioned; similar Beaver-Rats are, however, universally spread over the whole of the southern portion of Australia, including the eastern and western confines of that continent; but the animals from each of these localities appear to me to be distinct species; and a most complete series of the whole of them being now before me, I think I shall be able to point out, in my account of each of them, characters of sufficient importance to be regarded as specific.

The present species may be distinguished from all its congeners by the bright golden colouring of the sides of the face, lips, throat, shoulders, flanks and belly, the darker colouring being confined to the crown of the head and the upper part of the back only, whereas in two of the other species this dark colouring occupies so much of the upper part of the body as to include the shoulders and part of the fore arm; and, however near the whole of the species may assimilate in size, the present is the largest, as well as the one in which the colouring is the most contrasted and brilliant.

The native habitat of the *Hydromys chrysogaster* is New South Wales and Van Diemen's Land. It is strictly fluviatile in its habits, frequenting the muddy sides of creeks and water-holes, and the banks of the larger rivers and inlets of the sea. Rather shy in its disposition and nocturnal in its movements, it is not so often seen as might be supposed; at the same time it is by no means difficult to be procured when such an object is desirable. As might be inferred from the structure of its hind feet, the water is its native element; it swims and dives with the greatest facility, and easily secludes itself from view amidst the sedges lining the water's edge, or by descending to its hole after the manner of the common Water Vole of Europe. Like many other of the Australian Mammals, it reposes much on its hinder legs, in which position it may frequently be seen on large stones, snags of wood, or any other prominence near the water's edge.

Head, ears, back, outer surface of the hinder limbs, the portion of the body posterior to them and the base of the tail mingled black and buff, the former hue predominating; sides of the face, of the body, all the under surface and the inner side of all the limbs rich deep reddish orange; outer surface of the arms deep brown; upper surface of hinder feet pale glaucous buff, passing into brown on the tips of the toes; basal half of the tail black, apical half white.

The figures are somewhat smaller than life.

HYDROMYS FULVOLAVATUS, *Gould.*

Fulvous Beaver-Rat.

I POSSESS specimens of this animal from three different localities—some obtained near the River Murray by E. J. Eyre, Esq., now Lieut.-Governor of New Zealand, others procured at Lake Albert by Mr. Strange, and others shot by myself in the pools of the upper part of the River Torrens, all of which closely resemble each other, but differ very considerably from either of the foregoing species; I have therefore been induced to regard them as specifically distinct. To the Western Australian animal (*H. fuliginosus*) they are allied in the extreme tip of the tail only being white, and to *H. chrysogaster* in the colouring of the under surface, but in no other respect so far as colour goes. As the specific name implies, the whole of the body is washed with golden orange, a tint only relieved by the interspersion of numerous black hairs over the upper surface, giving that part a darker hue, without any decided line of demarcation separating the colouring of the upper from that of the under surface. The habits and economy of this species offer a close resemblance to those of *H. chrysogaster*. I usually found it on the muddy banks of the water-holes of South Australia, where, like the European Water Vole, it lived upon vegetables, mollusks, and other lacustrine animals common to such situations.

The feet of this species are somewhat darker coloured than those of *H. chrysogaster*. The general hue of the fur orange buff, but the numerous black hairs which are dispersed over the head and upper surface give those parts a dusky hue; the whiskers, which in the other species are entirely black, are here mingled black and white; outer surface of the limbs dark brown; upper surface of hinder feet pale brown, deepening into a darker hue on the toes; nails white; tail black, except at the extreme tip, which is white.

The figures are rather under the natural size.

HYDROMYS LEUCOGASTER, *Geoff.*

Drawn and W.J.Erchter del.et lith.

Hullmandel & Walton, Imp.

HYDROMYS LEUCOGASTER, *Geoff.*

White-bellied Beaver-Rat.

Hydromys leucogaster, Geoff. Ann. Mus., tom. vi. p. 81. tab. 36. figs. B, C, D ?

M. Geoffroy St. Hilaire has given the name of *leucogaster* to an animal of this genus, and I believe the subject of the accompanying Plate to be the one to which it was applied. The several specimens contained in my collection were obtained on the banks of the Hunter, Clarence, and other rivers traversing the districts lying between the mountain ranges and the sea. They are all similarly marked, and, as will be seen on reference to the Plate, differ very considerably from *H. chrysogaster*, the tawny white of the under surface occupying the belly only, while the shoulders and upper part of the fore arm are included in the darker colouring of the upper surface. I mention this latter point more particularly, because, were the colouring of the under surface the only difference between the two animals, some persons might suppose that difference to be due to the action of light, which, having abstracted the rich orange colouring, had left the parts thus coloured of a dull or tawny white.

The hinder feet of the two animals also differ, those of *H. leucogaster* being smaller and of a darker colour than those of *H. chrysogaster*, and having the toes of the fore feet for half their length from the nails white, a feature I never observed in the latter species. The only character in which they are alike consists in the extent of the white on the tail, which occupies the terminal half in both species.

Head, all the upper surface, shoulders, sides, outer surface of all the limbs, and the portion of the body posterior to them, mingled black and buffy grey, the former hue predominating; face, all the under surface of the body and the inner side of the limbs buffy white; upper surface of the hinder feet deep purplish buffy white; basal half of the tail black, apical half white.

The figures are somewhat less than the size of life.

HYDROMYS FULIGINOSUS, *Gould.*

HYDROMYS FULIGINOSUS, *Gould.*

Sooty Beaver-Rat.

Ngoŏr-joo, Aborigines of Perth, Western Australia.
Ngw'ir-ri-gin, Aborigines of King George's Sound.

THE specimens of this animal in my collection were procured in the neighbourhood of the lakes near Perth and at King George's Sound in Western Australia, by the late Mr. Gilbert, who in his letter to me on the subject expressed his opinion that they were quite different from any other species he had seen ; and surely an animal so different from all its congeners in the colouring of the body, in the darker colour of the hinder, and the greyish white hue of the fore feet, may with propriety be considered as specifically distinct. Independently of these differences, I may mention, that the uniformity of the body tint is all but unbroken, the face, the centre of the back, and the basal portion of the tail being simply a trifle darker than the rest of the fur. Mr. Gilbert hints that the *H. chrysogaster* also inhabits Western Australia, but in this I believe he was mistaken ; in all probability the South Australian species, to which I have given the name of *fulvolavatus,* is the animal he saw, but failed to procure.

Fur of the upper surface mingled buffy brown and black, the latter hue predominating and producing a deep sooty appearance, especially along the back, whence the specific name ; whiskers and fur of the face black ; that of the tail is also black, except the apical inch and a half, which is white ; fur of the under surface pale greyish brown ; fur of the outer surface of the limbs dark brown, of the fingers white ; nails white.

The figures are rather less than the size of life.

PTEROPUS POLIOCEPHALUS, *Temm.*

J. Gould and H.C. Richter del et lith.   Hullmandel y Walton Imp.

PTEROPUS POLIOCEPHALUS, *Temm.*

Grey-headed Vampire.

Pteropus poliocephalus, Temm. Monog., tom. i. p. 179, tom. ii. p. 66.—Gray, List of Mamm. in Brit. Mus., p. 36.

NEW SOUTH WALES is the true and probably the restricted habitat of this large species of Bat; for I have never seen a specimen from any other part of the Australian continent, and it certainly does not inhabit Van Diemen's Land as stated by M. Temminck : the situations in which I met with it were the dense and luxuriant brushes which fringe the south-eastern portion of Australia, such as those at Illawarra, in the neighbourhood of the Hunter, the Manning and the Clarence; I possess, however, a specimen said to have been killed at Bathurst, which, although of much smaller size, I believe to be the same. Like all other Bats, the Grey-headed Vampire is strictly nocturnal in its habits, and remains during the day suspended from the branches of the larger trees clothing the gullies and mountain sides; at nightfall it sallies forth in search of its natural food, which principally consists of the fruits and berries peculiar to the brushes, the small wild fig when ripe being a favourite article. The enormous numbers that may be seen sleeping pendent from the trees in the more secluded parts of the forest are beyond conception; it is not surprising therefore that the settlers whose abodes may be in the neighbourhood of one of these colonies, should find their peach orchards entirely devastated in a single night. Indeed no one of the native animals is more troublesome to the settlers than this large Bat, which, resorting to the fruit-grounds by night, when it is impossible to protect them from its attacks, commits the most fearful havoc. Many pages might doubtless be written respecting the habits and economy of these great Bats, but this can only be done by those who, having been long resident in the country, have had ample opportunities of observing them, which the rapidity of my explorations and the brevity of my stay did not admit. In describing the habits of a nearly allied species (the *Pteropus Javanicus*) Dr. Horsfield states, that " they congregate in companies, and selecting a large tree for their resort, suspend themselves by the claws of their hind limbs to the naked branches, affording to the stranger a very singular spectacle; in short, to a person unaccustomed to their habits, they might be readily mistaken for fruit of a large size, suspended from the branches. They thus pass the greater portion of the day in sleep; but soon after sunset they gradually quit their hold, and pursue their nocturnal flight in quest of food. They direct their course, by an unerring instinct, to the forests, villages and plantations, occasioning incalculable mischief, attacking and devouring indiscriminately every kind of fruit, from the abundant and useful cocoa-nut, which surrounds the dwelling of the meanest peasantry, to the rare and most delicate productions which are cultivated with care by princes and chiefs of distinction. Their flight is slow and steady, pursued in a straight line, and capable of long continuance." This interesting account of the habits of the Javan species doubtless applies in an equal degree to those of the present animal, since we may reasonably infer that the economy of two species so nearly allied is very similar.

Its flesh forms one of the multitudinous articles partaken of as food by the aborigines.

The entire head brown, grisled with grey; round the neck and advancing on to the back a very broad collar of deep rust-red; upper surface and the clothing of the arms glossy black, grisled with greyish olive, the olive hue becoming more apparent on the hind quarters; under surface brownish black, many of the hairs pointed with olive-yellow; down each flank a patch of rufous; ears and wing-membranes naked and of a deep purplish black; claws black, becoming horny at the tip.

The figures are of the natural size.

PTEROPUS CONSPICILLATUS.

PTEROPUS CONSPICILLATUS, *Gould.*

Spectacled Vampyre.

Pteropus conspicillatus, Gould in Proc. of Zool. Soc., 1849, p. 109.

THE native habitat of this fine species of Vampyre is Fitzroy Island, lying off the eastern coast of Australia, where it was discovered by Mr. John MacGillivray during the recent surveying voyage of H.M.S. Rattlesnake, under the command of the late Capt. Owen Stanley. It is about the same size as the *P. poliocephalus*, but has a somewhat larger head and much larger and more powerful teeth; it may moreover be distinguished from that species by the nuchal band being of a deep sandy buff, instead of deep rust-red, and not continuous round the neck; by the crown of the head and the back being almost jet-black; and by the eyes being conspicuously encircled with deep buff, whence the specific name, and which at once distinguishes it from every other known species.

I am indebted to Mr. MacGillivray for the following brief notes, which comprise all that is at present known respecting it:—

"On the wooded slope of a hill on Fitzroy Island, I one day fell in with this Bat in prodigious numbers, looking while flying along the bright sunshine, so unusual for a nocturnal animal, like a large flock of rooks; on close approach, a strong musky odour became apparent, and a loud incessant chattering was heard; many of the branches were bending under their load of bats, some in a state of inactivity suspended by their hind claws, others scrambling along among the boughs and taking to wing when disturbed. In a very short time I procured as many specimens as I wished, three and four at a shot, for they hung in clusters, but unless killed outright they remained suspended for some time; when wounded they are handled with difficulty, as they bite severely, and on such occasions their cry reminds one of the squalling of a child."

Crown of the head black, slightly grizzled with buff; round each eye a large oval patch of deep brownish buff, which advances on the sides of the face and shows very conspicuously; at the nape a broad crescent-shaped band of deep sandy buff, which extends down the sides of the neck and nearly meets on the breast; centre of the back glossy black, slightly grizzled with grey; cheeks, chin, all the under surface and rump black, slightly grizzled with buff; ears and wing-membranes naked, and of a deep purplish black; claws black.

The Plate represents a male of the natural size.

PTEROPUS FUNEREUS, *Temm.*

PTEROPUS FUNEREUS, *Temm.*

Funereal Vampire.

Pteropus funereus, Temm. Monog. tom. ii. p. 63. tab. 35. fig. 4.—Gray, List of Mamm. in Brit. Mus. p. 37.
Al-wo-re, of the Aborigines at Port Essington.

THIS species appears to be as exclusively confined to the northern portions of Australia as the *Pteropus poliocephalus* is to the south-eastern. M. Temminck gives the animal a very wide range, for he states that he has positive evidence of its existence on four other islands, namely Timor, Amboyna, Borneo and Sumatra. Mr. Gilbert's notes inform me that " it is extremely abundant in all parts of the Cobourg Peninsula; during the day it may be seen in great numbers suspended from the upper branches of the mangroves overhanging the creeks: while living it emits a very strong and disagreeable odour which is perceptible at a considerable distance; it becomes very active at night, and while flying about in search of food utters a loud, trembling, but shrill whistle." Frequent mention is made of this species in Dr. Leichardt's Journal of his Expedition from Moreton Bay to Port Essington, and despite of its disagreeable odour, it often formed for himself and party an excellent and welcome article of food. Like the other species, it feeds upon fruit and the honey of the various flowers; in one instance Dr. Leichardt found them feeding upon the blossoms of the tea-tree, and remarked that they were then more than usually fat and delicate, while those that had been revelling among the blossoms of the gum-trees were not so fat, and had a strong, unpleasant odour. So numerous did they become towards the latter part of the journey, that " twelve were brought in for luncheon, thirty more were procured during the afternoon, and at least fifty were left wounded and hanging to the trees; upon another occasion they were seen clustering in such numbers, that the branches of the low trees drooped with their weight so near the ground, that they could easily be killed with cudgels. In the neighbourhood of the river Roper, myriads were suspended in thick clusters on the highest trees, in the most shady and rather moist parts of the valley; they started as the travellers passed, and the flapping of their large membranous wings produced a sound like that of a hail-storm." Dr. Leichardt went the next day with two of his party to the spot where they had seen the greatest number, and while he was examining the neighbouring trees his companions shot sixty-seven, of which fifty-five were brought to the camp, and served for dinner, breakfast and luncheon, each of the party receiving eight: the animal here lived upon a small, blue, oval stone-fruit, of an acid taste, with a bitter kernel, growing on a tree of moderate size.

Considerable difference is found to exist in the colouring of this animal, but whether this difference is due to sex or age is at present unknown: the following are the descriptions of the two specimens now before me :—

In the one, the head, upper part of the body, the rump, and all the under surface is clothed with a thick, loose black fur, with a wash of deep chestnut between the shoulders; the centre of the back, and the arms, clothed with thin, shining, closely pressed black hairs. In the other, there is a wash of rufous round the eyes, and a broad collar of rich deep chestnut across the nape of the neck. In both the wing-membranes are deep purplish black; and the claws are black.

The figures are of the natural size.

MOLOSSUS AUSTRALIS, Gray

J. Gould and H. C. Richter del. et lith.   Hullmandel & Walton Imp.

MOLOSSUS AUSTRALIS, *Gray.*

Australian Molossus.

Molossus australis, Gray in Mag. of Zool. and Bot., vol. ii. p. 501.—Gould in Proc. of Zool. Soc., July 27, 1858.

THIS large and truly singular species of Bat was described by Dr. Gray just twenty years ago, his description having appeared in the Number of the "Magazine of Zoology and Botany" for December 1838. The specimen from which Dr. Gray took his characters was then and still is the property of the United Service Institution, to whose museum it was presented by Major M'Arthur, with the word "Australia" written on the label attached to it; beyond this nothing was known respecting it, and up to this time it remains the only example in Europe. My thanks are especially due to the President and Council of this Institution for permission to take this rare and valuable specimen out of their cases for the purpose of figuring in the present work; this had been done and my drawing made, when, just on the eve of publication, an unexpected letter arrived from a friend in Australia, of which the following is a copy :—

"Melbourne, Victoria, May 12, 1858.

"DEAR SIR,—A few days ago I saw five Bats together in a hollow tree near this place. Not having seen them figured in your own or any other work, I thought it likely they might prove to be a new species. I therefore made a sketch of one of them for you, and if you think proper to publish it you are quite welcome so to do. I also leave it to you to name the animal. A short description I have the honour to send you.

"Dentition : incisors $\frac{1 \cdot 1}{2 \cdot 2}$, canines $\frac{1 \cdot 1}{1 \cdot 1}$, false molars $\frac{2 \cdot 2}{2 \cdot 2}$, true molars $\frac{3 \cdot 3}{3 \cdot 3} = \frac{14}{16} = 30$.

"For the admeasurements of the animal I refer you to the sketch, which is of the size of life. Colour of the body sepia-brown, the belly somewhat lighter; wings greyish brown; hand or hind-foot and the extremities, including the tail, black; the under side of the fore-arms whitish flesh-colour; the palm and wrist of the hand black, as if covered with gloves; between the elbows and the knees a pure white streak stretching towards the root of the tail; irregular white spots occur on the neck and chest in some of the specimens, in others the neck and belly are covered with large white patches; tragus and the ears, where free from hairs, black, the remainder clothed with dark rust-coloured hairs; on the upper margin of the ear a row of diminutive tubercles; eyes black; space surrounding the nostrils naked and black; under lip nude and of a blackish brown; thumb and fourth finger of the hind-foot thicker than the three middle ones, while a sort of fine brush covers the former; the thumb is shorter than the fingers, but all have on the top of the nail a small tuft of fine hairs; the tail is prolonged for more than an inch beyond the intra-femoral membrane; on the throat a sort of pouch stretching inwards and downwards nearly three-quarters of an inch, covered with two distinct tufts of stiff brown hairs growing on the bottom of the pouch, and resembling a couple of artist's brushes; when the pouch is not distended, only the extremity of the brushes are visible, and are scarcely distinguishable from the other hairs; the naked portion of the pouch is flesh-coloured; there is no communication between the pouch and the inner portion of the neck. The upper incisor teeth are rather large, and resemble canines; the lower ones are very minute; the upper first and second true molars are nearly equal, and have three sharp tubercles externally; the third is smaller, and has two pointed tubercles; the first and second of the lower true molars have five points on their crowns.

"I shall be very much pleased to receive a few lines stating your opinion of the Bat, and if I can serve you in any way connected with natural history.

"With all due respect,
"Yours obediently,
"LUDWIG BECKER."

The drawing which accompanied this letter accorded so nearly with the specimen in the United Service Museum as not to leave a shadow of doubt as to the animals discovered by Dr. Becker being identical with it; the tufts of hair and the pouch are, however, almost obsolete in the specimen, which is probably due to a difference in sex, or of the season at which it was killed.

On submitting Dr. Becker's letter and drawing to R. F. Tomes, Esq., who has paid great attention to the *Vespertilionidæ*, that gentleman favoured me with the following remarks :—

"Welford, Stratford-on-Avon, June 1858.

"MY DEAR SIR,—I have compared the drawing and description of the *Molossus Australis* with a great many species of that genus in spirit and prepared skeletons, and conclude most certainly that it is a

Molossus. The only doubt I had was whether the genus *Nyctinomus* of Isidore Geoffroy should properly be separated from the genus *Molossus* established by his father. In order to determine this, if possible, I have cleaned a number of specimens and instituted a minute comparison. The results scarcely justify a *sub-generic* difference. With respect to the gular pouch, that appears, so far as my specimens in spirit inform me, to be peculiar to the Australian species; but I strongly suspect that it will be found to be a peculiarity due to sex and age, perhaps even in some measure to season, rather than of generic value. In the genus *Taphozous* this gular sac is entirely dependent on the age and sex of the animal, and its absence or presence has been made use of as a specific character very improperly.

"It is not a little remarkable that this Australian species should possess characters (with the exception of the pouch) which are as similar to the European ones as to any other species. I have often been surprised that Australia does not furnish a single form among the Bats that are not common to nearly all the world besides; indeed, many of the species are found in the Indian islands, and, curiously enough, in *China*. A collection of Chinese Bats which I have lately examined consisted of Indian and Australian species. I have taken very great pains in the examination of the drawing, and have prepared skeletons of several species on purpose to compare them. I would strongly advise you to continue the name of *Molossus Australis*, and you may, if you like, add that the animal belongs to that division of the genus which has been designated *Nyctinomus*.
"Ever yours truly,
"R. F. Tomes."

This, then, is all that I am at present able to communicate respecting the species, and it affords me much pleasure to furnish even this meagre account, for a more interesting animal I have not seen for some time.

The following is a description of the animal in the United Service Museum :—

Fur soft and dense, that of the upper surface of a deep reddish brown, and extending on to the basal parts of the limbs; fur of the under surface of a similar but much lighter hue, bordered on each side of the body by a broad fringe of white, which extends in a lesser degree on to the base of the lower limbs and the posterior part of the body; wing-membranes extremely thin, and presenting a silvery appearance; ears and face dark purplish brown.

Total length 6¼ inches ; extent of the wings from tip to tip 15¼.

The figure is of the natural size.

TAPHOZOUS AUSTRALIS, Gould.

J. Gould and H.C. Richter, del et lith.   Hullmandel & Walton, Imp.

TAPHOZOUS AUSTRALIS, *Gould.*

Australian Taphozous.

For a knowledge of this fine species of Bat, we are indebted to the researches of Mr. John MacGillivray, the Naturalist attached to H.M.S. Rattlesnake, during the recent survey of the northern coasts of Australia. After carefully comparing it with the various species contained in the unrivalled collection of the Vespertilionidæ in the Museum at Leyden, I have come to the conclusion that it is quite distinct, and have, therefore, given it the name of *Taphozous Australis*, it being, I believe, the first species of the form found in that country.

Mr. MacGillivray having kindly furnished me with a copy of his notes respecting this species, I cannot, perhaps, do better than give them in his own words ;—notes taken on the spot being of infinitely greater value than any that can be elaborated from the most careful examination of the dried specimen.

" Dentition :—incisors $\frac{1 \cdot 1}{2 \cdot 2}$; canines $\frac{1 \cdot 1}{1 \cdot 1}$; false molars $\frac{2 \cdot 2}{2 \cdot 2}$; true molars $\frac{3 \cdot 3}{3 \cdot 3} = \frac{14}{16} = 30$.

" Length 3 inches ; tail, 1 ; fore-arm, 2·5 ; hind-arm, 1 ; ear, 1 ; tragus, 0·25.

" Colour (there are two varieties) : above, ferruginous brown ; light brown in the centre of the back and across the abdomen, or entirely brownish grey ; basal half of the fur white ; below, ash-grey, with sometimes a slight reddish tinge ; muzzle black.

" Nostrils : simple, terminal, bordering the upper lip.

" Ears : not connected at the base ; rather large ; somewhat triangular, the two exterior angles rounded ; within thinly covered with scattered hairs ; upper margin with a row of small tubercles ; tragus one-fourth the length of the ear, irregularly quadrate ; narrowed at the base ; the exterior angles rounded, the inferior projecting most.

" Wings : above with a few hairs at the elbow, and rather thickly covered at the base of the intra-femoral membrane ; below with a band of scattered whitish hairs along the arm and fore-arm as far as the fifth digital phalange ; second phalange single-jointed ; the third, fourth and fifth three-jointed.

" Tail not reaching to the posterior margin of the intra-femoral membrane ; free for two-tenths of an inch.

" Incisors : upper very minute, simple ; lower small, three-lobed.

" Canines : upper strong, curved, acute, with a small basal lobe projecting behind, and internally forming a sharp point ; lower more slender, with projecting collar in front and on the sides.

" False molars : above, first very minute, second long, acute : below, acute, the second rather the longest.

" True molars : above, first and second nearly equal, with two sharp tubercles externally, and a low, sharp ridge internally ; third very small, transverse, with two pointed tubercles and a sharp ridge ; below, the first and second with five, and the third, which is the smallest, with four points on the crown.

" Habitat : the maritime caves in the sandstone cliffs of Albany Island, Cape York. In great numbers in three of the caves. Specimens obtained October 1848."

The animal is represented of the natural size.

RHINOLOPHUS MEGAPHYLLUS, Gray.

J. Gould and H.C. Richter, del. et lith.   Hullmandel & Walton, Imp.

RHINOLOPHUS MEGAPHYLLUS, *Gray.*

Great-leaved Horse-shoe Bat.

Rhinolophus megaphyllus, Gray in Proc. of Zool. Soc., part ii. p. 52.—Ib. List of Mamm. in Coll. Brit. Mus., p. 22.

This species, the largest of the Horse-shoe Bats that has yet been received from Australia, was described by Dr. Gray as long back as 1834, from a specimen collected in the caverns in the neighbourhood of the river Morumbidgee, in New South Wales. The example from which my figure was taken was obtained by Mr. Leycester near the Richmond river. I mention these localities particularly, because the animal is one of those which did not come under my notice during my explorations in Australia, although it was almost a never-failing practice with me to devote the last hour of the day to the study and collection of these curious little mammals.

Dr. Gray states, that " this Bat is very nearly allied to the true European *Rhinolophi*, and agrees with them in having four cells at the base of the hinder nose-leaf, and distant pectoral teats, but differs from them in having a much broader nose-leaf.

" The hinder nose-leaf is bristly, ovate-lanceolate, nearly as broad at the base as the face, with a rather produced tip; the septum of the nose is grooved; and the front leaf expanded with a quite free membranaceous edge. The head is elongated; the face depressed; the muzzle rounded; the ears are large, reaching when bent down rather beyond the tip of the nose. The fur is soft and of a pale mouse-colour. The membranes are dark and naked, with rather distant whitish hair on the under side, near the sides of the body."

The figures are of the size of life.

RHINOLOPHUS CERVINUS, Gould.

J.Gould and H.C.Richter del.et lith.

Hullmandel & Walton, Imp.

RHINOLOPHUS? CERVINUS, *Gould.*

Fawn-coloured Bat.

I have figured this species as a *Rhinolophus* with a mark of doubt, being somewhat uncertain as to whether I am correct in placing it in that genus; probably it ought to have been assigned to that of *Phyllorhina.* Mr. MacGillivray, to whom we are indebted for its discovery, was inclined to think it identical with *Rhinolophus aurantius*, but upon comparing it with that species, I am convinced it is distinct; I have therefore assigned it a specific appellation, and have selected that of *cervinus*, in reference to the colouring of the fur. The following notes respecting the animal were communicated to me by Mr. MacGillivray, and as they were made at the time he procured the specimens from which my figures are taken, it will be well perhaps to give them in his own words :—

"Dentition : incisors $\frac{1 \cdot 1}{2 \cdot 2}$; canines $\frac{1 \cdot 1}{1 \cdot 1}$; false molars $\frac{2 \cdot 2}{2 \cdot 2}$; true molars $\frac{3 \cdot 3}{3 \cdot 3} = \frac{14}{16} = 30.$

"Length : body, exclusive of the tail, 2 inches ; fore-arm, 1·7 ; hind-arm, 0·7 ; tail, 1 ; ears, 0·5 long, 0·45 wide ; extent of wings, 11 inches.

"Colour : above tawny brown, darkest on the face, head and shoulders ; below paler, and tinged on the belly with grey.

"Nose-leaf simple, long, straight-edged, 0·25 across.

"Ears : connected by a hairy fold of skin, large, broadly ovate, pointed ; posterior margin slightly sinuated near the tip, then rounded ; internally with anterior one-third thickly clothed with hair ; tragus obsolete, being indicated merely by a slight internal fold of the auricle.

"Wings naked ; index one-jointed, the others three-jointed.

"Tail continued 0·1 beyond the intra-femoral membrane.

"Incisors : above very minute ; below larger and three-lobed.

"Canines : strong, hooked, sharp, the upper ones the largest.

"False molars : above, first very minute, second large and pointed ; below, simple, pointed, the second the largest.

"True molars : first and second in each jaw with five, and the third with four sharp points.

"Habitat : Cape York ; also in the sandstone caves on Albany Island, where it occurs in great numbers. The two species do not associate together. Procured October 1848."

The figures are of the natural size.

RHINOLOPHUS AURANTIUS.

J. Gould and H.C.Richter del. et lith.

Hullmandel & Walton, Imp.

RHINOLOPHUS AURANTIUS, *Gray.*

Orange Horse-shoe Bat.

The Orange Horse-shoe Bat (Rhinolophus aurantius), Gray, App. to Eyre's Journ. of Exp. of Disc. into Central
Australia, vol. i. p. 405. tab. 1. fig. 1.

THE only information we possess respecting this beautiful Bat, is that it is abundant on the Cobourg Peninsula in Northern Australia; that it retires during the day-time to the hollow spouts and boles of the
various species of *Eucalypti*; and that it sallies forth on the approach of evening in search of its insect
food: its general habits and manners in fact so closely resemble those of the other members of the genus,
that a separate description of them is quite unnecessary.

Mr. Gray, who characterized the animal in the Appendix to Mr. Eyre's "Travels," above referred to, from
a specimen procured while flying near the Hospital at Port Essington, by Dr. Sibbald, R.N., remarks that it
is "peculiar for the brightness and beauty of its colour, the male being nearly as bright as the Cock of the
Rock (*Rupicola aurantia*) of South America."

The following is Mr. Gray's description of this pretty animal:—

"Ears moderate, naked, rather pointed at the end; nose-leaf large, central process small, scarcely lobed,
blunt at the tip; fur elongate, soft, bright orange; the hairs of the back with short brown tips, of the under
side rather paler, of the face rather darker; membranes brown, nakedish; tail rather produced beyond the
membrane at the tip; feet small and quite free from the wings.

"The female pale yellow, with brown tips to the hair of the upper parts."

The figures are of the natural size.

NYCTOPHILUS GEOFFROYI, *Leach.*

J. Gould and H. C. Richter del. et lith.   Hullmandel & Walton, Imp.

NYCTOPHILUS GEOFFROYI, *Leach**.

Geoffroy's Nyctophilus.

Nyctophilus Geoffroyi, Leach in Linn. Trans., vol. xiii. p. 73, 1820–22.—Less. Man. de Mamm., p. 86, 1827.—
Fisch. Synops. Mamm., p. 135, 1829.—Temm Mon., tom. ii. p. 47, 1835–41.—Wagn. Supp. Schreib.
Säugeth., tom. ii. p. 442, 1840.—Less. Nouv. Tab. Règn. Anim., p. 33, 1842.—Schinz, Synops. Mamm.,
tom. i. p. 217, 1844.—Tomes in Proc. of Zool. Soc., part xxvi. p. 29.

Since my plate and description of the animal I have called *Nyctophilus Geoffroyi* were printed, Mr. Tomes has very minutely investigated this group of bats, and published a monograph of the genus, and he now considers that the hint I there gave as to the probability of the species from Western Australia and Tasmania being distinct is a correct view of the case, and has come to the conclusion that the Western Australian species is the true *N. Geoffroyi*, and consequently that the animal from New South Wales, formerly figured by me under that name, should receive a new appellation; and he has accordingly named it after myself, *N. Gouldi*. It is much to be regretted that this conclusion should not have been arrived at before my plate and description were printed, as the synonymy of the New South Wales *N. Gouldi* has reference to the animal here represented, which is a native of Western Australia; however as Mr. Tomes's opinions are of value, and entitled to be recorded in this or any other work comprising an account of any of the members of the family *Vespertilionidæ*, I will quote his own words:—

"This, from its size," says Mr. Tomes, "is unquestionably the species on which Dr. Leach established the genus. The original description in the Linnean Transactions is much too vague to discriminate the exact species with certainty; but M. Temminck having become possessed of the original specimen, and given a more detailed description of it, I am enabled to determine with certainty which is the true *N. Geoffroyi*."

Mr. Gilbert states that this species is called *Bar-ba-lon* by the aborigines of King George's Sound, and *Bämbe* by the natives of Perth, and that it is the most abundant species in the colony of Western Australia. It is sometimes met with by the wood-cutters in the hollow spouts of the gum-trees in great numbers; from these retreats they emerge at twilight and flit about the shrubs and lower trees in search of insects.

The following is Mr. Tomes's description:—

"The face is moderately hairy, the hairs being pretty regularly scattered, but a little thicker on the upper lips and on the second nose-leaf than elsewhere; immediately over the eye is a small tuft of bristle-like black hairs, and a similar one near the hinder corner of the eye; at the angle of the mouth a few similar hairs may be observed; the fur of the back extends to a very trifling extent on to the interfemoral membrane, but all the other membranes are perfectly naked and of a dark brown colour, as are also all the other naked parts, with the exception of the tragus and the contiguous parts of the inside of the ear, which are brownish yellow. The fur of the body is rather long, thick, and very soft; on all the upper parts it is conspicuously bicoloured, black for nearly two-thirds of its length, the remainder being olive brown, of which the extreme tips are rather the darker portion; on the membranes uniting the ears the fur is uniform yellowish brown; the fur of the throat and flanks is uniform brownish white, that of the latter being sometimes more strongly tinted with brown; all the remaining underparts have the fur markedly bicoloured black at the base, with the terminal third brownish white, varying considerably in purity of colour in different individuals."

"This description," says Mr. Tomes, "was taken from a specimen kindly lent to me by Mr. Gould, and which is labelled 'Albany, King George's Sound, May 19, 1843.'"

The figures are of the natural size.

NYCTOPHILUS GEOFFROYI, Leach

J.Gould and H.C.Richter, del et lith.   Hullmandel & Walton, Imp

NYCTOPHILUS GEOFFROYI.

Geoffroy's Nyctophilus.

Nyctophilus Geoffroyi, Leach, in Linn. Trans., vol. xiii. p. 78.—Temm. Monog., vol. ii. pl. 34.— Gray, in Mag. of
Zool. and Bot., vol. ii. p. 12.—*Id.*, List of Mamm. in Coll. Brit. Mus., p. 25.—*Id.*, in Grey's Journ.
of Two Exp. in N.W. and W. Australia, vol. ii. p. 400.
Barbastellus Pacificus, Gray, Zool. Misc., vol. i. p. 38.
Nyctinomus ——, Benn. Cat. of Australian Museum, Sydney, p. 1. no. 2.

THE figures on the accompanying Plate are taken from specimens captured in New South Wales, and I
consider it necessary to state this particularly, because the long-eared Bat of Western Australia, though
very nearly allied, may prove to be distinct : the specimens I possess from the latter country are larger, and
have much more powerful teeth than any examples I have seen from the eastern parts of the continent. I
shall therefore speak of the present animal as an inhabitant of New South Wales and Van Diemen's
Land, with a slight doubt as to whether the Tasmanian animal may not be different also, all the specimens
I have yet examined being smaller and darker than those from New South Wales or Western Australia.
Every mammalogist is aware how closely the *Vespertilionidæ* are allied, and how difficult it is to obtain
correct information respecting the species inhabiting our own country. I may therefore be readily excused
for not coming to a hasty conclusion on the subject of those of the antipodes : one thing is certain, namely,
that the animal figured is identical with the specimens in the British Museum which were received from
New South Wales, and to which I find the name of *Geoffroyi* attached.

I frequently saw this animal during my sojourn in New South Wales, and remarked that it was a high
flier and extremely active in the air ; in other respects, as may be supposed, it closely assimilated in its
actions and economy to the nearly allied species in Europe. As the figures in the accompanying Plate are
the size of life, it will be unnecessary to give the admeasurements.

The face is fleshy brown, deepening into dark brown on the nose and laterally expanded nose-leaf ; fur
clothing the upper surface brown, that of the under surface greyish brown, washed with rufous on the sides ;
ears and wing-membranes purplish brown.

NYCTOPHILUS UNICOLOR, *Tomes.*

J.Gould and H.C.Richter, del. et lith.   Hullmandel & Walton, Imp.

NYCTOPHILUS UNICOLOR, *Tomes.*

Tasmanian Nyctophilus.

Nyctophilus unicolor, Tomes in Proc. of Zool. Soc., part xxvi. p. 33.

―――――――――

" ALL the specimens of this genus I have yet seen from Van Diemen's Land," says Mr. Tomes, "differ remarkably from those of the mainland of Australia, in having the fur everywhere short and cottony, perfectly devoid of lustre, and unicoloured; that of the upper parts is of a dark olive-brown without any variation of tint, excepting that it is perhaps a little darker along the middle of the back than elsewhere; beneath the fur is similar but paler in colour, with the tips of the hairs a little tinged with ash-colour; this is the colour of the whole of the under parts, with the exception of a patch on the throat, which is whitish brown, dirty white, and occasionally pure white.

"Immature examples often have the fur above and beneath of a very dark olive-brown, almost black. One specimen of this dark colour which I have examined has the spot on the throat almost pure white.

"So far as I have been able to ascertain, this species is subject to very trifling variations, either in colour or size in the adult state; and the size agrees so closely with that of the species which I have called *N. Gouldi,* that I at first thought the great difference in the texture and colour of the fur was due to the difference of locality."

To this description I have nothing to add. The specimens in my collection were transmitted from Tasmania to this country by Ronald C. Gunn, Esq., a gentleman who has done much to enrich our knowledge of natural history.

The upper figure is of the natural size, the lower one somewhat reduced.

NYCTOPHILUS TIMORIENSIS.

J Gould and HC Richter del. et lith.   Hullmandel & Walton Imp.

NYCTOPHILUS TIMORIENSIS.

Western Nyctophilus.

Vespertilio Timoriensis, Geoff. Ann. du Mus., tom. viii. p. 200. tab. 47, 1806.—Desm. Mamm., p. 146, 1820.—
    Fisch. Synop. Mamm., p. 118, 1829.—Temm. Mon., tom. ii. p. 253, 1835–41.—Wagn. Supp. Schreib.
    Saugth., tom. i. p. 520, 1840.—Schinz, Synop. Mamm., tom. i. p. 175, 1844.
———— ————?, Temm. Mus. Leyd.
Plecotus Timoriensis, Less. Man. de Mamm., p. 97, 1827.—Is. Geoff. in Guérin, Mag. de Zool. 1832.—Less. Nouv.
    Tab. Règn. Anim., p. 23, 1842.
Nyctophilus Timoriensis, Tomes in Proc. of Zool. Soc., part xxvi. p. 30.
Bam-ba, Aborigines of Perth in Western Australia.

It is believed by Mr. Tomes that this species of Bat, although bearing the name of *Timoriensis*, is never found in Timor, but that its true habitat is Western Australia; certain it is that it was there found by Mr. Gilbert, who states that it is very abundant in the neighbourhood of Perth, that it often flies into the houses, doubtless attracted by the light, and that its flight is extremely rapid.

"Although the original specimen of this species," says Mr. Tomes, "is reported to have been received from Timor, I am inclined to believe that there may have been some mistake respecting its locality. Among a great number of Bats from that island, contained in our museums and in that of Leyden, representatives of this genus do not appear; but specimens absolutely identical with the original in the Paris Collection have been obtained by Mr. Gould from Western Australia, and I have noted one in the Leyden Museum also from Australia, but without any precise indication of locality.

"The forms of this species are so similar to those of *N. Geoffroyi*, that it is needless to enter at greater length into details of description than is necessary to point out the differences between the two.

"In all the specimens I have been able to examine, viz. the original one in the Paris Museum, and three others collected in Australia by Mr. Gould, the ears are strongly sulcated, even more so than is observable in the *Plecotus auritus*, whilst in the *N. Geoffroyi* they are very faintly if at all marked; and instead of the small tufts of bristle-like hairs about the eyes, the present species has a tolerably regular series of similar ones fringing the eyelids.

"But the great difference in the size of the two animals is alone sufficient to distinguish them, the one being only nine inches in expanse, whilst the other attains fully thirteen inches; nearly as great a difference as exists between the *Pipistrelle* and the *Noctule* Bats.

"The fur of the upper parts is bicoloured, nearly black at the base, with the terminal half dark sepia-brown; that on the top of the head and on the membrane uniting the ears, unicoloured and paler; beneath, the fur has the basal half nearly black, the remainder being light brown, palest on the throat, on the middle of the belly, and on the pubes; on the shoulder of one example from Perth, Western Australia, is a patch of brownish rust-colour, but it does not occur in the other examples.

"This animal has been repeatedly described as a *Vespertilio*—*V. Timoriensis*; but it is strictly a *Nyctophilus*, as I have ascertained by the examination of the original specimen in the Paris Museum."

The figure is of the size of life.

SCOTOPHILUS GOULDI, Gray

J. Gould and H.C. Richter del et lith.   Hullmandel & Walton, Imp.

SCOTOPHILUS GOULDI, *Gray.*

Gould's Bat.

Scotophilus Gouldii, Gray in Grey's Journ. of Discoveries in Australia, App. vol. ii. p. 405.—Ib. List of Mamm. in Coll. Brit. Mus., p. 30.

———

THIS fine species of Bat is very generally dispersed over New South Wales, and, I believe, South Australia; but, as yet, I have only seen examples from the districts of the former country lying between the mountain ranges and the sea, where it frequents the outskirts of the brushes and the wooded borders of the great rivers. It may be readily distinguished by the upper half of the body being black, while the lower is suffused with brown; and by the hairs of the latter hue on the under surface being lengthened, and extending on to the arms and wing-membranes. It appears, however, to be subject to considerable variation in colour, some being parti-coloured as described, while in others the black predominates; others again, from Flinders' Range in South Australia, have the brown tint reaching nearly to the nape on the upper surface and to the chest on the under surface. I have some specimens also from this locality with a good deal of brown on the chin and throat. I was for some time inclined to consider the Flinders' Range specimens to be distinct; but, on submitting them to the inspection of Mr. Tomes, who has paid the most minute attention to this group of animals, that gentleman states that he considers them to be identical, and that the mere variation in colour, unaccompanied by a difference in structure, is not sufficient to warrant their separation.

The anterior half of the body, both above and beneath, is sooty-black; the posterior half of the upper surface brown; sides and abdomen brownish fawn-colour; wing-membranes purplish-brown.

The figures are of the natural size.

SCOTOPHILUS MORIO, Gray

SCOTOPHILUS MORIO, *Gray*.

Chocolate Bat.

Scotophilus morio, Gray in Grey's Journ. of Discoveries in Australia, App. vol. ii. p. 405.—Ib. List of Mamm. in
Coll. Brit. Mus., p. 29.

THIS species is about the size of *Scotophilus Gouldi*, but differs in having larger ears, and in the colouring of the entire body being of a uniform chocolate-brown. It is very common in New South Wales, between Moreton Bay and Sydney, and Mr. Gilbert states that it also inhabits Western Australia. I have not, however, his specimens to compare with those from New South Wales; its inhabiting the western coast must therefore rest upon his authority; if his assertion be correct, its range will probably be found to extend over the whole of the southern portion of the country. The animal Mr. Gilbert describes is called by the natives *Bam-be*, and in his notes he says that "it is rather uncommon, but may be readily recognized by its habit of flying at a great elevation, and generally around the branches of the loftiest *Eucalypti*."

The whole of the fur of both the upper and under surface of a uniform chocolate-brown, becoming somewhat darker or nearly black on the cheeks; wing-membranes purplish-brown.

The figures are of the natural size.

SCOTOPHILUS MICRODON, *Tomes*

Gould and H.C.Richter. del. et lith.

Hullmandel & Walton. imp.

SCOTOPHILUS MICRODON, *Tomes.*

Small-toothed Bat.

Scotophilus microdon, Tomes in Proc. of Zool. Soc., part xxvii. p. 68.—Ann. and Mag. Nat. Hist., 3rd ser. vol. v. p. 50.

MR. TOMES has very kindly favoured me with the loan of a specimen of the Bat represented in the accompanying Plate, for the purpose of enriching the 'Mammals of Australia.' This gentleman, believing the species to be entirely new to science, has characterized it in the 'Proceedings of the Zoological Society of London' for the year 1859, and I cannot perhaps do better than reproduce Mr. Tomes's account of the species, a course which I feel assured will be approved of by every mammalogist, from the confidence we all place in the investigations made by that gentleman.

"The present species is one having the same subgeneric characters as the common *Pipistrelle* of Europe and the *Scot. Greyii* and *S. pumilus* of Australia. To the latter species it is, by the form of its head and ears, most nearly affined, but may at once be distinguished from it by its greater size and by its smaller teeth.

"The crown is but little elevated above the facial line; but the muzzle, although short, is more pointed than is usual in the flat-crowned species. The ears are very small, nearly as broad as high, with the outer margin slightly hollowed out about the middle, below which is a faintly developed lobe, and immediately above which is the tip of the ear,—the latter being obtusely angular, and directed outwards. The inner margin is very much rounded, especially at two-thirds of the distance from the base, where the convexity is so prominent as to be quite as high as the tip itself, the portion between this prominence and the tip being nearly horizontal. Altogether the ear bears some resemblance to that of *Miniopteris*. *Scot. pumilus* is the only species which has ears of form similar to those of the present species; but they are, although the species is smaller, rather larger, relatively longer, and have their tips less outwardly directed and more rounded. The tragus, as in all others of this group, is curved inwards, and rounded at the end; but it differs from that of some others in being rather widest in the middle.

"In relation to the size of the animal, the wings are rather ample, and rather broad for their length, the fourth finger (that which determines the breadth of the wing) being longer than the two basal phalanges of the longest finger. All the wing-bones are somewhat slender. The thumb is rather long, not quite half enveloped in the membrane.

"The legs are rather long and slender, the tibiæ being quite as long as in *S. Gouldii*, a species of greater size than the present; they are just twice the length of those of *S. pumilus*. The feet are large, about the length of those of *S. Leisleri* of Europe, the toes taking up half their entire length, and the wing-membranes extending to half the distance between the extremity of the tibia and the base of the toes. Tip of the tail enclosed in the membrane.

"The fur of the head extends to rather near the end of the nose; and the upper lips are furnished with moustaches; so that the only naked space is around and in front of the eye. The fur of the back does not extend on to the interfemoral membrane, and only to a very limited extent on those of the wings; but that of the under parts encroaches on the membranes all round the body, especially beneath the arms, where it reaches nearly to the elbow. A straight line from that joint to the knee would pretty accurately define the hairy portions of the wing-membranes.

"In quality the fur is soft, and rather long, bicoloured above and beneath. That of the back of a specimen from South Australia is dark brown at the root, with the terminal half of the hairs reddish brown, uniformly of the latter colour around the rump and on the flanks; beneath, dark brown at the root, with the terminal third light cinnamon-brown, that on the membranes paler and unicoloured. Membranes lightish brown.

"Another specimen from Van Diemen's Land differs only from the last in being much darker in colour; the fur of the upper parts black at the root, tipped with sepia-brown; beneath, the same, but the brown tips lighter and more tinged with rufous, especially that on the membranes and around the pubal region, where it is unicoloured and reddish brown.

"The teeth of this species, although not sufficiently examined to furnish a comparative description, are nevertheless seen at a glance to be of very small size, not only in reference to the size of the animal, but also actually smaller than those of several other species of much less size, such as *S. trilatitius*, *S. lobatus*, and *S. abramis*; hence the specific name of *microdon* here bestowed upon it."

The figure is of the natural size.

SCOTOPHILUS PICATUS, *Gould.*

Gould & Richter del. et lith.

Hullmandel & Walton, Imp.

SCOTOPHILUS PICATUS, *Gould.*

Pied Scotophilus.

Vespertilio—Little Black Bat, Sturt, Exp. into Central Australia, vol. ii. App., p. 9.
Scotophilus picatus, Gould in Proc. of Zool. Soc., 1852.

———————

This pretty little Bat, which is the smallest and one of the most interesting of the true *Scotophili* inhabiting Australia, is so rare, that the single specimen, procured by my friend Captain Sturt, during his late hazardous journey towards the centre of that country, is the only one that has come under my notice; and all the information at present known respecting it is contained in the following note, given in the Appendix to the second volume of Captain Sturt's valuable account of his expedition quoted above :—

"This diminutive little animal flew into my tent at the depôt, attracted by the light. It is not common in that locality, or any other that we visited. It was of a deep black in colour, and had smaller ears than usual."

The whole of the fur both of the upper and under surface deep glossy black, with the exception of a crescentic mark of white which bounds the sides and the lower part of the abdomen; wing and tail membranes purplish brown.

The figures are of the natural size.

SCOTOPHILUS NIGROGRISEUS, Gould.

SCOTOPHILUS NIGROGRISEUS, *Gould.*

Blackish-grey Scotophilus.

A VERY fine specimen of this Bat was sent to me by Mr. Strange, who collected it in the neighbourhood of Moreton Bay. In size it is about equal to the *Scotophilus picatus*, to which it bears a close resemblance, but from which it is quite distinct. The *S. picatus* is an inhabitant of the distant interior, where it was collected by Captain Sturt in the neighbourhood of his farthest encampment, when he endeavoured to reach the centre of the continent from South Australia; the present animal, on the other hand, inhabits the country near the coast; it will be seen therefore that the two species affect very different localities.

The specimen from which my figure was taken will hereafter be deposited in the British Museum, where it may be examined by any mammalogist who may be desirous of investigating the singular group of animals to which it pertains. I may add, that Mr. Tomes, who has paid much attention to this group, coincides with me in considering it to be a new and distinct species from any previously described.

Fur soft and velvety to the touch, the general hue greyish-black, becoming somewhat paler on the posterior part of the upper surface; abdomen washed with brown, and fading into very light brown on the vent; wing- and tail-membranes purplish-brown.

The figures are of the natural size.

SCOTOPHILUS GREYI, *Gray*

J. Gould and H. C. Richter del. et lith.   Hullmandel & Walton, Imp.

SCOTOPHILUS GREYI, *Gray*.

Grey's Scotophilus.

Scotophilus Greyii, Gray, List. of Mamm. in Coll. Brit. Mus., p. 30.—Ib. Zool. of Voy. of Erebus and Terror, pl. 20, fig. 2.

THIS diminutive species was named by Dr. Gray in his " List of the Specimens of Mammalia in the Collection of the British Museum," and was also figured, as above stated, in the " Voyage of the Erebus and Terror." It is said to be a very rare species, and not to have been hitherto found in any other part of Australia than Port Essington. Dr. Gray has given it the above appellation in honour of Sir George Grey, the present Governor of the Cape of Good Hope, to whom such a tribute is justly due for his devotion to the natural sciences generally: from his enlightened views much good has already accrued to every community over which Sir George has had influence, and, as a traveller, he must be considered one of the most intrepid of England's sons.

The fur of this little animal is of a light reddish brown, somewhat paler on the under than on the upper surface; the nose is reddish flesh colour, and the wing-membranes of the usual purplish brown, as seen in other members of the genus.

The figures are of the natural size.

SCOTOPHILUS PUMILUS, Gray

J. Gould and H.C. Richter del. et lith.

Hullmandel & Walton, Imp.

SCOTOPHILUS PUMILUS, *Gray*.

Little Bat.

Scotophilus pumilus, Gray in App. to Grey's Two Exp. in N.W. Australia, vol. ii. p. 406.—Ib. List of Mamm. in Brit. Mus., p. 30.

ALTHOUGH the *Vespertilionidæ* are fairly represented in Australia, the species inhabiting that country are not very numerous. The Bat here represented is certainly one of the very least of the Australian members of the family, for it scarcely exceeds in size the European Pipistrelle. It was my usual practice when travelling in Australia to look around me during the last half-hour of daylight for Bats, at which to discharge the contents of my gun before retiring to my tent, and by this means several species were collected, which might otherwise even now be unknown in Europe. It was not, however, always necessary to shoot this little animal, for it is very tame, and my black attendants often amused themselves by cutting it down with a switch as it passed before them, or rapidly skimmed over the water, a frequent habit with it. I found it especially abundant on the upper part of the River Hunter, particularly on the banks of the rivulets descending from the mountain ranges.

I have never heard of the *Scotophilus pumilus* being collected by any one but myself, and I regret to say that I am unable to give any details as to its habits and economy.

Fur of the upper surface greyish-brown, and of a darker or blackish hue at the base; under surface paler; cheeks blackish; wing-membranes purplish-brown.

The figures are of the natural size.

VESPERTILIO MACROPUS, *Gould*

J. Gould and H. C. Richter del. et lith.   Hullmandel & Walton, Imp.

VESPERTILIO MACROPUS, *Gould*.

Great-footed Bat.

Mr. Tomes having carefully examined my collection of Bats, and come to the conclusion that this animal has not been described, I have, in accordance with his views, characterized it as distinct. It is a native of South Australia, in every respect a true *Vespertilio*, and remarkable for having rather lengthened and elegantly-formed ears, a delicately-constructed body, large wings, and very large hind feet, whence its specific name; besides these peculiarities it is also distinguished from every other Australian Bat by the hoary colouring of its fur, particularly on the lower part of the abdomen, where it is nearly white; it appears, however, subject to some variation in this respect, as in one of my specimens the hoary tint gives place to a pale reddish hue; but I believe hoary to be the prevailing colour.

General tint of the fur greyish-brown, becoming hoary on the posterior parts of the body, especially on the lower part of the abdomen, whence it gradually becomes paler, and fades into buffy-white on the vent; wing-membranes light brown.

The figures are of the size of life.

VESPERTILIO TASMANENSIS.

J. Gould and H. C. Richter del. & lith.

Hullmandel & Walton, Imp.

VESPERTILIO TASMANIENSIS.

Tasmanian Bat.

Noctulina Tasmanensis, Gray, List of Mamm. in Coll. Brit. Mus., p. 194.

It would appear that this species enjoys an unusually wide range of habitat; for not only does it inhabit Tasmania, but, according to Mr. Tomes, it is also found in the Philippines, and even on the continent of India. Had I not known that Mr. Tomes had closely investigated the Vespertilionidæ, and that from his intimate knowledge of the subject he is considered an authority in such matters, I should have hesitated to make this statement.

On submitting my drawings to Mr. Tomes, he suggested that the ears should be a little more indented on the lower side, after the manner of the Notch-eared Bat of Europe; but the Plate having been printed, this could not be attended to.

The specimen from which my figure was taken is in the British Museum.

The fur of this species is of a light brown hue, with a slight tinge of olive, and is lighter on the under than on the upper surface; the wing-membranes and the interior of the ear are of the usual purplish-brown hue; the nose and lips reddish flesh colour.

The figure is of the natural size.

ARCTOCEPHALUS LOBATUS.

ARCTOCEPHALUS LOBATUS.

Cowled Seal.

Otaria cinerea, Gray in King's Narrat. Australia, vol. ii. p. 413.—Id. in Griff. Anim. Kingd., vol. v. p. 183 (not Péron?), 1827.
Arctocephalus lobatus, Gray, Spic. Zool., i. t. (skull).—Bull. Sci. Nat., vol. xvi. p. 113.—J. Brookes's Cat. Mus., p. 37, 1828.—Gray, Zool. of Ereb. and Terror, Mamm., pl. 16, p. 4.—Id. Cat. of Spec. of Mamm. in Coll. Brit. Mus., part ii., Seals, p. 44.—Id. Proc. of Zool. Soc., part xxvii. p. 110.
Phoca lobata, Fisch. Syn., vol. ii. p. 574.
Otaria Lamairii, J. Müll. Wieg. Archiv, 1843, p. 334?
Otaria stelleri, (Mus. Leyden, 1845) Faun. Japon., t. 21, 22, 23 (animal), t. 22. fig. 3 (skull).
Otaria jubata, part, Gray, Cat. of Osteol. Coll. of Brit. Mus., p. 33.

THERE is perhaps no one group of the Mammals of Australia so little understood as the Seals; hence it is very gratifying when we are able to obtain any reliable information respecting the species that visit the rocky shores of that continent and the adjacent islands. As I did not see many of these animals during my visit to Australia, I must content myself with letting those who have say what they know of the subject, taking care that the animals are correctly figured, and that the passages quoted are correctly applied. I would also remark that the list of synonyms are given on the authority of Dr. Gray's 'List of the Seals contained in the Collection of the British Museum;' and as this gentleman has paid much attention to the Seals of the Southern Ocean, I have no doubt that they may be depended upon.

The specimens spoken of by Mr. Gilbert, in the note from his MSS. given below, as having been procured by him on the Houtmann's Abrolhos, as well as the one which Mr. Angus mentions as killed by Sir George Grey in Rivoli Bay, are all in the British Museum; and it is from these specimens that my figures are taken. There is but little doubt in Dr. Gray's mind that Mr. Gilbert's specimens from the Houtmann's Abrolhos are the female or young of the much larger male shot by Sir George Grey in Rivoli Bay, although the latter is twice the size of the former, being fully ten feet in length and as large in girth as a moderate-sized horse. No great length of time has elapsed since the islands in Bass's Straits and the south coast of Australia were first visited by the sealers; but in that comparatively short interval they have dealt out destruction among these inoffensive animals to such an extent that they are now all but exterminated. Collins (in 1798, when his account of New South Wales was published) mentions that "The rocks towards the sea were covered with Fur-Seals of great beauty, of a species which seemed to approach nearest to that known to naturalists as the Falkland Island's Seal." Few, if any, are now to be seen there.

"In the collection I now send you," says Mr. Gilbert, "you will receive eight Seals, of various sizes, the largest of which is a mature male, though it is not so large, by a third, as the very old ones, of which I saw several, but could not obtain either of them. Among them is a half-grown male and a full-grown female; the others are young animals, and the smallest a suckling.

"This animal is extremely numerous on all the low islands of the Houtmann's Abrolhos, particularly those having sandy beaches; but it does not confine itself to such places, being often found on the ridges of coral and madrepores, over which we found it very painful walking, but over which the Seals often outran us. On many of the islands they have been so seldom (perhaps, indeed, never before) disturbed, that I frequently came upon several females and their young in a group under the shade of the mangroves; and so little were they alarmed, that they allowed me to approach almost within the reach of my gun, when the young would play about the old ones, and bark and growl at us in the most amusing manner; and it was only when we struck at them with clubs that they showed any disposition to attack us, or defend their young. The males, however, would generally attack the men when attempting to escape: but, generally speaking, the animal may be considered harmless; for even after being disturbed they seldom attempt to do more than take to the water as quickly as possible. They differ much in colour, the males being considerably darker than the females."

I am indebted to Mr. G. F. Angus for a drawing of this animal, taken from the specimen killed by Sir George Grey, as mentioned above.

"I send you," says Mr. Angus, "a sketch of the Seal killed by Sir George Grey, while Governor of South Australia, in Rivoli Bay, on the south-east coast of that colony. I was with Sir George when it was shot and afterwards clubbed, and made my sketch, and took its admeasurements on the spot after death."

Dr. Gray states that this species and the *A. Hookeri* "are called Hair-Seals by the sealers because they are destitute of any under fur; but this appears to be the case only with the older specimens, for the young of *A. lobatus* is said to be covered with soft fur, which falls off when the next coat of hair is developed. The under fur is entirely absent in the half-grown *A. lobatus* in the British Museum collection."

The adult has the face, front and sides of the neck, all the under surface, sides, and back dark or blackish brown, passing into dark slaty grey on the extremities of the limbs; the hinder half of the crown, the nape and back of the neck rich deep fawn-colour; eyes black.

In the young a reverse of this colouring occurs, the upper surface being dark, and the face and under surface buff.

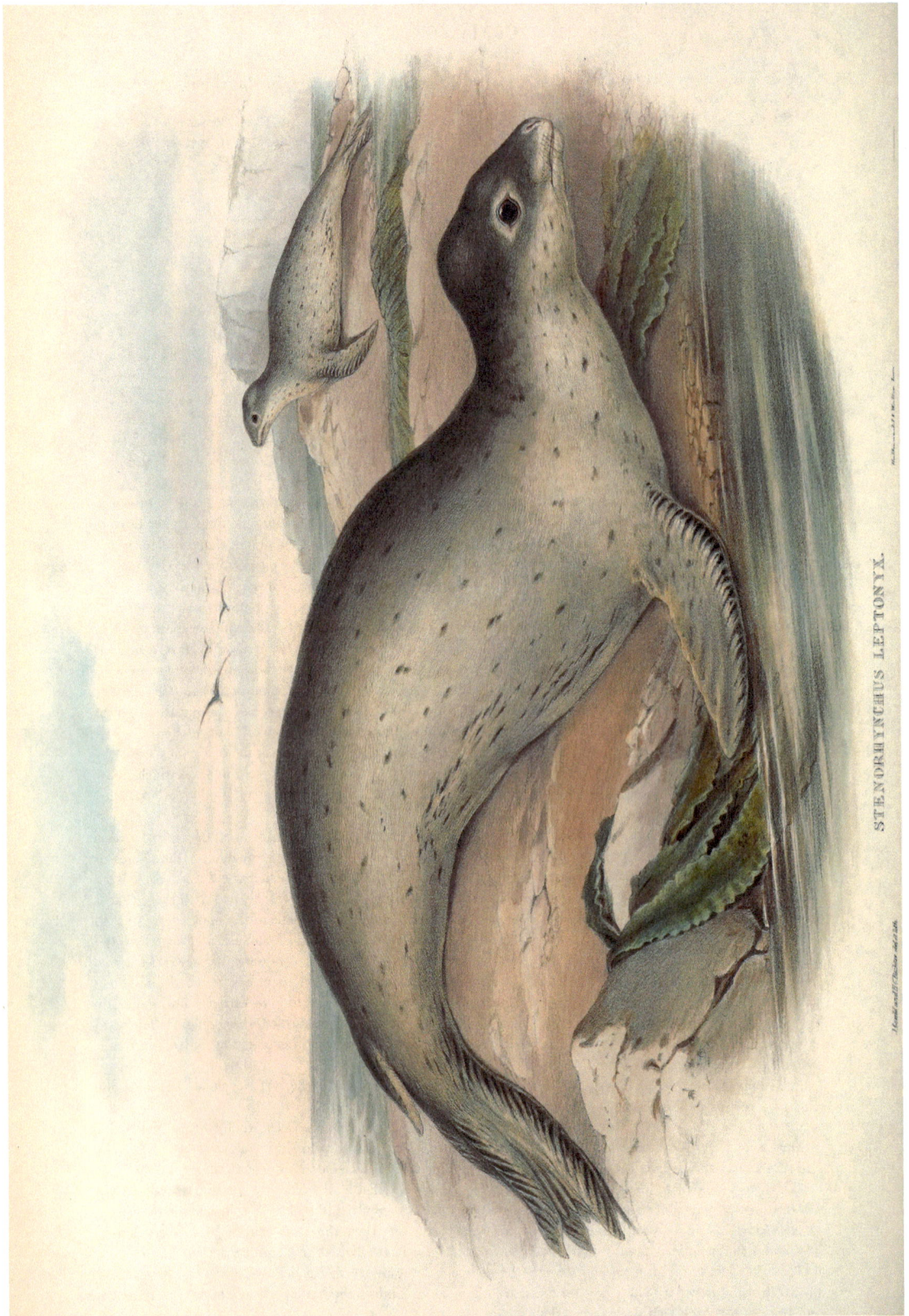

STENORHYNCHUS LEPTONYX.

STENORHYNCHUS LEPTONYX.

Sea Leopard.

Phoca Leptonyx, Blainv. Journ. Phys., vol. xci. p. 288, 1820.—Desm. Mamm., p. 247.—Cuv. Oss. Foss., vol. v. p. 208.
t. 18. fig. 2.—Gray in Griff. Anim. Kingd., vol. v. p. 178.—Blainv. Ostéogr., Phoca, t. 1, and t. 4. fig.
(skull).—F. Cuv. Dent. Mamm., p. 118. t. 38 A.

Seal from New Georgia, Home, Phil. Trans. 1822, p. 240. t. 29 (skull).

Phoque quatrième, Blainv. in Desm. Mamm., p. 243 (note).—Cuv. Oss. Foss., vol. v. p. 207.

Stenorhynchus Leptonyx, F. Cuv. Dict. Sci. Nat., vol. xxxix. p. 549. t. 44.—Ib. Mém. Mus., vol. xi. p. 190. t. 13. fig. 1.
—Ib. Dent. Mamm., p. 118. t. 38 A.—Nilsson, Wieg. Archiv, vol. vii. p. 307.—Ib. Scand. Faun., t.   .—
Gray, Zool. of Ereb. and Terror, Mamm., t. 3 (animal), t. 4 (skull) p. 4.—Ib. Cat. of Osteol. Spec. in
Brit. Mus., p. 31.—Blainv. Ostéogr., Phoca, t. 5. fig. 9 (teeth and skull).—Gray, Cat. of Spec. of Mamm.
in Brit. Mus., part ii., Seals, p. 13.

Phoca Homei, Less. Dict. Class. Nat. Hist., vol. xiii. p. 417.

The small-nailed Seal, Hamilton Smith in Jard. Nat. Hist. Mamm., vol. viii. p. 180. t. 11.

Stenorhynchus aux petits ongles, Homb. et Jacq. Voy. à Pole Sud, t. 9.

Phoca ursina, or *Sea Bear*, Polack.

Sea Leopard of the Whalers.

Upon landing on the sandy beach of one of the quiet bays of Port Arthur, Tasmania, I found myself between the salt water and a huge specimen of the Seal figured on the accompanying Plate; of course, as I had never seen the animal before, it was not to be lost without a struggle; and, after a slight resistance on the part of the animal, a strong cord was fastened round its neck, with the view of towing it after my boat and killing it by drowning, that the specimen might not be injured; but the attempt at dispatching the animal by this means proved futile, as the more it was towed through the water, the more it appeared to gain strength, and other means of depriving it of life had to be resorted to. I have notices of two other specimens having been taken on the south coast of Australia, almost in the immediate neighbourhood of Sydney. For the particulars of their capture, as well as for a very fine drawing of the species, I am indebted to Mr. G. F. Angas, who made the latter immediately after the death of one of them. I mention these solitary instances of its occurrence, because I have reason to believe that the animal is not common in the localities mentioned.

The note accompanying Mr. Angas's drawing is somewhat interesting, inasmuch as it informs us that the stomach of the Seal contained a specimen of that remarkable animal the Ornithorhynchus.

"We have lately added to our Museum Collection," says Mr. Angas, "a fine specimen of an adult Sea Leopard (*Stenorhynchus Leptonyx*), killed some miles above the salt water in the Shoalhaven River; it had an Ornithorhynchus in its stomach when captured; it is much larger than one killed on Newcastle Beach. The dentition is exactly the same as that of the animal figured under the name above-mentioned in the 'Zoology of the Voyage of the Erebus and Terror.'"

I am again obliged to remark that the above list of synonyms is given on the authority of Dr. Gray. For my own part, I have not been able to give sufficient attention to the subject to vouch for their correctness, but Dr. Gray's well-won reputation will be a sufficient guarantee in this respect.

This species of Seal is of a more lengthened or slender form than the *Arctocephalus lobatus*; its length is about ten feet, and its weight probably four hundred pounds.

The general colouring of the animal is greenish creamy white, becoming of a dark slaty hue on the head and back, and speckled with the same dark hue on the sides.

CANIS DINGO, *Blumenb.*

J.Gould and H.C.Richter, del. et lith.   Hullmandel & Walton, Imp.

CANIS DINGO, *Blumenb.*

The Dingo.

HEAD, OF THE SIZE OF LIFE.

THE opposite life-sized head of the Dingo, or native Australian Dog, is portrayed so faithfully, through the talent of Messrs. Richter and Krefft, that I am certain no one will regret my giving two plates of this animal: whether the head be viewed as a zoological illustration or as a work of art, it must be equally acceptable.

The natural history of the Dingo is so fully entered into in the letter-press accompanying the reduced figures, both from my own observation of the animal in a state of nature and from the writings of previous authors, that to recapitulate them here would be superfluous; I therefore refer my readers to that account.

It will be seen that the animal is subject to much variety of colour; I might therefore have multiplied the plates to almost any extent; but such a measure would have been of very questionable utility: I have therefore confined myself to one representing the normal style of colouring.

CANIS DINGO, *Blumenb.*

CANIS DINGO, *Blumenb.*

The Dingo.

Canis Dingo, Blumenb.—Shaw, vol. i pl. 76.—Gray, List of Spec. of Mamm. in Coll. Brit. Mus., p. 57.
—— *familiaris*, var. *Australasiæ*, Desm.—Benn. Gard. and Menag. of Zool. Soc. del., vol. i. p. 51, with fig.
Chrysæus Australiæ, Lieut.-Col. Hamilton Smith in Jard. Nat. Lib. *Dogs*, vol. i. p. 188. pl. 10.

"WHETHER the numberless breeds of dogs, which are the companions of the human race in every region of the globe, were originally descended from one common stock, and owe their infinite varieties solely to their complete domestication, the modifications by which they are distinguished having been gradually produced by the influence of circumstances,—whether, on the contrary, they are derived from the intermixture of different species, now so completely blended together as to render it impossible to trace out the line of their descent,—and whether on either supposition the primæval race or races still exist in a state of nature, are questions which have baffled the ingenuity of the most celebrated naturalists. Theory after theory has been advanced, and the problem is still as eagerly debated as ever, and with as little probability of arriving at a satisfactory conclusion. In the investigation of this difficult subject, however, as in the search after the philosopher's stone, many curious facts have been brought to light which would otherwise in all probability have remained buried in obscurity; and the causes which are continually operating to produce a gradual change of character, both in outward form and in intellectual capacity, among the brute creation, have received considerable elucidation. It is thus that theories, however erroneous in themselves, are frequently made subservient to the advancement of science, by the important facts which are incidentally developed by their authors in the ardour of their zeal for the establishment of a favourite hypothesis."

Such are the words of the late Edward Turner Bennett at the commencement of his paper on the history of the Dingo in "The Gardens and Menagerie of the Zoological Society delineated." Agreeing with Mr. Bennett in the impossibility of arriving at a satisfactory conclusion on the subject, I feel that I cannot close the present work without giving a figure and description of an animal which forms so prominent a feature in the fauna of Australia. It may be expected also that I should myself have formed some opinion as to its claim to be regarded as indigenous or otherwise; and if this opinion should be at variance with those of some Australian zoologists who have lately written on the subject, I may state that it has not been formed without due consideration. Without going into the probable origin of this particular race of dogs, or offering reasons why it should not be considered as indigenous, I may briefly state that I believe it has followed the black man in his wanderings from Northern Asia through the Indian Islands to Australia, the southern portion of which country appears to be its boundary in this direction; for I believe it has never been found in Van Diemen's Land in the wild or semi-wild state in which it occurs on the Australian continent. From what I saw of the animal in a state of nature, I could not but regard it in the light of a variety to which the course of ages had given a wildness of air and disposition; indeed it appeared to have all the habits of a skulking low-bred dog, and none of the determined air and ferocity of disposition of the wolf or jackal: in confirmation of this opinion, I may cite the facility with which the natives bring it under subjection, and the parti-colouring of its hairy coat; for although the normal colouring is red or reddish sand-colour; black, or black and white, individuals are not unfrequently seen; and that this variation in the colouring is not due to crossing with the domesticated races introduced when the country was first discovered, is proved by the following passage in the Appendix to "Collins's Voyage," a work published soon after the colonization of New South Wales, where he says, "the dogs of this country are of the jackal species; they never bark, are of two colours, the one red, with some white about it, the other black: some of them were very handsome." The existence of parti-coloured Dingos is still further confirmed by Mr. Gilbert's note on the animal, as observed by him in Western Australia: "The Dingo is very common over all parts of this colony. There are a very great number of varieties, varying from reddish brown to black, white, light brown, and black and white." Now, on the other hand, it may be affirmed that late geological discoveries will set aside the idea of its being a mere variety and tend to prove that this dog existed in Australia even prior to the aborigines; for it is said that a skeleton of a Dingo has been discovered at Warnamborl, beneath a bed of volcanic ash; but I believe no fossil remains have yet reached this country. The following letter on the subject has been kindly transmitted to me by Mr. Gerard Krefft, a gentleman to whom I am indebted for a beautiful drawing of the head, and an entire figure of the animal sketched either from life or immediately after it was killed:—

"In reply to your inquiry about the Australian Native Dog, I beg to state that it is proved without a doubt, as far as my own judgment goes, that the Dingo is an original inhabitant of the Australian continent.

"There is now, at the Museum in Melbourne, a fossil skull, found with other animal remains in a cave at

Mount Macedon, by Mr. Selwyn, the Geological Surveyor of Victoria. This skull, according to the authority of Professor M'Coy, is identical with that of the Dingo of the present day.

"An article to this effect was published by the learned Professor in the 'Argus' of 1857; but as it is not in my power to consult a file of this Journal, I am unable to furnish any further particulars.

"All the specimens of the Dingo procured by me during my stay at the Lower Murray were distinguished by a white tip at the extremity of the tail, and among the 'trophies' which so generally ornament shepherds' huts in Australia, I do not recollect to have seen a single tail without the white tip.

"The black variety is more scarce; the single specimen which I secured was a young bitch, quite black, except the inside of the fore legs and paws and the outside of the hind legs and paws, which were of a tan-colour. The head was more pointed than in the yellow variety, and had a distinct patch of white, about the size of a shilling, on each cheek.

"I made a drawing of the animal on the spot, and another one of the head, life size; both sketches are now, I trust, in the hands of Professor M'Coy. This dog had been prowling about Jamieson's Station for several nights; it fell at last a victim to strychnine, and I secured its skin."

During my wanderings in Australia I saw much of the Dingo in a state of nature, and can bear testimony to its great tenacity of life and the consequent difficulty of destroying it. I also witnessed the destructive nature of its habits in various ways, particularly its mode of "rushing" the sheep-fold, when it not only wantonly kills great numbers, but scatters the remainder to such an extent as almost to occasion the loss of the entire flock. It is not altogether for the purpose of supplying the cravings of hunger that the Dingo visits the sheep-pen, but in mere wantonness, dealing out his vengeance right and left with a single bite, which, although not fatal at the moment, the sheep seldom recovers, but lingers and soon dies. Mr. Gilbert states that its more usual mode of attack is to follow a flock of sheep, and when a lamb drops behind to immediately pounce upon and carry it off; and Collins mentions that such is its invincible predilection for poultry, that not even the severest beatings can repress it.

"The Dingos, or native dogs, 'Warragal' of the Aborigines," says Dr. Bennett, "are the wolves of the colony, and are perhaps unequalled for cunning. These animals breed in the holes of rocks: a litter was found near Yas Plains, which the discoverer failed to destroy, thinking to return and catch the mother also, and thus exterminate the whole family; but the 'old lady' must have been watching him, for on his returning a short time after, he found all the little dingos had been carried away, and he was never able, although diligent search was made in the vicinity, to discover their place of removal. The cunning displayed by these animals, and the agony they can endure without evincing the usual effects of pain, would seem almost incredible, had it not been related by those on whose testimony every dependence can be placed. The following are a few among a number of extraordinary instances. One had been beaten so severely, that it was supposed all the bones were broken, and it was left for dead. Upon the person accidentally looking back, after having walked some distance, his surprise was much excited by seeing 'master dingo rise, shake himself, and march into the bush, evading all pursuit. One supposed to be dead was brought into a hut, for the purpose of undergoing 'decortication;' at the commencement of the skinning process upon the face, the only perceptible movement was a slight quivering of the lips, which was regarded at the time as merely muscular irritability: the man, after skinning a very small portion, left the hut to sharpen his knife, and returning, found the animal sitting up, with the flayed integument hanging over one side of the face. Another instance was that of a settler, who, returning from a sporting exhibition with six kangaroo dogs, met with a dingo which was attacked by the dogs and worried to such a degree, that finding matters becoming serious, and that the worst of the sport came to his share, the cunning dingo pretended to be dead; thinking he had departed the way of all dogs, they gave him a parting shake and left him. Unfortunately for the poor dingo, he was of an impatient disposition, and was consequently premature in his resurrection; for before the settler and his dogs had gone any distance, he was seen to rise and skulk away, but at a slow pace, on account of the rough treatment he had received; the dogs soon re-attacked him, when he was handled in a manner that must have effectually prevented any resuscitation taking place a second time. The Dingo, like all dogs in a state of nature, never barks, but simply whines, howls, and growls, the explosive noise being only found among the dogs which are domesticated."

I cannot conclude this paper without stating that the Dingo affords considerable exercise and amusement to the Nimrods of Australia, who hunt it precisely as the fox is hunted in England, and for which it forms no mean substitute.

The size of the Dingo is about that of the English Fox-hound, but it is much lower on the legs. The accompanying Plates represent the head of the natural size, and the whole animal reduced.